Explosives Detection Using Magnetic and Nuclear Resonance Techniques

NATO Science for Peace and Security Series

This Series presents the results of scientific meetings supported under the NATO Programme: Science for Peace and Security (SPS).

The NATO SPS Programme supports meetings in the following Key Priority areas: (1) Defence Against Terrorism; (2) Countering other Threats to Security and (3) NATO, Partner and Mediterranean Dialogue Country Priorities. The types of meeting supported are generally "Advanced Study Institutes" and "Advanced Research Workshops". The NATO SPS Series collects together the results of these meetings. The meetings are co-organized by scientists from NATO countries and scientists from NATO's "Partner" or "Mediterranean Dialogue" countries. The observations and recommendations made at the meetings, as well as the contents of the volumes in the Series, reflect those of participants and contributors only; they should not necessarily be regarded as reflecting NATO views or policy.

Advanced Study Institutes (ASI) are high-level tutorial courses intended to convey the latest developments in a subject to an advanced-level audience

Advanced Research Workshops (ARW) are expert meetings where an intense but informal exchange of views at the frontiers of a subject aims at identifying directions for future action

Following a transformation of the programme in 2006 the Series has been re-named and re-organised. Recent volumes on topics not related to security, which result from meetings supported under the programme earlier, may be found in the NATO Science Series.

The Series is published by IOS Press, Amsterdam, and Springer, Dordrecht, in conjunction with the NATO Public Diplomacy Division.

Sub-Series

A.	Chemistry and Biology	Springer
B.	Physics and Biophysics	Springer
C.	Environmental Security	Springer
D.	Information and Communication Security	IOS Press
E.	Human and Societal Dynamics	IOS Press

http://www.nato.int/science
http://www.springer.com
http://www.iospress.nl

Series B: Physics and Biophysics

Explosives Detection Using Magnetic and Nuclear Resonance Techniques

edited by

Jacques Fraissard
Université Pierre et Marie Curie
Ecole de Physique et Chimie Industrielles
Paris, France

and

Olga Lapina
Boreskov Institute on Catalysis
Russian Academy of Sciences
Novosibirsk, Russia

Springer

Published in cooperation with NATO Public Diplomacy Division

Proceedings of the NATO Advanced Research Workshop on
Explosives Detection Using Magnetic and Nuclear Resonance Techniques
St. Petersburg, Russia
7–9 July 2008

Library of Congress Control Number: 2009929398

ISBN 978-90-481-3061-0
ISBN 978-90-481-3060-3
ISBN 978-90-481-3062-7 (eBook)

Published by Springer,
P.O. Box 17, 3300 AA Dordrecht, The Netherlands.

www.springer.com

Printed on acid-free paper

CONTENTS

PREFACE

A week, or even a day without a report on bombing in the news has become rare. As a consequence, detecting a threat before it causes harm is a prominent public interest. However, there is a frightening variety of threats of which one has to be aware: those that have already been encountered and further ones of which nobody except the initiator(s) may have the slightest idea. Consequently, a systematic approach is assumed to be helpful to identify carefully all explosives.

Weapons and any kind of bomb are obviously the most common form of threat. Most weapons are detectable by metal detectors, but the existence of ceramic knifes, for example, should be kept in mind in this context. Bombs include not only grenades, ordnance, mines and other ammunition but also improvised explosive devices which do not have any blue-print for their construction. Figure 1 shows the vast range of possibilities for designing a bomb and what it might contain in addition to the explosive and fuse. Even the range of possible explosives exceeds that which usually constitutes the military arsenal. In particular, peroxides not containing any nitric oxide, which is typical of many explosives particularly in the military environment, have very often played a tragic role (e.g., the London attack on 7 July 2005). One reason for the occurrence of such substances is the fact that they can be easily home-made with ingredients readily available in any chemist's shop. Therefore, while few chemicals find use as military explosives, these can be combined with platiscizers and other materials to create a plethora of formulations. Several explosives based on nitrogen are listed in Table 1 along with their abbreviations. High explosives consist of an intimate mixture of oxidant and reductant, either within a single molecule, such as nitroglycerin, pentaerythritol tetranitrate (PETN), trinitrotoline (TNT), or triacetone triperoxide (TATP), or within an ionic solid, such as ammonium nitrate, when mixed with fuel oil. Mixtures of high explosives are frequently used. For example, Semtex is a blend of cyclomethylenetrinitramine (RDX) and PETN.

In theory the problem of mine detection should be the easiest one because, despite the fact that there are hundreds of mine types, only a few explosive types are used. However, many "high-tech" approaches have proven not totally satisfactory. (Because the explosives are often hermetically sealed in the mine, or because of other nitrogen-containing materials in the ground?). Figure 2 provides a useful overview, albeit non-exhaustive, of current Bulk and Trace explosive detection technologies of interest. The necessity to provide definitive "yes or no" answers about the presence of explosives has led to interest in using tools that offer chemical-specific identification of explosive materials. To this end, researchers have turned their attention to spectroscopic techniques.

TABLE 1. Some representative common high explosives and their compositions in wt %

Explosives based on nitrogen	Formula	C	H	N	O	N + O
Ammonium nitrate (AN)	$H_4N_2O_3$	0	5.04	35.01	59.97	94.98
Ammonium picrate (Expl D)	$C_6H_6N_4O_7$	29.28	2.46	22.76	45.5	68.26
Cyclonite (RDX)	$C_3H_6N_6O_6$	16.22	2.72	37.84	43.22	81.06
Ethylenediamine dinitrate	$C_2H_{10}N_4O_6$	12.91	5.42	30.1	51.58	81.68
Guanidine nitrate	$CH_6N_4O_3$	9.84	4.95	45.89	39.32	85.21
Hexamethylenetriperoxide diamine (HMTD)	$C_6H_{12}N_2O_6$	34.62	5.81	13.46	46.11	59.57
Hexanitrohexaazaisowurtzitane (HNIW or CL20)	$C_6H_6N_{12}O_{12}$	16.45	1.38	38.36	43.82	82.18
Hydrazine nitrate	$H_5N_3O_3$		5.3	44.2	50.09	94.29
Mannitol hexanitrate	$C_6H_8N_6O_{18}$	15.94	1.78	18.59	63.69	82.28
Monomethylamine nitrate	$CH_4N_2O_3$	13.05	4.38	30.43	52.14	82.57
Nitrocellulose	$C_6H_7N_3O_{11}$	24.24	2.37	14.14	59.23	73.37
Nitroglycerin (NG)	$C_3H_5N_3O_9$	15.87	2.22	18.5	63.41	81.91
Nitrotriazolone (NTO)	$C_2H_2N_4O_3$	18.47	1.55	43.08	36.9	79.98
Octogen (HMX)	$C_4H_8N_8O_8$	16.22	2.72	37.84	43.22	81.06
Pentaerythritol tetranitrate (PETN)	$C_5H_8N_4O_{12}$	19	2.55	17.72	60.73	78.45
Picric acid	$C_6H_3N_3O_7$	31.46	1.32	18.34	48.88	67.22
Tetrazene	$C_2H_8N_{10}O$	12.77	4.29	74.44	8.5	82.94
Tetryl	$C_7H_5N_5O_8$	29.28	1.76	24.39	44.58	68.97

Spectrometry is predominantly a tool for characterizing chemical substances. Some techniques are widely used in laboratories: Fourier transform infrared (FTIR), nuclear magnetic resonance (NMR). Others are already available for outdoor applications: Raman, laser-induced breakdown spectrometry (LIBS), γ-spectrometry; further ones are still in development, for example terahertz (THz) and nuclear quadrupole resonance (NQR). This latter seems to be the most promising bulk chemical identification technique being developed for explosives detection.

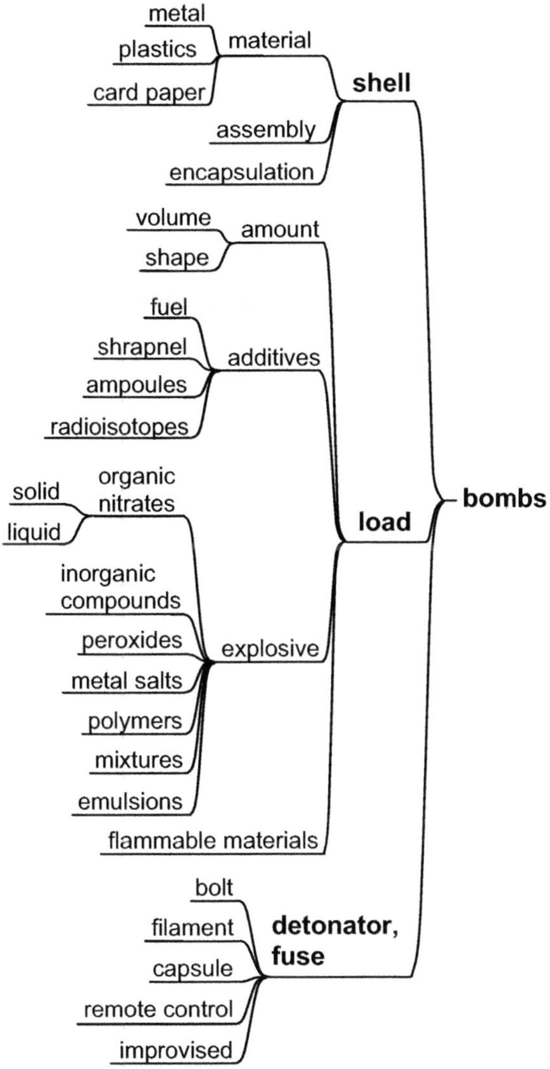

Figure 1. Parts of bombs.

What are the theoretical advantages of NQR?

This technique can be compared to NMR, but the main difference is that a magnet is not required. Therefore the NQR technique is relatively inexpensive and more compact. In NQR, as in NMR, a radio-frequency pulse excites transitions between polarized nuclear spin states. However, unlike NMR, these polarized states are not created via an externally applied magnetic field, but rather they are a result of purely intramolecular electronic forces.

Specifically, the nuclear quadrupole moment, which only arises in nuclei that are non-spherical (spin quantum number > 1/2), interacts with the electric field gradient from the molecule's electron cloud, forming non-degenerate spin orientations. Since nitrogen-14 is a quadrupole-active nucleus, this technique is feasible for explosives detection applications.

The selectivity of NQR is considerable since the electric field gradient depends exclusively on the sight-symmetry the nucleus finds itself in with respect to the valence electrons. Therefore, the exact resonance frequency is extremely dependent on molecular geometry and, thus, peaks can occur across a very wide range. Even chemically equivalent nuclei can have distinct peak positions due to crystal packing geometries. For instance, the three ring nitrogens in RDX show three distinct NQR frequencies corresponding to ν_+, ν_- and ν_0 transitions (about 100 kHz apart from each other).

The NQR frequency is insensitive to long range variations in composition, therefore it does not change in physical mixture of the explosive with other materials such as the plasticizers in plastic explosives. The NQR signal strength varies linearly with the amount of explosive present, and is independent of the explosive distribution within the detection volume.

The NQR method can provide automated detection of explosives types in configurations missed by the imaging x-ray method. Therefore combining NQR with existing x-ray systems can provide improved scope for the checkpoint detection of explosive threats.

If NQR is so good, why it is not used everywhere?

The main limitation is the signal-to-noise ratio, particularly with the interference that exists in a field environment, because of the fact that NQR polarization is much weaker than that from an external magnetic field. The distinctive signatures are there, but the sensitivity is so poor that the signatures are difficult to pull out of the noise. The typical NQR signal is weaker than the thermal noise in the detector circuit. External sources of radio frequency interference also make detection difficult. Fortunately, methods to coherently manipulate the NQR signal exist, and these methods can be used to increase the signal amplitude and reduce interference.

In addition the high selectivity is partly a disadvantage, as it is not easy to build a multichannel system necessary to cover a wide range of target substances. Also it is difficult to detect substances fully screened by metallic enclosures, etc.

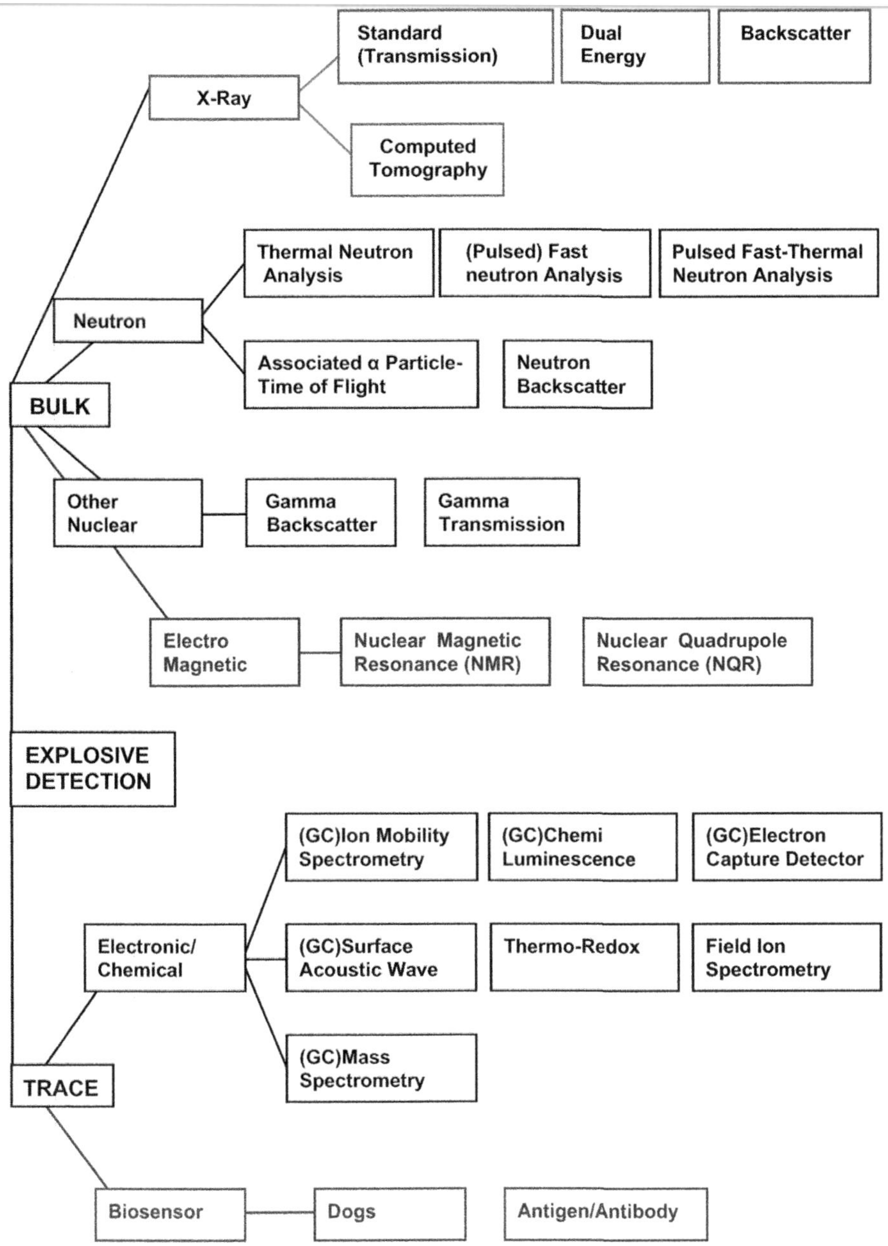

Figure 2. Organization of current explosives detection technologies. (Adapted from NAV97, Fig 2.1-1).

The Naval Research Laboratory in the USA and the Kaliningrad Institute in Russia are the most important research centres in this field. However, there are many other laboratories working to improve the sensitivity of

NQR for the detection of explosives, albeit in a completely independent fashion. The aim of this workshop was to bring together everybody interested in this question in an attempt to make NQR the universal technique for the detection of bombs regardless of type, even in the form of mixtures of explosives.

The choice of Russia as the venue for this meeting is not arbitrary. This country is one of those particularly subject to terrorist attacks. However, due to the economic situation in Russia, communication with the researchers from partner countries has been quite limited. Such a workshop can help to foster constructive interaction to the benefit of this specifically crucial field.

The conference took place in the University of Saint Petersburg, situated in the heart of old city, on the bank of the Neva river, in front of the Hermitage and the Saint Isaac Cathedral (1783–1789). We must thank the authorities of this university which gave us the magnificent Petrosvsky hall, the only hall in the "Twelve Colleges" which is still in its original Baroque style, practically intact since its construction under Peter the Great in 1724.

Many people have contributed to the success of the ARW on which this volume is based. We thank of course all participants for contributing to the intellectual dialogue. It is also a great pleasure to acknowledge the main financial support provided by the Scientific Affairs Division of the North Atlantic Treaty Organization (NATO).

We thank also the financial support provided by:

→ University of Pierre and Marie Curie (Paris, France)
→ Boreskov Institute of Catalysis (Novosibirsk, Russia)
→ Bruker Analytik GMBH (Karlsruhe, Germany)

Many individual whose great help with the organization of the ARW we acknowledge with gratitude include A.A. Shubin, D.F. Khabibulin, T.V. Zamulina, L.Ya. Startseva, from Novosibirsk, and V.I. Chizhik, A.G. Gribanov from St. Petersburg.

Jacques Fraissard
Olga Lapina

FUNDAMENTALS OF PULSED NITROGEN-14 QUADRUPOLE RESONANCE

DANIEL CANET*, MAUDE FERRARI
Nancy-Université, Méthodologie RMN (UMR 7565 CNRS-UHP Nancy 1), B.P. 239 54506-Vandoeuvre-les-Nancy (Cedex) France

Abstract The essential features of nitrogen-14 Nuclear Quadrupole Resonance, a new tool based on density matrix calculations is proposed. After a brief review of the density matrix theory, it is demonstrated that, for each of the three NQR transitions, the (3,3) density matrix can be reduced to a (2,2) matrix, evidently easier to handle. (2,2) rotation matrices are defined for predicting, in a straightforward manner, the system evolution under a rf pulse. The first example treated by this methodology concerns nutation experiments (evolution of the signal amplitude as a function of the pulse length) and it is shown that the NMR (Nuclear Magnetic Resonance) flip angle, in the case of powder samples, should be substituted by a pseudo flip angle which is no longer proportional to the pulse length. Still for powder samples, it is demonstrated that, in NQR, data averaging continuously improves when shortening the repetition time. Finally it has been possible to define proper phase cycles in view of measuring relaxation times (T_1 and T_2) by a two-pulse sequence. In all cases, experimental verifications were performed in order to assess this methodology.

Keywords: NQR, nitrogen-14, pulsed QR, density matrix, nutation, polarization, relaxation time, flip angle

1. Introduction

The theory of spin 1 NQR of powder samples involve serious difficulties associated with (i) the special form of the Hamiltonian, (ii) the lack of a preferred direction (as the static magnetic field in NMR), (iii) the orientation

* E-mail: Daniel.Canet@rmn.uhp-nancy.fr

J. Fraissard and O. Lapina (eds.), *Explosives Detection Using Magnetic and Nuclear Resonance Techniques*,

of microcrystallites with respect to the direction of the radio-frequency field. This article will deal with a general method based on the density matrix formalism for treating the action of a radio-frequency (rf) pulse whatever is the state of the system prior to the application of the rf pulse. The method allows us to account as well for evolution periods. After an outline of the density matrix methodology, it will be shown that each of the three NQR lines of a spin 1 can be treated with a simple (2,2) density matrix thus amenable to simple calculations. Several fields of application will be considered, including nutation experiments, data averaging, two-pulse sequences for the measurement of relaxation times... Before developing density matrix calculations, we recall the simple theory based on the eigenstates of the quadrupolar Hamiltonian and on the concept of transition probability.[1]

The quadrupolar Hamiltonian is well-known and can be written as

$$\hat{H}_Q = (1/3)\omega_Q[(3\hat{I}_z^2 - \hat{I}^2) + \eta(\hat{I}_x^2 - \hat{I}_y^2)] \tag{1}$$

where $\hat{I}_x$, $\hat{I}_y$ and $\hat{I}_z$ are the classical spin operators (in the case of a spin 1, $\hat{I}^2$ is simply equal to $2\hat{E}$, $\hat{E}$ being the identity operator) and where ω_Q (quadrupolar interaction) and η (asymmetry parameter) are related to the electric field gradient (efg) tensor in its diagonal form (V_{xx}, V_{yy}, V_{zz}), e being the electron electric charge and Q the quadrupolar moment of the considered nucleus:

$$\omega_Q = eQV_{zz}/2 \tag{2}$$

$$\eta = (V_{xx} - V_{yy})/V_{zz} \tag{3}$$

The molecular axes x, y and z constitute what is generally called the Principal Axis System (PAS).

The three eigenvectors of the Hamiltonian $\hat{H}_Q$ can be expressed as follows on the basis of the three eigenvectors of the $\hat{I}_z$ operator, namely $|0\rangle$, $|1\rangle$ and $|-1\rangle$, corresponding to its three eigenvalues 0, 1 and −1:

$$\begin{aligned} \phi_1 &= |0\rangle \\ \phi_2 &= (1/\sqrt{2})(|1\rangle + |-1\rangle) \\ \phi_3 &= (1/\sqrt{2})(|1\rangle - |-1\rangle) \end{aligned} \tag{4}$$

Notice that in the case of axial symmetry ($\eta = 0$), the energy level corresponding to the eigenstates $|1\rangle$ and $|-1\rangle$ is degenerate and that $|1\rangle$ and $|-1\rangle$ are convenient as eigenvectors.

In order to determine the allowed transitions, one has to define the time-dependent Hamiltonian by which transitions are induced. It is the radio-frequency (rf) Hamiltonian, $\hat{H}_{rf}(t)$, which represents the action of the rf field of amplitude B_1 and which can be expressed as

$$\hat{H}_{rf} = \gamma B_1 \cos\omega t(\sin\theta\cos\varphi\hat{I}_x + \sin\theta\sin\varphi\hat{I}_y + \cos\theta\hat{I}_z) \tag{5}$$

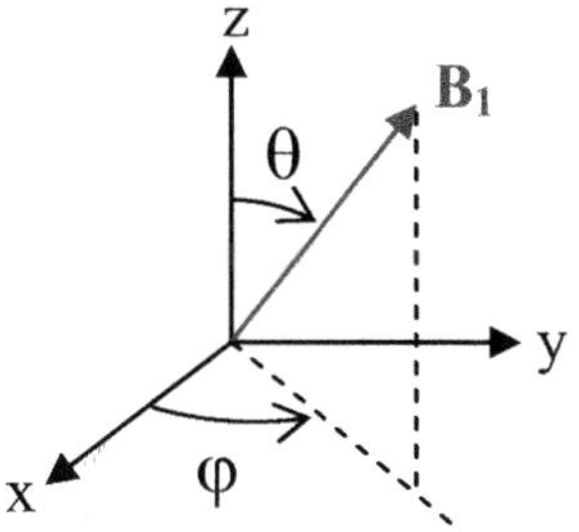

Figure 1. The polar angles defining the rf field orientation in the PAS system of a given microcrystallite.

γ is the nitrogen-14 gyromagnetic constant and the polar angles θ and φ define the orientation of the rf field in the PAS (Figure 1). Now, a transition is possible between two states represented by the eigenvectors Ψ and Φ provided that the scalar product $\langle\Psi|\hat{H}_{rf}|\Phi\rangle$ is non zero (the square of the modulus of this latter quantity provides the so-called transition probability). It is a simple matter to show that only the following scalar products are non zero: $(\langle 0|\hat{I}_x|1\rangle + \langle 0|\hat{I}_x|-1\rangle)$, $(\langle 0|\hat{I}_y|1\rangle - \langle 0|\hat{I}_y|-1\rangle)$, $(\langle 1|\hat{I}_z|1\rangle + \langle -1|\hat{I}_z|-1\rangle)$. Hence, the transition denoted below by ω_x is made possible by the first term of (5), the transition ω_y by the second term and the transition ω_z by the third term. These three possible transitions are sketched in Figure 2, along with experimental spectra. The latter, as all experimental results shown in this article, have been obtained with a home-made spectrometer described elsewhere.[2] Notice that, when the asymmetry parameter η is zero (electric field gradient tensor of axial symmetry), only one line exists at the frequency ω_Q. In the general case ($\eta \neq 0$), three lines are observed at frequencies

$$\begin{aligned} \omega_x &= \omega_Q(1+\eta/3) \\ \omega_y &= \omega_Q(1-\eta/3) \\ \omega_z &= 2\omega_Q\eta/3 \end{aligned} \tag{6}$$

The problem we have to deal with is the effect of a radio-frequency pulse at one of these three frequencies. It turns out that, because of the reasons invoked above, the only means is to rely on a density matrix (or density operator) treatment. We therefore start with a survey of the density matrix methodology.

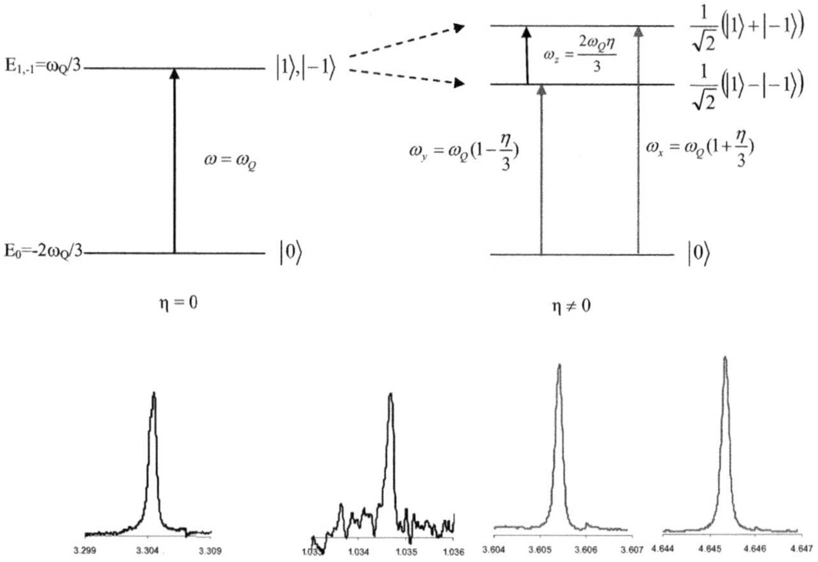

Figure 2. The three nitrogen-14 NQR lines in the general case (right) and in the case of an axially symmetric efg tensor (left). Below are shown experimental spectra: the single line of Hexamethylene-tetramine (left) and the three lines of sodium nitrite (right).

2. A survey of the density matrix (density operator) approach

When a system is time-dependent, its wave function $\Psi(t)$ obeys the Schrödinger equation[3]:

$$\hat{H}(t)\Psi(t) = -\frac{\hbar}{i}\frac{\partial\Psi}{\partial t} \tag{7}$$

where $\hat{H}(t)$ is the time dependent Hamiltonian, $\hbar$ is the Planck constant divided by 2π and $i^2 = -1$. Here

$$\hat{H}(t) = \hat{H}_Q + \hat{H}_{rf}(t) \tag{8}$$

The wave function (vector) $\Psi(t)$ can be expanded upon any basis. We shall choose here the basis of the eigenvectors of $\hat{H}_Q$, $\{\phi_k\}$, which, in our case, are given in (4):

$$\Psi(t) = \sum_{k=1}^{n} c_k(t)\phi_k \tag{9}$$

n is the dimension of the Hilbert space (space of the wave functions) and, here, is equal to 3. Let us define a (n,n) matrix by its kl element

$$\sigma_{kl} = \overline{c_k c_l^*} \qquad (10)$$

The star indicates the complex conjugate while the bar represents an ensemble average. The denomination of density matrix arises from the following features: diagonal elements $\sigma_{kk} = \overline{|c_k|^2}$ just represent the population of the kth energy level while off-diagonal elements represent the so-called coherences induced by the application of an rf field (and therefore zero at equilibrium). This is the first interest of the density matrix which indicates how populations evolve and also the existence of coherences, thus possibly detectable signals. The other interest lies in the capability to calculate (or to predict) the value of any quantity G, to which is associated the operator $\hat{G}$

$$\overline{\langle G \rangle} = Tr(\hat{\sigma}\hat{G}) \qquad (11)$$

$\hat{\sigma}$ is called the *density operator*, associated with the *density matrix*. In (11), the brackets indicate an expectation value (in the sense of quantum mechanics) while the bar (omitted in the rest of this article) indicates an ensemble average. This means that the density matrix (operator) leads effectively to the measured quantity.

The evolution of the density operator is governed by the Liouville–von Neumann equation

$$\frac{d\hat{\sigma}}{dt} = i[\hat{\sigma}, \hat{H}(t)] \qquad (12)$$

where the notation [,] refers to the commutator of the operators between brackets.

It can be noted that, although equations (11) and (12) are derived very easily, the very problem will be to solve equation (12). This will be the subject of the forthcoming sections. Nevertheless, the starting point is the density matrix (operator) at equilibrium. All off-diagonal elements are zero and the kth diagonal element, in the limit of high temperature (T being the absolute temperature), can be written as

$$\sigma_{kk}^{eq} \propto (1 - E_k / k_B T) \qquad (13)$$

where E_k is the energy of the kth energy level and k_B the Boltzmann factor.

3. The effect of an rf pulse on the density matrix

In order to solve the Liouville–von Neumann equation, we can always expand the density operator on a basis of nine orthonormalized operators $\hat{A}_k$ (nine being the dimension of the Hilbert space) thus such that $\left\langle \hat{A}_k \middle| \hat{A}_l \right\rangle = Tr(A_k^\dagger A_l))$ (*Tr*: trace; $A^\dagger$: adjoint of *A*). The choice of these operators is not unique[4] and one approach is to choose those which are eigenvectors of the Liouvillian super-operator. Let us recall that the super-operator $\hat{\hat{L}}$ (applied to operators and not to wave functions) is defined as[5]

$$\hat{\hat{L}}\hat{A} = [\hat{A}, \hat{H}] \tag{14}$$

We shall therefore look for operators such that

$$\hat{\hat{L}}_Q \hat{A}_k = \omega_k \hat{A}_k \tag{15}$$

where $\hat{\hat{L}}_Q$ is the Liouvillian associated with the quadrupolar Hamitonian $\hat{H}_Q$. We start from the basis of the operator products (i.e. $\hat{I}_x$, $\hat{I}_y$, $\hat{I}_z$, $\hat{I}_x\hat{I}_y$, etc.) and we calculate their commutators with $\hat{H}_Q$. Then, by inspection, we determine their linear combinations which satisfy equation (15). After normalization, we obtain

$$\hat{A}_1 = \frac{1}{2}(\hat{I}_x - i\hat{I}_y\hat{I}_z - i\hat{I}_z\hat{I}_y) \qquad \hat{\hat{L}}_Q \hat{A}_1 = \omega_x \hat{A}_1 \tag{16.1}$$

$$\hat{A}_2 = \frac{1}{2}(i\hat{I}_y + \hat{I}_x\hat{I}_z + \hat{I}_z\hat{I}_x) \qquad \hat{\hat{L}}_Q \hat{A}_2 = -\omega_y \hat{A}_2 \tag{16.2}$$

$$\hat{A}_3 = (\hat{I}_z + i\hat{I}_x\hat{I}_y) \qquad \hat{\hat{L}}_Q \hat{A}_3 = \omega_z \hat{A}_3 \tag{16.3}$$

$$\hat{A}_4 = \frac{1}{\sqrt{6}}(3\hat{I}_z^2 - 2) \qquad \hat{\hat{L}}_Q \hat{A}_4 = 0 \tag{16.4}$$

$$\hat{A}_5 = \frac{1}{\sqrt{2}}(\hat{I}_x^2 - \hat{I}_y^2) \qquad \hat{\hat{L}}_Q \hat{A}_5 = 0 \tag{16.5}$$

$$\hat{A}_6 = \frac{1}{\sqrt{3}}(\hat{E}) \qquad \hat{\hat{L}}_Q \hat{A}_6 = 0 \tag{16.6}$$

$$\hat{A}_7 = \frac{1}{2}(\hat{I}_x + i\hat{I}_y\hat{I}_z + i\hat{I}_z\hat{I}_y) \qquad \hat{\hat{L}}_Q \hat{A}_7 = -\omega_x \hat{A}_7 \tag{16.7}$$

$$\hat{A}_8 = \frac{1}{2}(i\hat{I}_y - \hat{I}_x\hat{I}_z - \hat{I}_z\hat{I}_x) \qquad \hat{\hat{L}}_Q\hat{A}_8 = \omega_y\hat{A}_2 \tag{16.8}$$

$$\hat{A}_3 = (\hat{I}_z - i\hat{I}_x\hat{I}_y) \qquad \hat{\hat{L}}_Q\hat{A}_9 = -\omega_z\hat{A}_9 \tag{16.9}$$

$\hat{E}$ is the identity operator. ω_x, ω_y and ω_z are the frequencies given in equation (6)

Now, we can always expand the density operator on the basis (16.1)–(16.9)

$$\hat{\sigma}(t) = \sum_{k=1}^{9} a_k(t)\hat{A}_k e^{i\omega_k t} \tag{17}$$

The factor $e^{i\omega_k t}$ has been introduced by anticipation in order to simplify the forthcoming differential equations. The next step is to expand the Hamiltonian on the $\{\hat{A}_k\}$ basis and to use this expansion along with the one of equation (17) for obtaining, from (14), the differential equations which govern the evolution of coefficients a_k. These calculations are detailed in reference[6] and we shall consider only the transition ω_x anticipating that conclusions regarding this transition can be transposed to the other two transitions. The relevant part of the Hamiltonian given in (5) is

$$\hat{H}_{rf} = \Omega_1^{eff} \cos\omega t\hat{I}_x \tag{18}$$

With ω close to ω_x and

$$\Omega_1^{eff} = \gamma B_1 \sin\theta\cos\varphi \tag{19}$$

The nine differential equations can then be written as

$$\frac{da_1}{dt} = \frac{i\Omega_1^{eff}}{2\sqrt{2}}(\sqrt{3}a_4 + a_5)$$

$$\frac{da_2}{dt} \approx 0$$

$$\frac{da_3}{dt} \approx 0$$

$$\frac{da_4}{dt} = \frac{i\Omega_1^{eff}}{2\sqrt{2}}(\sqrt{3}a_1) - \frac{i\Omega_1^{eff}}{2\sqrt{2}}(\sqrt{3}a_7)$$

$$\frac{da_5}{dt} = \frac{i\Omega_1^{eff}}{2\sqrt{2}}(a_1) - \frac{i\Omega_1^{eff}}{2\sqrt{2}}(a_7)$$

$$\frac{da_6}{dt} \approx 0$$

$$\frac{da_7}{dt} = -\frac{i\Omega_1^{eff}}{2\sqrt{2}}(\sqrt{3}a_4 + a_5)$$

$$\frac{da_8}{dt} \approx 0$$

$$\frac{da_9}{dt} \approx 0$$

They are seen to reduce to four equations which can still be reduced to two equations by defining the following quantities

$$\begin{aligned} b_1 &= \sqrt{2}(a_1 - a_7) \\ b_2 &= \sqrt{3}a_4 + a_5 \\ b_3 &= a_1 + a_7 \\ b_4 &= a_4 - \sqrt{3}a_5 \end{aligned} \tag{20}$$

which lead to

$$\begin{aligned} \frac{db_1}{dt} &= i\Omega_1^{eff} b_2 \\ \frac{db_2}{dt} &= i\Omega_1^{eff} b_1 \\ \frac{db_3}{dt} &\approx 0 \\ \frac{db_4}{dt} &= 0 \end{aligned} \tag{21}$$

Therefore, only two operators are required (those corresponding to b_1 and b_2) for the expansion of the density operator, in view of determining its evolution

$$\hat{\sigma} = b_1 \hat{B}_1 + b_2 \hat{B}_2 \tag{22}$$

with

$$\hat{B}_1 = \frac{1}{i\sqrt{2}}(\hat{A}_1 - \hat{A}_7) = \frac{-1}{\sqrt{2}}(\hat{I}_y\hat{I}_z + \hat{I}_z\hat{I}_y)$$
$$\hat{B}_2 = \frac{1}{2}\left(\sqrt{3}\,\hat{A}_4 + \hat{A}_5\right) = \frac{1}{\sqrt{2}}(\hat{I}_z^2 - \hat{I}_y^2) \quad (23)$$

In the matricial form, on the basis of $\hat{I}_z$ eigenvectors, this yields

$$\boldsymbol{B}_1 = \frac{1}{2}\begin{bmatrix} 0 & i & 0 \\ -i & 0 & -i \\ 0 & i & 0 \end{bmatrix}$$
$$\boldsymbol{B}_2 = \frac{1}{2\sqrt{2}}\begin{bmatrix} 1 & 0 & 1 \\ 0 & -2 & 0 \\ 1 & 0 & 1 \end{bmatrix} \quad (24)$$

On the basis of the quadrupolar Hamiltonian eigenvectors $((1/\sqrt{2})(|1\rangle + |-1\rangle);\ |0\rangle;\ (1/\sqrt{2})(|1\rangle - |-1\rangle))$, we obtain

$$\boldsymbol{B}_1 = \frac{1}{\sqrt{2}}\begin{bmatrix} 0 & i & 0 \\ -i & 0 & 0 \\ 0 & 0 & 0 \end{bmatrix}$$
$$\boldsymbol{B}_2 = \frac{1}{\sqrt{2}}\begin{bmatrix} 1 & 0 & 0 \\ 0 & -1 & 0 \\ 0 & 0 & 0 \end{bmatrix} \quad (25)$$

Because we are dealing with the transition $|0\rangle \to (1/\sqrt{2})(|1\rangle + |-1\rangle)$, the useful part of the density matrix is the (2,2) matrix occupying the left upper corner, just as for the matrices (25). Therefore, from now on, we can proceed only with (2,2) matrices

$$\boldsymbol{B}_1 = \frac{1}{\sqrt{2}}\begin{bmatrix} 0 & i \\ -i & 0 \end{bmatrix}$$
$$\boldsymbol{B}_2 = \frac{1}{\sqrt{2}}\begin{bmatrix} 1 & 0 \\ 0 & -1 \end{bmatrix} \quad (26)$$
$$\boldsymbol{\sigma} = \begin{bmatrix} \sigma_{11} & \sigma_{12} \\ \sigma_{21} & \sigma_{22} \end{bmatrix}$$

σ_{11} and σ_{22} represent the populations of energy levels associated with $(1/\sqrt{2})(|1\rangle + |-1\rangle)$ and $|0\rangle$, respectively. The off-diagonal elements indicate, as usual, coherences, which here correspond to detectable signals. Also, it can be noticed that, except the $\frac{1}{\sqrt{2}}$ factor (which would be replaced by $\frac{1}{2}$), the matrices $\boldsymbol{B}_1$ and $\boldsymbol{B}_2$ are identical to the matrices associated with the operators $-\hat{I}_y$ and $\hat{I}_z$ of a spin ½. The same remark holds for the density matrix or the density operator which can be expanded as indicated by equation (22), whereas the density matrix can be written as

$$\boldsymbol{\sigma} = \frac{1}{\sqrt{2}}\begin{bmatrix} b_2 & ib_1 \\ -ib_1 & -b_2 \end{bmatrix} \tag{27}$$

It is important to determine the polarization of the NQR signal. The latter corresponds to the off-diagonal elements in (27) and we have to look for the contribution of ib_1 to the observables $<I_x>$, $<I_y>$ or $<I_z>$. As an example, we consider the case of $<I_x>$. This quantity can be calculated by

$$<I_x> = Tr(\hat{\sigma}\hat{I}_x) = Tr[(\sum_k a_k e^{i\omega_k t}\hat{A}_k)(\hat{A}_1 + \hat{A}_7)]$$

This leads to

$$<I_x> = (a_1 + a_7)\cos\omega_x t - i(a_1 - a_7)\sin\omega_x t =$$

$$= b_3 \cos\omega_x t - \frac{i}{\sqrt{2}} b_1 \sin\omega_x t$$

As b_3 is not involved in the ω_x transition, we are left with the second term in the above expression and

$$<I_x> \propto b_1 \sin\omega_x t \tag{28}$$

It can be demonstrated that neither b_1 nor b_2 are involved in $<I_y>$ or $<I_z>$ (see reference[6] for further details). Thus, the NQR signal at the frequency ω_x is *polarized in the x direction*. The same conclusions hold for signals at ω_y and ω_z which are polarized along the *y* and *z* directions, respectively.

In order to solve the two differential equations in (21), let us define by β the equivalent of the flip angle in NMR

$$\beta = \Omega_1^{eff}\delta \tag{29}$$

where δ is the duration of the rf pulse. We shall use the following notations: b_1^- and b_2^- the values of b_1 and b_2 before the rf pulse and b_1^+ and b_2^+ these values after the rf pulse. One has

$$\begin{aligned} b_1^+ &= \cos\beta b_1^- + \sin\beta b_2^- \\ b_2^+ &= -\sin\beta b_1^- + \cos\beta b_2^- \end{aligned} \tag{30}$$

which can also be written

$$\begin{bmatrix} b_1^+ \\ b_2^+ \end{bmatrix} = \begin{bmatrix} \cos\beta & \sin\beta \\ -\sin\beta & \cos\beta \end{bmatrix} \begin{bmatrix} b_1^- \\ b_2^- \end{bmatrix} \tag{30bis}$$

Now, using equations (27) and (30), we are able to calculate σ^+, as a function of the density matrix before the pulse, σ^-, or rather as a function of b_1^- and b_2^-:

$$\boldsymbol{\sigma}^+ = \frac{1}{\sqrt{2}} \begin{bmatrix} -\sin\beta b_1^- + \cos\beta b_2^- & i(\cos\beta b_1^- + \sin\beta b_2^-) \\ -i(\cos\beta b_1^- + \sin\beta b_2^-) & \sin\beta x_1^- - \cos\beta b_2^- \end{bmatrix} \tag{31}$$

It turns out that the same result can be obtained by writing

$$\hat{\sigma}^+ = \hat{R}_{0^\circ}^\dagger(\beta/2)\hat{\sigma}^-\hat{R}_{0^\circ}(\beta/2) \tag{32}$$

that is in matricial form

$$\boldsymbol{\sigma}^+ = \frac{1}{\sqrt{2}} \begin{bmatrix} \cos\frac{\beta}{2} & -i\sin\frac{\beta}{2} \\ -i\sin\frac{\beta}{2} & \cos\frac{\beta}{2} \end{bmatrix} \begin{bmatrix} b_2^- & ib_1^- \\ -ib_1^- & -b_2^- \end{bmatrix} \begin{bmatrix} \cos\frac{\beta}{2} & i\sin\frac{\beta}{2} \\ i\sin\frac{\beta}{2} & \cos\frac{\beta}{2} \end{bmatrix} \tag{33}$$

Or alternatively

$$\boldsymbol{\sigma}^+ = \frac{1}{\sqrt{2}} \begin{bmatrix} \cos\frac{\beta}{2} & -i\sin\frac{\beta}{2} \\ -i\sin\frac{\beta}{2} & \cos\frac{\beta}{2} \end{bmatrix} \begin{bmatrix} \sigma_{11}^- & \sigma_{12}^- \\ \sigma_{21}^- & \sigma_{22}^- \end{bmatrix} \begin{bmatrix} \cos\frac{\beta}{2} & i\sin\frac{\beta}{2} \\ i\sin\frac{\beta}{2} & \cos\frac{\beta}{2} \end{bmatrix} \tag{33bis}$$

We have introduced here a rotation matrix

$$\boldsymbol{R}_{0^\circ}(\beta/2) = \begin{bmatrix} \cos\frac{\beta}{2} & i\sin\frac{\beta}{2} \\ i\sin\frac{\beta}{2} & \cos\frac{\beta}{2} \end{bmatrix} \tag{34}$$

The subscript 0° refers to the phase of the rf pulse (see below for the rotation matrix associated with a 90° rf phase change).

(33) is the key equation for calculating the effect of an rf pulse on any spin 1 NQR transition. The analogy with a spin ½ can be noticed but cannot further be employed because we are unable in NQR to define a rotating frame, which is an invaluable tool in NMR. Density matrix calculations are therefore compulsory but this not a real penalty because we are dealing with (2,2) matrices. One may be puzzled by the comparison of (30bis) and (33bis): β vs. $\beta/2$. In (30bis), the rotation concerns one vector while in (33bis) it is a matrix which is rotated and this amounts to rotate two vectors, hence the ½ factor. Anyhow, (33) is shown to be equivalent to (31) on pure mathematical grounds. In fact the above developments could justify the methodology of fictitious spins ½ .[7,8] More physically, we can resort to the analogy with a spin ½, thus to NMR. It is well known that, in NMR, rf pulses rotate the nuclear magnetization components, but this occurs in the rotating frame (which implies a preferential direction) and is applied to nuclear magnetization (which implies again a preferential direction). Moreover the rotating frame implies that B_1 is decomposed into two components of equal amplitude, one rotating clockwise (and which will be active), the other rotating anticlockwise (inactive). Half the B_1 amplitude is therefore used and, as a consequence, the "flip angle" is $\beta/2$ in NMR vs. β in NQR. *Thus, all things being equal, a NQR experiment requires half the rf power of an homologous NMR experiment*! This is verified by the comparison of NMR and NQR experiments, the results of which are shown in Figure 3. The NMR signal maximum (π/2 pulse) is obtained with a rf pulse length roughly twice larger than the one used in NQR for reaching a similar result (see below for the concept of pseudo π/2 pulse in NQR).

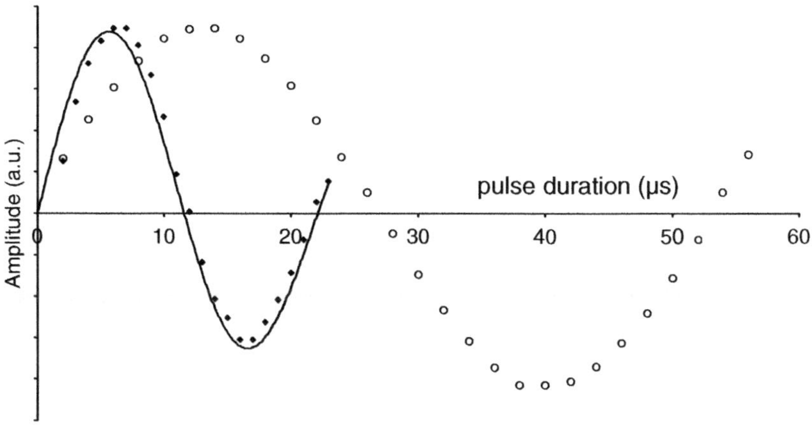

Figure 3. Comparison of NMR (empty circles) and NQR (filled lozenges) signal amplitudes as a function of the rf pulse length (nutation experiments), demonstrating that half the NMR power is required by an NQR experiment. NQR experiments were performed at 4.64 MHz (ω_x line of sodium nitrite). Proton NMR (water sample) experiments were also performed at 4.64 MHz (thanks to a variable field electromagnet) with the same equipment and with a rf field amplitude corrected by the ratio $\gamma(^{14}N)/\gamma(^{1}H)$ of gyromagnetic constants.

In order to proceed with (33bis), we need the form of the density matrix at equilibrium. Owing to the basis on which it has been constructed (the eigenvectors of the time independent Hamitonian), it has the simple form

$$\boldsymbol{\sigma}_{eq} = \begin{bmatrix} p_0 - \Delta & 0 \\ 0 & p_0 + \Delta \end{bmatrix} \tag{35}$$

where p_0 is a population and 2Δ the population difference between the two energy levels. This density matrix can be decomposed as follows

$$\boldsymbol{\sigma}_{eq} = p_0 \begin{bmatrix} 1 & 0 \\ 0 & 1 \end{bmatrix} - \Delta \begin{bmatrix} 1 & 0 \\ 0 & 1 \end{bmatrix} + \begin{bmatrix} 0 & 0 \\ 0 & 2\Delta \end{bmatrix} \tag{36}$$

Because the unity matrix is invariant under any rotation, we can disregard the first two terms of (36) and write the density matrix at equilibrium as

$$\boldsymbol{\sigma}_{eq} = \begin{bmatrix} 0 & 0 \\ 0 & m_0 \end{bmatrix} \tag{37}$$

with $m_0 = 2\Delta$. (37) will be the working form of the equilibrium density matrix.

The last point concerns the phase of rf field. By calculations similar as those developed above, it can be shown that for a 90° phase shift, the rotation matrix becomes

$$\boldsymbol{R}_{90^\circ}(\beta/2) = \begin{bmatrix} \cos\frac{\beta}{2} & -\sin\frac{\beta}{2} \\ \sin\frac{\beta}{2} & \cos\frac{\beta}{2} \end{bmatrix} \tag{38}$$

whereas a 180° phase shift is simply accounted for by a sign change of β. Of course, the phase of the first pulse in any pulse sequence is totally arbitrary. The importance is the transmitter phase of the forthcoming pulses. Let ϕ the phase of a given rf pulse in any sequence; we have just to look at the transmitter polarization for rotation matrices such as (34) and (38), recognizing that it concerns only the off-diagonal elements. Since the "polarization vector" can be expressed as a linear combination of the two polarization vectors corresponding to the 0° and 90° phases, the (1,2) diagonal element can be written as $\cos\phi(i\sin\frac{\beta}{2}) + \sin\phi(\sin\frac{\beta}{2})$, the (2,1) element being the complex conjugate of the latter.

4. The powder average

The preceding developments concern a given microcrystallite, the orientation of which is defined by the angles θ and φ of Figure 1. As all orientations can be found in a powder sample, an average must be calculated for any observable quantity. As an example, let us consider the transition ω_x in the general case ($\eta \neq 0$) and let us assume that the system is at equilibrium prior to the rf pulse. As indicated by equation (33bis), the effect of a rf pulse on the density matrix can be accounted for by rotation matrices involving the angle $\beta/2$ with $\beta = \gamma B_1 \delta \sin\theta \cos\varphi$. Inserting (37) in (32) in place of σ^-, we obtain

$$\hat{\sigma}^+ = \hat{R}^\dagger_{0^\circ}(\beta/2)\hat{\sigma}_{eq}\hat{R}_{0^\circ}(\beta/2) = m_0 \begin{bmatrix} \sin^2\frac{\beta}{2} & \frac{i}{2}\sin\beta \\ -\frac{i}{2}\sin\beta & \cos^2\frac{\beta}{2} \end{bmatrix} \tag{39}$$

The observed signal arises from the off-diagonal elements of (39) and its value is obtained by the following average

$$S_{\omega_x} = \frac{1}{4\pi}\int_0^{\pi}\int_0^{2\pi} \sin(\gamma B_1 \delta \sin\theta \cos\varphi)\sin\theta\cos\varphi \sin\theta \, d\theta \, d\varphi \tag{40}$$

Because the signal is polarized along the x direction (see equation (28)), the coil receptivity must be accounted for by the factor $\sin\theta\cos\varphi$, while $\sin\theta \, d\theta \, d\varphi$ is the Jacobian. $\frac{1}{4\pi}$ is the normalization factor. Similar expressions prevail for the other two transitions

$$S_{\omega_y} = \frac{1}{4\pi}\int_0^{\pi}\int_0^{2\pi} \sin(\gamma B_1 \delta \sin\theta \sin\varphi)\sin\theta\sin\varphi \sin\theta \, d\theta \, d\varphi \tag{41}$$

$$S_{\omega_z} = \frac{1}{2}\int_0^{\pi} \sin(\gamma B_1 \delta \cos\theta)\cos\theta \sin\theta \, d\theta \tag{42}$$

Of course, because these expressions depend only on the orientation of the three axes with respect to the B_1 field direction and because the choice of the x, y, z axes is arbitrary (as far as these expressions are concerned), (40)–(42) yield the same result. In passing, it can be noted that (42) is the Bessel function of order 3/2.[9] This feature has been verified by numerical integration.

In the case of an efg tensor of axial symmetry ($\eta = 0$), the transition ω_z disappears while the two transitions ω_x and ω_y merge. Because of axial symmetry, the polarization of the resulting signal must be the same irrespective to any direction in the (x,y) plane; this means that it must not depend on φ. Because ω_x and ω_y are polarized along the x and y axes, respectively (see the factors $\sin\theta\cos\varphi$ and $\sin\theta\sin\varphi$ in (40) and (41)), the polarization of the resulting signal is obtained by combining geometrically these two polarizations, i.e.

$$\sqrt{x^2 + y^2} = \sqrt{\sin^2\theta\cos^2\varphi + \sin^2\theta\sin^2\varphi} = \sin\theta .$$

Hence the signal polarization depends only on θ, meaning that only an integration over θ has to be performed. β becomes $\gamma B_1 \delta \sin\theta$. The coil receptivity is accounted for by $\sin\theta$ and the additional $\sin\theta$ comes from the Jacobian. Altogether, one has

$$S_{\omega_0} = \frac{1}{2}\int_0^{\pi} \sin(\gamma B_1 \delta \sin\theta)\sin^2\theta \, d\theta \tag{43}$$

It can be noticed that this latter expression is no longer a Bessel function. We shall see the consequences of these mathematical formulations in the following section. For now, the merging of the two transitions ω_x and ω_y, 90° out of phase, can be appreciated from Figure 4.

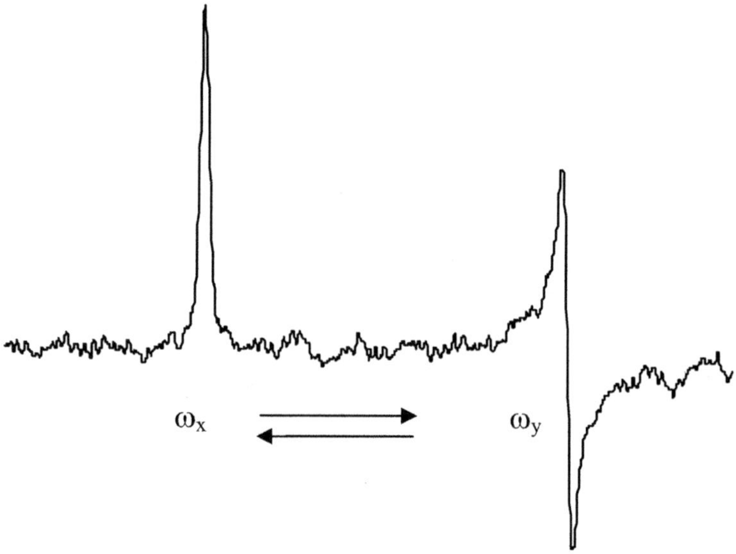

Figure 4. Sketch of a NQR spectrum when η tends to zero. The two lines corresponding to ω_x and ω_y are 90° out of phase and therefore do not combine in a simple additive process.

5. Nutation experiments

As a first application of the above theory, we shall consider nutation experiments which consist in measuring the NQR signal amplitude as a function of the rf pulse length. In fact; it is more instructive to look at their dependence upon the angle α that we define as

$$\alpha = \gamma B_1 \delta \tag{44}$$

This is just the flip angle in NMR (δ is, as before, the pulse duration) and α can be easily inserted in formulas (42)–(43). B_1 can be deduced from the results of a NMR experiment (see Figure 3). Because of powder average, we do not expect to observe the first maximum at $\alpha = 90°$ (as in NMR) but at a larger value since most microcrystallites experience a pulse characterized by the angle β (defined in (29)) which is necessarily smaller (or much smaller) than α. The shift will be more pronounced for the first zero which, contrary to NMR and for the same reasons as above, should occur at a value much larger than 180°. A typical experimental result is shown in Figure 5; the theoretical curve has been plotted according to equation (40). Other examples can be found in the literature.[7–11] Another consequence of the powder average is the damping of the nutation curve which is by no means due to relaxation phenomena during the pulse application. In order to have a more accurate overview of these peculiar nutation curves, we have plotted in Figure 6 the "universal" nutation curve for the case $\eta \neq 0$ (valid for the three transitions in accord with the considerations of the previous section) along with the "universal" nutation curve for the case $\eta = 0$. One may be puzzled by the fact that the curve corresponding to $\eta = 0$ is drastically different from the curve corresponding to $\eta \neq 0$. Again, this is due to the fact that when $\eta \to 0$, the two lines ω_x and ω_y merge in a single line ω_0. Although they present the same behavior as a function of α, they are 90° out of phase (see Figure 4), and cannot be simply added. They may be thought as behaving like *x* and *y*, respectively, and the single line behaves therefore as $\sqrt{x^2 + y^2}$ yielding a particular nutation curve. It can be seen that besides the differentiation between these two cases ($\eta \neq 0$ and $\eta = 0$), these nutation curves are not informative concerning the asymmetry parameter. This is in contrast with chlorine NQR for which a single line is observed (spin 3/2) and for which, fortunately, nutation experiments provide the asymmetry parameter.[12] We have still to look at the α value corresponding to the first maximum and to the first zero in the nutation curves. Due to the lack of a preferred direction (as the B_0 field direction in NMR), the angle β varies from one microcrystallite to the other, as does the pulse efficiency (see the previous section), and these α values cannot correspond to true 90° or 180° pulses. Consequently, they will be dubbed "pseudo 90°" and "pseudo 180°" pulses". However, it is important to know the value of the pseudo 90° pulse if one wants to proceed, for sensitivity reasons, with the signal maximum and to realize that the corresponding pulse length is not exactly half the one of the pseudo 180° pulse. As a matter of fact, the value of 119°5 for the pseudo 90° pulse has been put forward most of the time without reference to the asymmetry parameter[13–16] and the correct value of 101°5[6,17] for the case $\eta = 0$ was quoted only once.[11] A last comment can be made about the values of the pseudo 180° pulse (208° in the case $\eta = 0$ and 257° in the case $\eta \neq 0$). They appear to be quite different for these two cases and to be far from the inversion NMR flip angle (180°). As explained in a forthcoming section, this will be of some importance when dealing with pulse sequences aimed at measuring relaxation times.

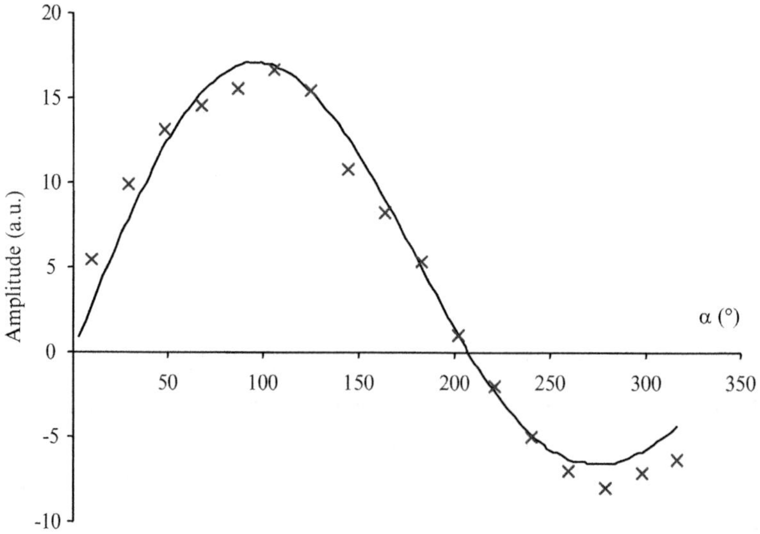

Figure 5. Experimental data (×) displayed with the corresponding theoretical nutation curve (ω_x transition of sodium nitrite at 4.64 MHz). The signal amplitude is plotted as a function of $\alpha = \gamma B_1 \delta$, δ being the pulse duration.

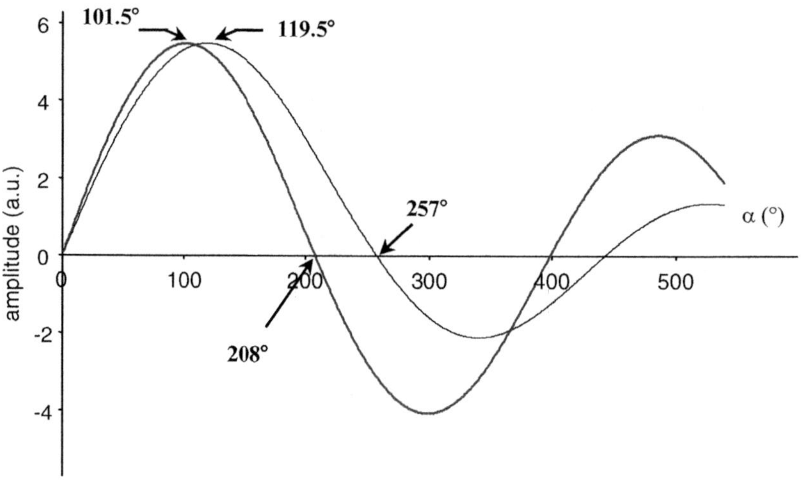

Figure 6. Universal nutation curves for the cases $\eta \neq 0$ (bold) and $\eta = 0$.

6. Signal averaging

As NQR has a poor sensitivity, data averaging is usually mandatory. If a steady state is assumed, an improvement of the S/N ratio by $\sqrt{n}$ occurs after accumulation of n transients. The first problem is to determine wether a steady state can really be reached in the case of a repetition time T smaller than five times the longitudinal relaxation time T_1 (which would allow the complete recovery to the thermodynamic equilibrium). With β having the same meaning as in equation (29) for a given microcrystallite, one has, from equations (32)–(33), for the signal occurring after m pulses[18]

$$s_m = s_0(1-e^{-T/T_1})\frac{1-e^{-mT/T_1}\cos^m\beta}{1-e^{-mT/T_1}\cos\beta} + s_0 e^{-T/T_1}\cos^m\beta \tag{45}$$

where s_0 is the unitary signal as obtained from an equilibrium situation. The quantity $e^{-mT/T_1}\cos^m\beta$ in equation (45) becomes smaller and smaller as m increases so that a steady state is actually reached when this quantity becomes negligible and one obtains for the corresponding signal

$$s_{st} = s_0\frac{1-e^{-T/T_1}}{1-\cos\beta e^{-T/T_1}} \tag{46}$$

Rather than looking for the number of transients necessary for attaining this steady state (this number of transients is larger when T/T_1 decreases; figures can be found in reference[18]), we shall focus here on the amplitude of the steady state signal. In the case $\eta \neq 0$, the powder averaging for the ω_x line is expressed as (with the same notations as in (40))

$$S_{\omega_x}^{st} = \frac{1}{4\pi}(1-e^{-T/T_1})\int_0^{\pi}\int_0^{2\pi}\frac{\sin(\gamma B_1\delta\sin\theta\cos\varphi)\sin^2\theta\cos\varphi}{1-\cos(\gamma B_1\delta\sin\theta\cos\varphi)e^{-T/T_1}}\,d\theta\,d\varphi \tag{47}$$

and similar expressions for ω_y and ω_z which anyway lead to the same result. In the case $\eta = 0$, still with the same notations as in (43), one has

$$S_{\omega_0}^{st} = \frac{1}{2}(1-e^{-T/T_1})\int_0^{\pi}\frac{\sin(\gamma B_1\delta\sin\theta)\sin^2\theta}{1-\cos(\gamma B_1\delta\sin\theta)e^{-T/T_1}}\,d\theta \tag{48}$$

Equations (47) and (48) have been plotted as a function of the ratio T/T_1 for different values of the pulse length δ. They qualitatively exhibit the same behavior and the curves shown in Figure 7 (calculated by numerical integration) correspond to the signal maximum, i.e. to pseudo 90° pulses. The unexpected feature of these plots is the absence of any maximum. In NMR, there is either an optimal repetition time T or an optimal pulse length (expressed as the flip angle α; see (44)); these two quantities are related by the well known Ernst angle α_{opt} such as $\cos\alpha_{opt} = e^{-T/T_1}$.[19] The monotonic decrease of the NQR signal intensity can be understood again by the lack of a preferential direction (or of a static magnetic field responsible for a dominant Zeeman term). In powder NQR, most microcrystallites do not experience a pulse of angle α but a pulse of much lower angle (β of equation (29)). Consequently they are less disturbed from their equilibrium state and therefore recover faster. It turns out that this property combined with the powder average of equations (47) and (48) lends to the monotonic decrease of Figure 7. This means that, in order to benefit from the maximum sensitivity, the shortest repetition rate (compatible with data acquisition) must be employed. This is illustrated and experimentally verified by the spectra of Figure 8 where, for a given measuring time, the maximum is effectively obtained for the shortest T/T_1 ratio.

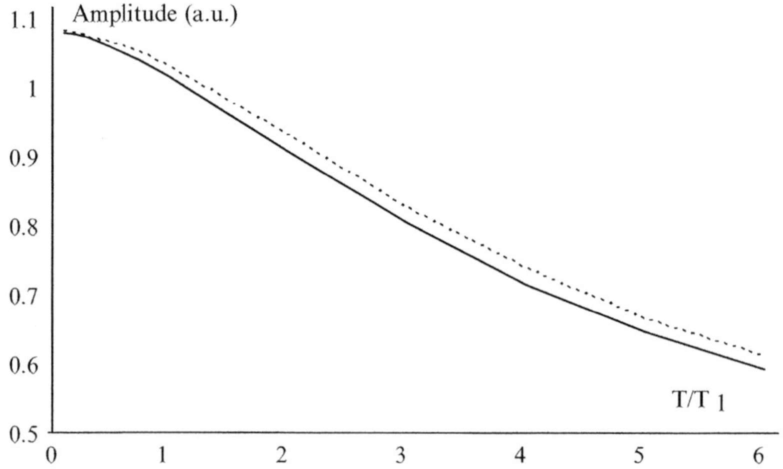

Figure 7. Plot of the steady state NQR signal (data averaging) as a function of T/T_1 (*T*: repetition time; T_1: longitudinal relaxation time). Solid line: $\eta = 0$; dotted line: $\eta \neq 0$.

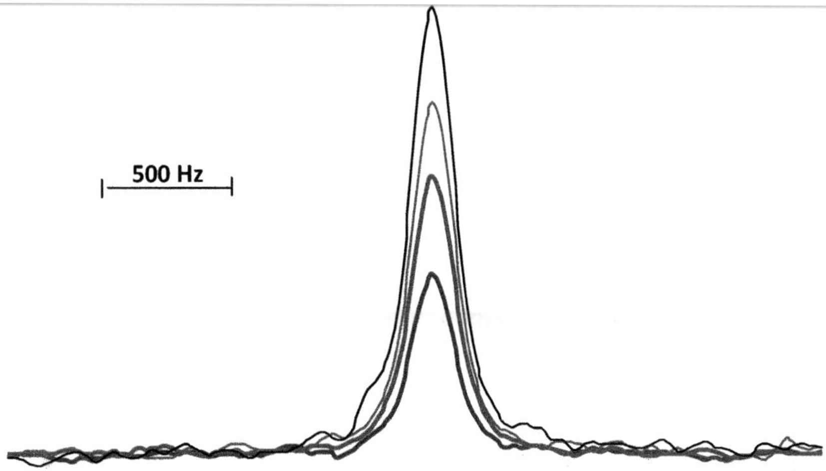

Figure 8. 4.6 MHz sodium nitrite spectra (resulting from data averaging in the time domain) obtained in 22 s for different values of the ratio T / T_1 : 1.3, 2, 3, 5 from top to bottom.

7. Two-pulse sequences for measuring relaxation times

The last application of the above theory will be devoted to the set-up of two-pulse sequences in view of determining either the longitudinal relaxation time T_1 (related to the populations, in other words to the diagonal elements of the density matrix) or to the transverse relaxation time T_2 (related to the coherences, in other words to the off-diagonal elements of the density matrix).[20]

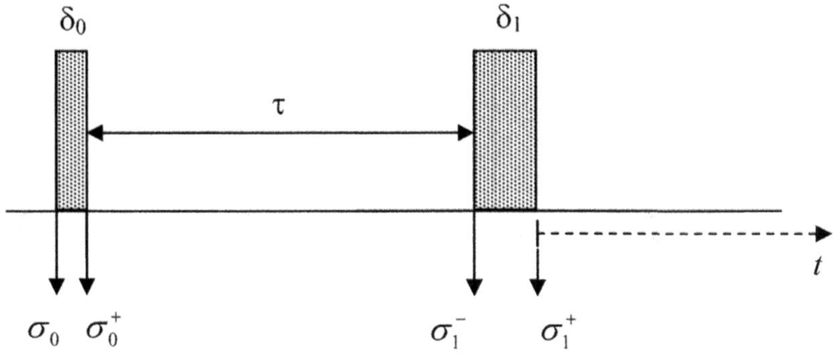

Figure 9. Sketch of a sequence made of two pulses of duration δ_0 and δ_1 with the notations used for the density matrix. The objective is to evaluate the NQR signal after the second pulse, i.e. as a function of the variable *t*.

The most general two-pulse sequence is sketched in Figure 9. As usual, we will consider the transition ω_x and use the angles β defined as (see (19) and (29))

$$\begin{aligned} \beta_0 &= \gamma B_1 \delta_0 \sin\theta \cos\varphi \\ \beta_1 &= \gamma B_1 \delta_1 \sin\theta \cos\varphi \end{aligned} \tag{49}$$

where δ_0 and δ_1 are the pulse durations. We recall that θ and φ are the polar angles defining the direction of the B_1 field in the principal axis system of a given microcrystallite. We start with the system at thermal equilibrium so that the working form of the density matrix is given by (37). The phase of the first rf pulse being arbitrary (see above), we can set it to 0° and thus use the rotation matrix given by (34). Applying (33bis), we obtain easily the density matrix at the end of the first pulse

$$\sigma_0^+ = m_0 \begin{pmatrix} \sin^2 \dfrac{\beta_0}{2} & \left(\dfrac{i}{2}\right) \sin\beta_0 \\ -\left(\dfrac{i}{2}\right) \sin\beta_0 & \cos^2 \dfrac{\beta_0}{2} \end{pmatrix} \tag{50}$$

We have now to account for relaxation phenomena during the interval τ separating the two pulses. Concerning the natural transverse relaxation, it is enough to multiply each diagonal element by $e^{\frac{-\tau}{T_2}}$ (T_2 being the transverse relaxation time). Extra broadening, which gives rise to what is generally denoted by T_2^* ($T_2^* < T_2$) and which is due to variations of the quadrupolar interaction, will be simulated by a factor $e^{i\omega^*\tau}$ in the element (1,2) and $e^{-i\omega^*\tau}$ in the element (2,1). ω^* can be thought as the difference between an extra resonance and the main resonance and, at the outcome, a distribution of ω^* should of course be considered. Longitudinal relaxation (defined by the relaxation time T_1) acts on diagonal elements for making them recover toward their equilibrium value (see (37)). Thus the element (1,1) has to be multiplied by $e^{\frac{-\tau}{T_1}}$ whereas the (2,2) element evolves as $m_0 + \left(m(0) - m_0\right) e^{\frac{-\tau}{T_1}}$, $m(0)$ being here equal to $m_0 \cos^2 \dfrac{\beta}{2}$ (see (50)). At this point, it is possible to write the density matrix just before the second pulse

$$\sigma_1^- = m_0 \begin{pmatrix} \sin^2 \dfrac{\beta_0}{2}\, e^{\frac{-\tau}{T_1}} & \left(\dfrac{i}{2}\right) \sin \beta_0\, e^{\frac{-\tau}{T_2}}\, e^{i\omega^* \tau} \\ -\left(\dfrac{i}{2}\right) \sin \beta_0\, e^{\frac{-\tau}{T_2}}\, e^{-i\omega^* \tau} & 1 - \sin^2 \dfrac{\beta_0}{2}\, e^{\frac{-\tau}{T_1}} \end{pmatrix} \tag{51}$$

The effect of the second pulse can be evaluated by applying again (33bis) with the appropriate rotation matrix, according to the phase of the second pulse. We give below the density matrix just after the second pulse for two phases: the same as for the first pulse, $(\beta_1)_{0^\circ}$; 90° out of phase with respect to the first pulse, $(\beta_1)_{90^\circ}$.

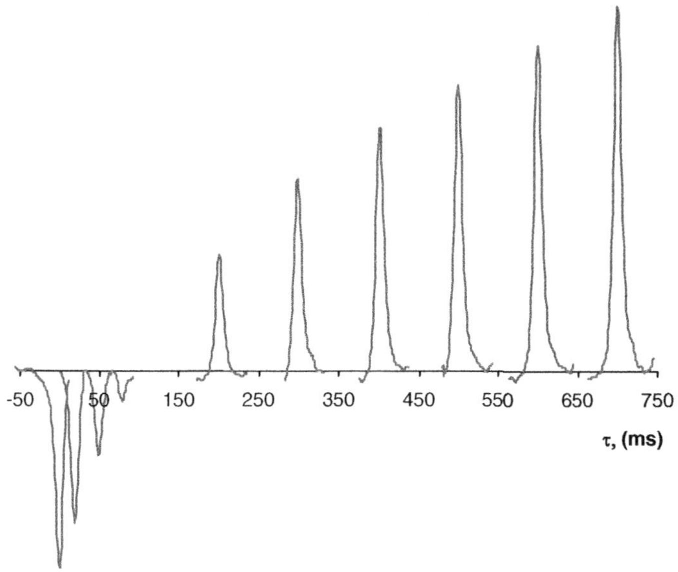

Figure 10. Example of a pseudo inversion-recovery experiment leading to the longitudinal relaxation time (ω_x transition of sodium nitrite at 4.64 MHz).

The NQR signal at $t = 0$ (see Figure 9) is represented by the off-diagonal elements of (52) or (53). The fist two terms depend on T_2 and ω^* while the last term depends on T_1.

$$\sigma_1^+[(\beta_1)_{0^\circ}] = m_0 \left(\begin{array}{l|l} \begin{array}{l} \sin^2\frac{\beta_0}{2}\cos^2\frac{\beta_1}{2}\, e^{\frac{-\tau}{T_1}} \\ +\left(1-\sin^2\frac{\beta_0}{2}\, e^{\frac{-\tau}{T_1}}\right)\sin^2\frac{\beta_1}{2} \\ +\frac{1}{2}\sin\beta_0 \sin\beta_1 \cos\omega^*\tau\, e^{\frac{-\tau}{T_2}} \end{array} & \begin{array}{l} \frac{i}{2}\sin\beta_0\cos^2\frac{\beta_1}{2}\, e^{\frac{-\tau}{T_2}}\, e^{i\omega^*\tau} \\ -\frac{i}{2}\sin\beta_0\sin^2\frac{\beta_1}{2}\, e^{\frac{-\tau}{T_2}}\, e^{-i\omega^*\tau} \\ +i\sin\beta_1\left(\frac{1}{2}-\sin^2\frac{\beta_0}{2}\, e^{\frac{-\tau}{T_1}}\right) \end{array} \\ \hline \begin{array}{l} \frac{i}{2}\sin\beta_0\sin^2\frac{\beta_1}{2}\, e^{\frac{-\tau}{T_2}}\, e^{i\omega^*\tau} \\ -\frac{i}{2}\sin\beta_0\cos^2\frac{\beta_1}{2}\, e^{\frac{-\tau}{T_2}}\, e^{-i\omega^*\tau} \\ -i\sin\beta_1\left(\frac{1}{2}-\sin^2\frac{\beta_0}{2}\, e^{\frac{-\tau}{T_1}}\right) \end{array} & \begin{array}{l} \sin^2\frac{\beta_0}{2}\sin^2\frac{\beta_1}{2}\, e^{\frac{-\tau}{T_1}} \\ +\left(1-\sin^2\frac{\beta_0}{2}\, e^{\frac{-\tau}{T_1}}\right)\cos^2\frac{\beta_1}{2} \\ -\frac{1}{2}\sin\beta_0 \sin\beta_1 \cos\omega^*\tau\, e^{\frac{-\tau}{T_2}} \end{array} \end{array}\right) \quad (52)$$

$$\sigma_1^+[(\beta_1)_{90^\circ}] = m_0 \left(\begin{array}{l|l} \begin{array}{l} \sin^2\frac{\beta_0}{2}\cos^2\frac{\beta_1}{2}\, e^{\frac{-\tau}{T_1}} \\ +\left(1-\sin^2\frac{\beta_0}{2}\, e^{\frac{-\tau}{T_1}}\right)\sin^2\frac{\beta_1}{2} \\ +\frac{1}{2}\sin\beta_0 \sin\beta_1 \sin\omega^*\tau\, e^{\frac{-\tau}{T_2}} \end{array} & \begin{array}{l} \frac{i}{2}\sin\beta_0\cos^2\frac{\beta_1}{2}\, e^{\frac{-\tau}{T_2}}\, e^{i\omega^*\tau} \\ +\frac{i}{2}\sin\beta_0\sin^2\frac{\beta_1}{2}\, e^{\frac{-\tau}{T_2}}\, e^{-i\omega^*\tau} \\ -\sin\beta_1\left(\frac{1}{2}-\sin^2\frac{\beta_0}{2}\, e^{\frac{-\tau}{T_1}}\right) \end{array} \\ \hline \begin{array}{l} -\frac{i}{2}\sin\beta_0\sin^2\frac{\beta_1}{2}\, e^{\frac{-\tau}{T_2}}\, e^{i\omega^*\tau} \\ -\frac{i}{2}\sin\beta_0\cos^2\frac{\beta_1}{2}\, e^{\frac{-\tau}{T_2}}\, e^{-i\omega^*\tau} \\ -\sin\beta_1\left(\frac{1}{2}-\sin^2\frac{\beta_0}{2}\, e^{\frac{-\tau}{T_1}}\right) \end{array} & \begin{array}{l} \sin^2\frac{\beta_0}{2}\sin^2\frac{\beta_1}{2}\, e^{\frac{-\tau}{T_1}} \\ +\left(1-\sin^2\frac{\beta_0}{2}\, e^{\frac{-\tau}{T_1}}\right)\cos^2\frac{\beta_1}{2} \\ -\frac{1}{2}\sin\beta_0 \sin\beta_1 \sin\omega^*\tau\, e^{\frac{-\tau}{T_2}} \end{array} \end{array}\right) \quad (53)$$

7.1. T_1 MEASUREMENTS

A very simple two-step phase cycling for the elimination of the first two terms in the off-diagonal elements of (52) and (53) is given below (other phase cyclings of the same type exist[20,21])

β_0	β_1	*Acq*
0°	0°	+
180°	0°	+

(The sign affected to the digitalized signals indicated in the column *Acq*).

Of course, the remaining term leads to a conventional NQR signal, the amplitude of which recovers according to T_1, as a function of τ. In fact for $\beta_0 = 180°$ and $\beta_1 = 90°$, this would be the conventional NMR inversion-recovery experiment. Here, we must perform a powder average over

$$\sin\beta_1\left(\frac{1}{2}-\sin^2\frac{\beta_0}{2}e^{\frac{-\tau}{T_1}}\right).$$

We have chosen an angle α_0 (see (44)) corresponding to the first zero of the nutation curve (pseudo 180°) pulse and an angle α_1 corresponding to the first maximum of the nutation curve (pseudo 90°). This experiment will therefore dubbed pseudo inversion-recovery. Theoretically (from the above form of the density matrix which involves a single longitudinal relaxation time) and experimentally (Figure 10), we arrive at a mono-exponential recovery which, however, does not start from the opposite of the equilibrium NQR signal (but rather from a smaller value, due to the powder average). It must be noted that, in principle, a bi-exponential behavior should be expected from the consideration of all probabilities transitions.[22]

7.2. T_2 MEASUREMENTS

We have now to remove the third term in the off-diagonal elements of (52) or (53). Again this can be accomplished by a very simple phase cycling:

β_0	β_1	*Acq*
0°	0°	+
0°	180°	+

We can notice that, contrary to the above sequence for measuring T_1 and as expected, the signal maximum would be obtained for $\beta_0 = 90°$ and $\beta_1 = 180°$ as in a NMR spin echo experiment. Of course, we will have, at the outcome, to perform a powder average. Taking into account this phase cycling, the off-diagonal element of the density matrix after the second pulse can be written as

$$(\sigma_1^+)_{12} = \frac{i}{2}\sin\beta_0 e^{-\tau/T_2}(\cos^2\frac{\beta_1}{2}e^{i\omega^*\tau} - \sin^2\frac{\beta_1}{2}e^{-i\omega^*\tau}) \tag{54}$$

Subsequently to this second pulse, the evolution of this element is simply obtained by multiplying (54) by $e^{-t/T_2}e^{i\omega^* t}$. Therefore, the first term appears as the continuation of the NQR signal produced by the first (β_0) pulse while the second one is responsible to the formation of an echo at $t=\tau$. As T_2 should be deduced from the decay of the echo amplitude, interferences with the first term can be a problem. Conversely, if the signal is processed in the magnitude mode, the conjunction of these two terms can yield a sensitivity improvement at the expense of a shift of the echo maximum (thus occurring at $t_{max} < \tau$), the latter nevertheless still decaying according to $e^{-(t+t_{max})/T_2}$.[20] Clearly, for accessing a clean measurement of T_2 values, we must get rid of the first term in equation (54). Referring to equation (53), this can be achieved through equation (53) by the following phase cycle

β_0	β_1	Acq
0°	90°	–
0°	270°	–

which yields for the density matrix off-diagonal element of interest

$$(\sigma_1^+)_{12} = -\frac{i}{2}\sin\beta_0 e^{-\tau/T_2}(\cos^2\frac{\beta_1}{2}e^{i\omega^*\tau} + \sin^2\frac{\beta_1}{2}e^{-i\omega^*\tau}) \tag{55}$$

Combining (54) and (55) in the following four-step phase cycling

β_0	β_1	Acq
0°	0°	+
0°	180°	+
0°	90°	–
0°	270°	–

we arrive at an expression for the element $(\sigma_1^+)_{12}$ which involves only the term leading to the echo formation

$$(\sigma_1^+)_{12} = -i\sin\beta_0 e^{-\tau/T_2}\sin^2\frac{\beta_1}{2}e^{-i\omega^*\tau} \tag{56}$$

An experimental example, obtained with this phase cycling, is shown in Figure 11. It can finally be noticed that the four-step phase cycling indicated above is just the well known EXORCYCLE procedure,[23] widely used in spin echo NMR.

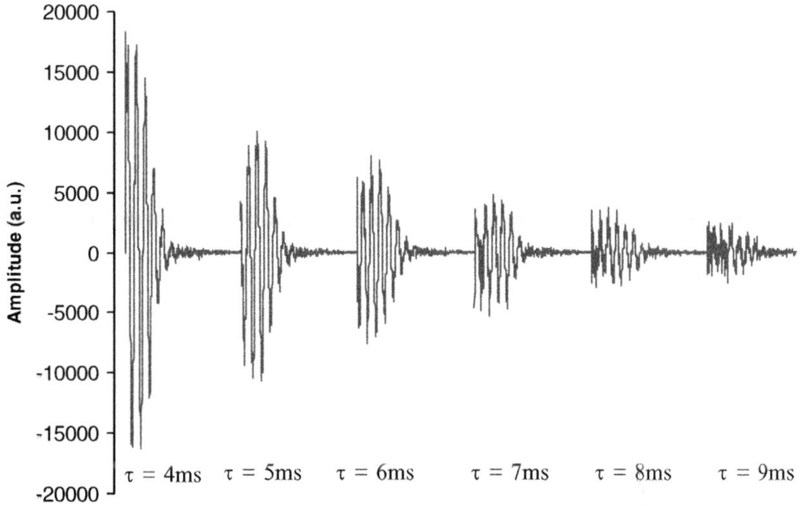

Figure 11. Example of a pseudo spin-echo experiment leading to the transverse relaxation time (ω_x transition of sodium nitrite at 4.64 MHz). Due to the insertion of an appropriate dead-time following the pseudo 180° pulse, the echo maximum is not visible for small τ values.

8. Conclusion

We have developed a tool, simple to use, by which one is able to predict spin 1 NQR dynamics. This tool only involves products of (2,2) matrices and can be adapted in a straightforward manner to each of the three NQR transitions. It can be thought as being similar to the fictitious spin ½ methodology but accounts for the specificities of NQR which arise from the lack of a preferential direction. In particular, the effect of a rf pulse cannot be deduced from the concept of flip angle (associated with the concept of rotating frame), widely used in NMR. We have shown that an efficient method was to apply (2,2) rotation matrices to the density matrix. Likewise, the concept of precession no longer holds in NQR and we have shown that, for each transition, the NQR signal is polarized along one direction of the electric field gradient tensor principal axis system (e.g. the x direction for the ω_x transition). By means of this methodology, we ware able to assess some particularities pertaining to nitrogen-14 NQR of powder samples: nutation experiments, data averaging, two-pulse sequences. We are confident that a lot of other experiments can be treated that way, especially those involving pulse trains and echo formation.

References

1. Smith JAS, Nuclear quadrupole resonance spectroscopy. General principles. J Chem Ed 48, 39–48 (1971)
2. Hiblot N, Cordier B, Ferrari M, Retournard A, Grandclaude D, Bedet J, Leclerc S, Canet D, A fully homemade ^{14}N quadrupole resonance spectrometer. C.R. Chimie 11, 568 (2008)
3. Slichter CP, Principles of Magnetic Resonance. Springer, Berlin (1978)
4. Bloom M, Burnell EE, de Lange CA, Density matrix methods in NMR. In: NMR of Ordered Liquids, edited by Burnell EE, de Lange CA, Kluwer, Dordrecht (2003)
5. Jeener J, Superoperators in magnetic resonance. Adv Magn Reson 10, 1–38 (1982)
6. Canet D, Merlat L, Cordier B, Grandclaude D, Retournard A, Ferrari M, ^{14}N pulsed nuclear quadrupolar resonance. 2. Effect of a single radio-frequency pulse in the general case. Molec Phys 104, 1391–1399 (2006)
7. Miller JB, Suits BN, Garroway AN, Circularly polarized RF magnetic fields for spin-1 NQR. J Magn Reson 151, 228–234 (2001)
8. Lee YK, Spin-1 Nuclear quadrupole resonance theory with comparisons to nuclear magnetic resonance. Conc Magn Reson 14, 155–171 (2002)
9. Prescott DW, Miller JB, Tourigny C, Sauer KL, Nuclear quadrupole resonance single-pulse spin echoes. J Magn Reson 194, 1–7 (2008)
10. Suits BH, Garroway AN, Miller JB, Sauer KL, ^{14}N magnetic resonance for materials detection in the field. Solid State Nucl Magn Reson 24, 123–136 (2003)
11. Youlin X, Chahohui Y, NQR spectroscopy of powder sample with I = 1 and 3/2 (I). Prog Natural Sci 6, 284–292 (1996)
12. Bloom M, Hahn EL, Herzog B, Free magnetic induction in nuclear quadrupole resonance. Phys Rev 97, 1699–1709 (1955)
13. Vega S, Theory of T_1 relaxation measurements in pure nuclear quadrupole resonance for spins I=1. J Chem Phys 61, 1093–1100 (1974)
14. Pratt JC, Watton A, Zero-field pulsed response and dipolar couplings in systems of spin I=1 nuclei. J Chem Soc Faraday Trans 2, 83, 2261–2269 (1987)
15. Krishnan MS, Temme FP, Sanctuary BC, Theory of pulses in nuclear quadrupole resonance spectroscopy. Molec Phys 78, 1385–1404 (1993)
16. Rudakov TN, Mikhaltsevitch, Flexman JH, Modified steady-state free precession pulse sequences for the detection of pure nuclear quadrupole resonance. Solid State Nucl Magn Reson 25, 94–98 (2004)
17. Cordier B, Grandclaude D, Retournard A, Merlat L, Canet D, ^{14}N pulsed nuclear quadrupole resonance 1 Nutation experiments in the case of an axially symmetric electric field gradient tensor. Molec Phys 103, 2593–2598 (2005)
18. Ferrari M, Hiblot N, Retournard A, Canet D, ^{14}N pulsed nuclear quadrupole resonance 3 Effect of a pulse train. Optimal conditions for data averaging. Molec Phys 105, 3005–3012 (2007)
19. Ernst RR, Sensitivity Enhancement in Magnetic Resonance. Adv Magn Reson 2, 1–135 (1966)
20. Ferrari M, Canet D, ^{14}N pulsed nuclear quadrupole resonance 4 Two-pulse sequences for the determination of T_1 and T_2 relaxation times. Molec Phys (submitted) (2008)
21. Rudakov TN, Mikhaltseitch VT, Spin locking spin echo in nitrogen-14 quadrupolar spin-system with axially symmetric electric field gradient tensor. Chem Phys Lett 363, 1–6 (2002)

22. Petersen G, Bray PJ, ^{14}N nuclear quadrupole resonance and relaxation measurements of sodium nitrite. J Chem Phys 64, 522–530 (1976)
23. Bodenhausen G, Freeman R, Turner DL, Suppression of artifacts in two-dimensional J spectroscopy. J Magn Reson 27, 511–514 (1977)

^{14}N NQR DETECTION OF EXPLOSIVES WITH HYBRID SENSORS

MYRIAM PANNETIER-LECOEUR, CLAUDE FERMON*,
HADRIEN DYVORNE, GREGORY CANNIES,
GÉRALD LE GOFF
Commissariat à l'Energie Atomique, DSM/IRAMIS/SPEC
91191 Gif-sur Yvette Cedex, France

Abstract Nuclear Quadrupolar Resonance of nitrogen allows selective detection of solid explosives containing Nitrogen. That technique has the advantage to be highly selective but its sensitivity is poor, in particular for some products presenting long relaxation time like Tri-Nitro-Toluene (TNT). The limitation is given by the noise of the detecting system, mainly a resonant coil tuned at the detection frequency. Here we present a new approach with mixed sensors having a potential sensitivity better than resonant coils for radio frequency detection and a wide band detection allowing the detection of several explosives with a rather simple system.

Keywords: NQR, explosive detection, spin electronics

1. Introduction

Nuclear Quadrupolar Resonance (NQR) is a solid state spectroscopic technique that applies to nuclei of spin $I \geq 1$ in a non cubic environment. Nuclei with an electric quadrupolar moment, such as ^{14}N, ^{35}Cl and ^{63}Cu, have their energies split by an electric field gradient, created by the electronic bonds in the local environment. The resonance frequencies are compound specific and allow detection of various kinds of objects (explosives, drugs, oil...) on the basis of their chemical composition, which is an important advantage over other methods that detect extrinsic properties such as metallic mine detectors. The technique requires that the substance is in a solid state and that the environment of the nuclei of interest is non cubic, otherwise the electric field gradient is zero because of symmetry consideration in the cubic case, or because of motion averaging in the case of a liquid substance.

* E-mail: claude.fermon@cea.fr

J. Fraissard and O. Lapina (eds.), *Explosives Detection Using Magnetic and Nuclear Resonance Techniques*,

Most solid explosives contain an important proportion of nitrogen which allows their detection by NQR. The resonance frequencies lie in the 500 kHz–5 MHz range.

The simplest NQR detection experiment consists on an excitation produced by a radiofrequency pulse which induces a rotation of the spin orientation of the nuclei having a resonance at the corresponding frequency. Then the excited nuclei process and relax to their equilibrium. In the classical scheme of detection, the precession of the nuclear magnetization at the resonance frequency after the pulse is detected through a voltage induced in a coil by the Faraday effect. Usually the same coil is used for excitation and detection.

The important parameters that govern the efficiency of detection for a given nucleus are its concentration, the resonance frequency and the relaxation times T_1, T_2* and T_2.

T_1 is the longitudinal relaxation time and it characterizes the return to equilibrium of the system.

T_2* is the loss of coherence of the spin precession including extrinsic factors such as inhomogeneities (temperature, resonance frequency…) inside the sample. It can be also seen as the time inverse of the resonance line width.

$T_2 > T_2^*$ is the intrinsic transversal relaxation time i.e. the loss of coherence of the spins.

For an optimal detection of a substance, the knowledge of the various relaxation times is important. It allows optimizing the simple detection sequence described above, but it also allows improving more sophisticated sequences. Depending on the NQR characteristics, various elaborate sequences have been proposed: the common point of these sequences is the use of multi-pulse excitations which permit to refocus the magnetization and to increase the effective observation time above the value T_2* which pertains to the simple one pulse excitation sequence.

NQR detection of explosives has until now be widely studied.[1–5] However, NQR systems have until now not really been implemented in real conditions due to the fact that the sensitivity is still rather low, especially for TNT which presents a low density of Nitrogen and long relaxation times. The limitation in sensitivity is given by the thermal noise of the detecting coil. For that reason, alternative very low noise sensors as SQUIDs[6] or atomic magnetometers[7] are explored. Here we present a third alternative solution based on superconducting-magnetoresistive hybrid sensors. These sensors developed in our laboratory[8] are achieving sensitivities in the femto-tesla range.

2. Superconducting-magneto-resistive hybrid sensors

2.1. PRINCIPLE

Hybrid sensor is made of a wide superconducting loop of typically few millimeters to several centimeters in diameter, containing a micron-sized constriction. A micron-sized magnetoresistive sensor, which is a field sensor, is placed on the top or below the constriction. When an external field is applied on the superconducting loop, a supercurrent is created in the loop to prevent the entrance of the magnetic flux in the loop. This supercurrent is flowing through the constriction where the current density becomes very high, creating locally a high magnetic field. This local high magnetic field is measured by the magneto-resistive sensor. The ratio between the field seen by the loop and the field experienced by the sensor can be several hundreds for a centimeter size loop.

In our present systems, we are using GMR sensors which present a typical MR ratio of 10–15% at 77 K with a yoke shape. Figure 1 gives an example of a mixed sensor with four GMRs are placed in a bridge configuration on a 5 μm wide constriction.

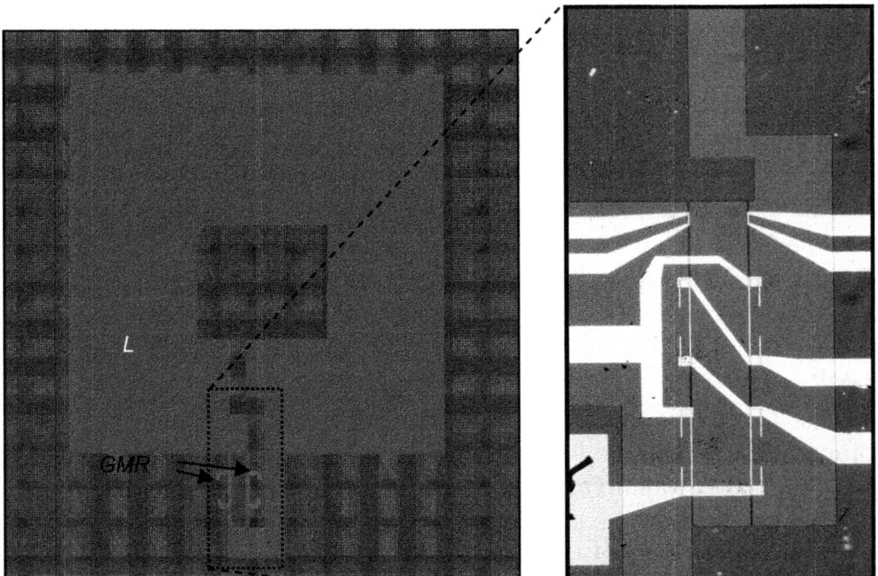

Figure 1. View of a hybrid sensor with four GMRs mounted in bridge configuration. Left: Global schematic view, Right: detailed view of the GMR bridge.

The superconducting flux-field transformer can be built in low-Tc or High-Tc material. Both mixed sensors have been tested for low frequency detection. High-Tc mixed sensors are more sensitive because the magnetoresistive sensor can accept a larger bias current and then can deliver a higher voltage variation for a field change. However, YBaCuO based mixed sensors are more difficult and expensive to produce.

For niobium based mixed sensors (low-Tc devices), the GMR stack is first grown on silicon or glass. The patterning of the sensor and its connections is performed. Then an insulating layer as Si_3N_4 is deposited and polished. At the end an evaporation of high purity niobium film is performed with a lift off process.

For YBCO based sensors (high Tc devices), it is necessary to grow the superconductor first in order to obtain good superconducting properties. Then an insulating layer is deposited and polished. The GMR stack can then be deposited with good magnetic properties. The etching of the complete system is then done with three steps: an etching of the complete stack with the superconducting loop shape, the selective etching of the GMR stack with the GMR sensor shape and passivation of the edges of the latter. Then contacts of the GMR sensors are processed.

2.2. NOISE PROPERTIES AND SENSITIVITY

The noise of the hybrid sensor is given by the thermal noise of the magnetoresistance, i.e. $\sqrt{4kTR}$ where R is the value of the resistance for frequencies larger than 10 kHz. Below this frequency, a 1/f noise appears which reduces the sensitivity of these devices. For NQR experiments, this 1/f noise is not a limitation. For a 1 kΩ resistance bridge, the noise at 77 K is 2 nV/√Hz and at 4 K is 0.47 nV/√Hz.

The GMR sensor have a resistance variation of about 3%/mT. The bias current injected is typically 5 mA for a 1 kΩ resistance. Hence the voltage variation is 150 mV/mT and in a bridge configuration 300 mV/mT. This gives a thermal noise of 6 pT/√Hz at 77 K and 1.5 pT/√Hz at 4 K in the GMR sensor.

The gain of the flux-field transformer can be modeled[9] and it is roughly proportional to the ratio between the size of the superconducting loop and the constriction width. The sensor used for the experiments has a 2 cm loop and a 5 µm constriction, leading to a gain of 700. This give a thermal noise of 5 fT/√Hz at 77 K and 1.4 fT/√Hz at 4 K with 5 mA bias current. A 6 cm loop and 5 µm constriction would have a gain of 2,000 and a 6 cm loop and 2 µm constriction, 5,000.

For comparison, the noise level of a single turn surface resonant coil has been calculated in reference.[10] It corresponds to a field detectivity of 1.5 fT/√Hz.

Hence at 800 kHz, present sensitivities of hybrid sensor detection and resonant coil detection are comparable and with the next generation of sensors, we plan to have a gain of 5–8 in terms of sensitivity.

3. NQR Experimental setup

In order to test the response of hybrid sensors, we have mounted a NQR system able to detect ^{14}N resonances in the range of 100 kHz to 10 MHz. The spectrometer has a 0.1° phase accuracy, 100 dB amplitude control and 10 ns pulse resolution. A double frequency change associated to very low noise preamplifiers in the reception part insures an optimized signal to noise ratio of the electronics. Then the effective noise is given mainly by the thermal noise of the resonant coil or the hybrid sensor + preamplifier. The two preamplifier chains for the resonant coil and the hybrid sensor are simultaneously operational and the acquisition can be switched between each channel. The coils used in the experiments have a solenoid shape (radius = 6 cm, length = 1 cm), and are well-fitted to the volume of the explosive tested, about 80 g. The working distance is comprised between 5 and 10 cm. The hybrid sensor has an additional distance of 2 cm from the explosives. The pulse power was at maximum 500 W. The pulse length was optimized for the best signal (Figure 2).

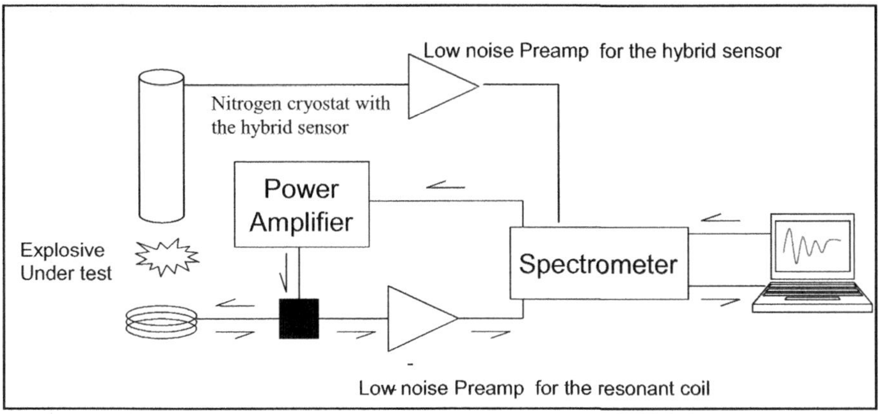

Figure 2. Schematics of the experimental setup. A surface coil is connected to a home-made spectrometer controlled by a PC. It acts as the pulse emitter and a RF receiver. A hybrid sensor is placed on top of the explosive and connected to its own preamplifier. The probe and the preamplifiers are separated from the remaining part by several meters.

4. Hybrid sensor response to an RF pulse

In the first experiments, we have tested the response of a Hybrid sensor to strong RF pulses.[11] The pulse has two effects: a heating of the full sensor and a magnetic saturation of the GMr element. Nb and YBaCuO based mixed sensors have then a very different behaviour. For Nb based mixed sensor, the heating of the constriction is sufficiently high to reach the critical temperature transition of the Nb and then the mixed sensor is reset. As soon as the pulse is finished the constriction cool down and the mixed sensor is ready to measure, back to the zero super current state. The sensor is fully operational after a time shorter than 0.3 μs. We have performed a simulation of the cooling of the constriction taking into account the heat capacity and the various thermal conductivities. We have obtained a cooling characteristic time of 50 ns. For YBaCuO based mixed sensors, the critical temperature is not reached. The super current oscillates at the RF frequency from minus the critical current of the constriction to plus the critical current of the constriction and when the pulse is finished, the mixed sensor is in state with a super current equal to the value at the end of the pulse. This case is less favorable because depending of the value of this super current, the GMR resistance can be different inducing a small dephasing of the NQR signal but this effect is rather negligible. Figure 3 shows the response of a YBCO based mixed sensor after a RF pulse compared to the response of a resonant coil at 400 kHz on a NMR proton signal.

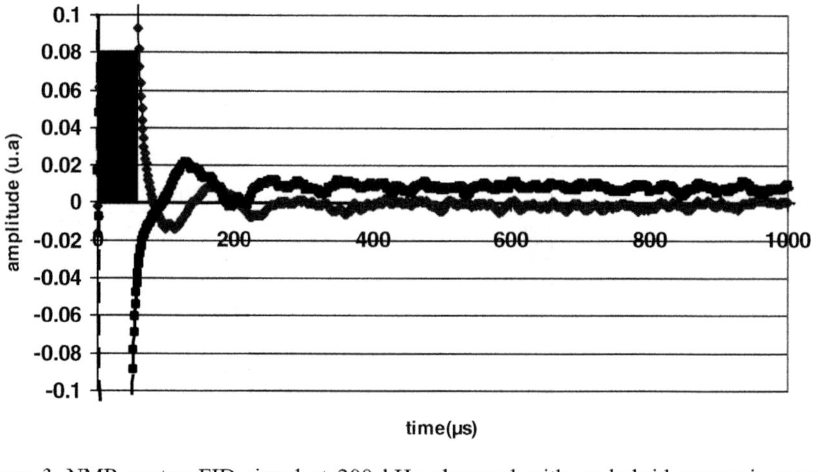

Figure 3. NMR proton FID signal at 200 kHz observed with an hybrid sensor in a very inhomogeneous field. The pulse width is 60 μs. We see that the NMR signal can be seen just after the RF pulse. This experiment highlights the non resonant character of mixed sensors detection which allows detecting broad or multiple resonance signals. Phase and quadrature signals are traced.

5. Experimental results

We have performed experiments on various explosives and we present here results obtained on RDX and TNT. The measured quantities were 100 g and the temperature was 5°C for the RDX and 20°C for the TNT.

RDX has a rather short T_1 of ~18 ms at room temperature and then can be easily and rapidly detected in a single pulse experiment. Figure 4 shows the spectrum around the 3.4 MHz line obtained with a Nb based hybrid sensor presenting a noise figure at 4 K of 10 fT/√Hz. Acquisition time was 10 s. This time can be easily shortened by using a SORC sequence.[12] We can observe that the line width with the hybrid sensor is small than the line width. This is presumably due to the coupling of the nuclear spin system with the resonant coil. This damping effect is well known at high fields with protons.

TNT is harder to detect because the relaxation time is longer (6 s), the resonance frequency is lower (0.8 MHz) and the ^{14}N density is smaller. TNT is the present limitation of existing systems as it requires minutes to be detected. Figure 5 shows the signal observed at 0.843 MHz with the coil and with the hybrid sensor. Here also the signal has been detected by a single pulse for more clarity but more sophisticated sequences like SLSE have been implemented allowing a faster discrimination of the presence of an explosive.

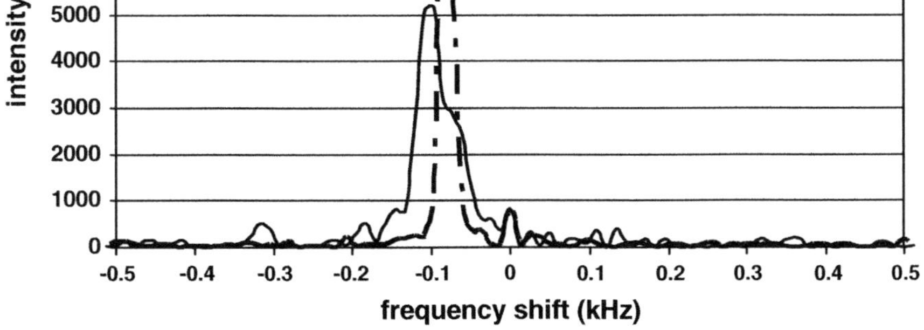

Figure 4. RDX signal detected at 3.4 MHz with a single pulse sequence with the resonant coil (continuous line) and the hybrid sensor (dotted line). This signal has been averaged during 10 s. The hybrid sensor used is a Nb-based sensor at 4 K. The signal is given in arbitrary units as function of frequency offset in kilohertz.

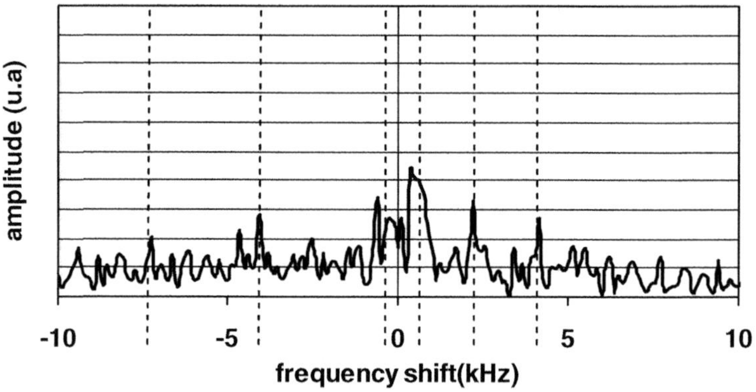

Figure 5. TNT spectrum observed with a single pulse sequence at 842 kHz with a hybrid YBCO sensor at 77 K with a noise of 8 fT/√Hz (including the preamplifier noise). Both phases monoclinic and orthorhombic are present. The lowest peak is at 834 kHz (monoclinic phase) and the highest seen is at 846 kHz. Other peaks have been detected at 854 and 868 kHz.

6. Conclusion and perspectives

Hybrid sensors exhibit a promising way to achieve high sensitivity detection of solid explosives by NQR. Present sensitivity of hybrid sensors give signal to noise comparable to optimized resonant coils but with a non resonant system allowing simultaneous detection of several explosives. Next generation of hybrid sensors with larger size loops and smaller constriction will give us a sensitivity which could be five to ten times better than existing systems.

References

1. ML Buess, AN Garroway, JB Miller: "Detection of Explosives by Nuclear Quadrupole Resonance", US Patent no 5206592 (April 1993)
2. JAS Smith: "Method and Apparatus for Detecting Certain Compounds", US Patent no 4887034 (December 1989)
3. J Barras, MJ Gaskell, N Hunt, RI Jenkinson, KR Mann, DAG Pedder, GN Shilstone and JAS Smith:"Detection of ammonium nitrate inside vehicles by nuclear quadrupole resonance", Appl Magn Reson 25 (3–4), 411–437 (2004)
4. VS Grechishkin: "NQR device for detecting plastic explosives, mines and drugs", Appl. Phys. A55, 505–507 (1992)
5. QRsciences, Port Melbourne, Australia; http://www.qrsciences.com
6. DF He, M Tachiki, H Itozaki, "N-14 NQR using a high-T_c rf SQUID with a normal metal transformer", Supercond Sci Technol 21 (1), 015023 (2008)
7. SK Lee, KL Sauer, SJ Seltzer, O Alem, MV Romalis, "Subfemtotesla radio-frequency atomic magnetometer for detection of nuclear quadrupole resonance" Appl Phys Lett 89, 214106 (2006)
8. M Pannetier, C Fermon, G Le Goff, J Simola, E Kerr, "Femtoteslta magnetic field measurement with magnetoresistive sensors", Science 304, 1648–1650 (2004)

9. M Pannetier, C Fermon, G Le Goff, E Kerr, "Ultra-sensitive mixed sensors – Design and performance", Sensors Actuators, A-Phys 129, 247–250 (2006)
10. YS Greenberg, "Application of superconducting quantum interference devices to nuclear magnetic resonance", Rev. Mod. Phys 70, 175 (1998)
11. M Pannetier-Lecoeur, C Fermon, N Biziere, J Scola, AJ Walliang, "RF response of superconducting-GMR mixed sensors: application to NQR" IEEE Trans Appl Supercond 17, 598–601 (2007)
12. SM Klainer, TB Hirschfeld, RA Marino, in "Fourier, Hadamard and Hilbert transform in chemistry" (New York, Plenum, 1982) p. 147.

POLARIZATION ENHANCED NQR DETECTION AT LOW FREQUENCIES

JANKO LUŽNIK[1*], JANEZ PIRNAT[1], VOJKO JAZBINŠEK[1], ZVONKO TRONTELJ[1], TOMAŽ APIH[2], ALAN GREGOROVIČ[2], ROBERT BLINC[2], JANEZ SELIGER[3]
[1]*Institute of Mathematics, Physics and Mechanics, Ljubljana, Slovenia*
[2]*J. Stefan Institute, Ljubljana, Slovenia*
[3]*Faculty of Mathematics and Physics, University of Ljubljana, Slovenia*

Abstract In this contribution we present our current research on polarization enhanced nuclear quadrupole resonance (NQR) detection at low frequencies with the emphasis on ^{14}N NQR trinitrotoluene (TNT) detection at room temperature. Combination of proton–nitrogen level crossing polarization transfer and pulsed spin-locking sequence makes ^{14}N NQR in TNT fast and sensitive enough to be used in routine detection of explosives. Enhancement factors for ^{14}N NQR lines in TNT were calculated and compared with experimental values. Good agreement between measured and calculated signal enhancement factors was observed. ^{14}N NQR signals in a 15 g trinitrotoluene sample of predominantly monoclinic modification were measured in 15 s in different polarization magnetic fields. The conditions for optimal polarization enhancement were determined.

Keywords: ^{14}N NQR, polarization enhanced nuclear quadrupole resonance, TNT

1. Introduction

Nuclear Quadrupole Resonance (NQR), with its ability to identify specific molecules, is potentially a powerful method in solid state physics, chemistry and pharmacy. In the last 10–15 years, several attempts have been made to improve the detection of military explosives, improvised explosive devices (IED) and other illicit materials – mainly narcotics – by ^{14}N NQR.[1–9]

* E-mail: janko.luznik@imfm.uni-lj.si

Unfortunately, many of these substances have ^{14}N NQR frequencies in the low frequency domain between 100 and 1,000 kHz, hence a rather low signal to noise (s/n) ratio. Therefore, the measuring times for the required signal averaging can be hours and they are thus too long for practical applications. With a combination of proton polarization transfer to nitrogen nuclei and multi-pulse spin-locking sequences the measuring time can be significantly reduced to an acceptable level of the order of 10 s, provided the proton and the nitrogen spin-lattice relaxation times (T_1) are suitable.

There are two ways to increase the s/n ratio by proton–nitrogen level crossing polarization transfer: (a) proton–nitrogen nuclear double resonance techniques[10–13] using changes in the proton NMR signal as an indirect indication of the ^{14}N NQR transitions; and (b) direct ^{14}N NQR detection where the signal is enhanced by proton polarization transfer via proton–nitrogen level crossing in a time variable magnetic field.[14–19] The first technique requires a homogeneous applied external magnetic field and is therefore not convenient for work in the field. For direct ^{14}N NQR detection there is no requirement for homogeneity of the polarizing magnetic field, because the signal enhancement depends on the average magnetic field, while pure NQR detection is performed in a zero magnetic field.

The aim of this study was to find a combination of polarization enhanced (PE) ^{14}N NQR detection together with a suitable pulse sequence in pulsed NQR spectroscopy to obtain low frequency ^{14}N NQR signals in TNT at room temperature in 10–20 s.

2. Theoretical

In NQR (NMR) spectroscopy signal intensities are limited by the difference in the occupation number ΔN between two energy levels.[20–23] In thermal equilibrium the occupation of different energy levels follows the Boltzmann distribution function

$$N \propto \exp\left(-\frac{E}{k_B T}\right) \tag{1}$$

Knowing that in rf spectroscopy the high temperature limit ($k_BT >> E$) is applicable, the population difference between two energy levels is:

$$\Delta N_Q \propto \frac{\Delta E}{k_B T} = \frac{h\nu_Q}{k_B T} \tag{2}$$

The population difference is thus proportional to the NQR transition frequency:

$$\Delta N_Q \propto \nu_Q$$

This population difference can be increased by applying cross polarization transfer. When the measured sample contains also protons, their spins can be polarised in a strong magnetic field. Consequently the population difference for protons is:

$$\Delta N_H \propto \gamma B \propto h\nu_H \; ; \qquad \Delta N_H \propto \nu_H \tag{3}$$

For strong polarizing magnetic fields where $\nu_H >> \nu_Q$ also $\Delta N_H >> \Delta N_Q$ (ν_H is the proton NMR transition frequency in the polarizing magnetic field B). After proton polarization is complete – the polarization time should be longer than T_1 of the protons – the polarizing magnetic field is adiabatically removed. This field can be obtained either with an electromagnet or with a suitable permanent magnet. During demagnetization the proton NMR transition frequency goes to zero, but the energy levels population difference for protons remain temporarily unbalanced and available for improvement of the ^{14}N NQR signal in a time short compared to the proton spin-lattice relaxation time T_1. In the adiabatic demagnetization process the proton NMR and the nitrogen NQR transition frequencies become equal (level crossing), and the relative population differences ΔN_H for protons and ΔN_Q for nitrogens equalize due to the proton–nitrogen dipolar coupling. The population differences ΔN_H and ΔN_Q are proportionally distributed between protons and nitrogens. A new average population difference can be defined as:

$$\Delta N = \frac{N_H \Delta N_H + N_N \Delta N_Q}{N_H + N_N} = (1-\rho)\Delta N_H + \rho \Delta N_Q \tag{4}$$

where N_H and N_N are the numbers of protons and nitrogen nuclei in the molecule of the sample and

$$\rho = \frac{N_N}{N_H + N_N} \tag{5}$$

Using a pulse NQR spectrometer in a zero magnetic field and triggering it immediately after demagnetization is completed, under optimal conditions this enhancement factor for the quadrupole signal intensity can be obtained

$$EF = \frac{\Delta N}{\Delta N_Q} = (1-\rho)\frac{\Delta N_H}{\Delta N_Q} + \rho = (1-\rho)\frac{\nu_H}{\nu_Q} + \rho \tag{6}$$

The case spin I = 1 ^{14}N nuclei with $\eta \neq 0$: The allowed transitions and energy levels for this case are shown in Figure 1. The corresponding frequencies are equal to[22]:

$$\nu^+ = \frac{3e^2qQ}{4h}\left(1+\frac{\eta}{3}\right), \quad \nu^- = \frac{3e^2qQ}{4h}\left(1-\frac{\eta}{3}\right) \quad \text{and} \quad \nu^0 = \frac{e^2qQ}{2h}\eta \qquad (7)$$

Here, e^2qQ/h is the quadrupole coupling constant where Q is the nuclear electric quadrupole moment, $q = q_{zz}$ is the maximal component of the electric field gradient tensor, e is the electron charge, h the Planck constant, and $\eta = (q_{xx} - q_{yy})/q$ is the asymmetry parameter.

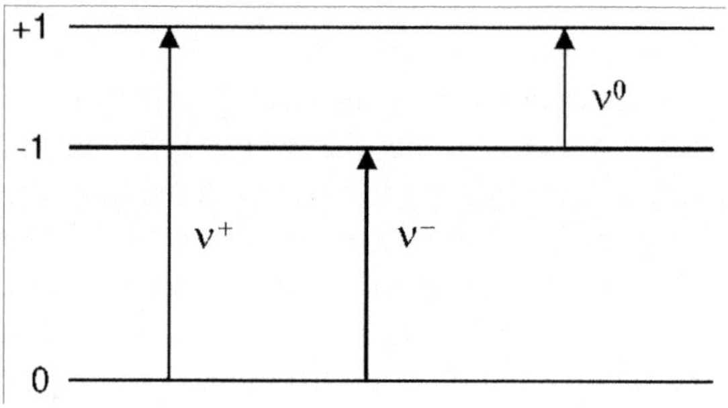

Figure 1. The quadrupole energy levels and allowed transitions for a spin 1 nucleus.

In the case I = 1 the polarization transfer is a three step process[18] with successive crossings of energy levels: ν^+ (step I), ν^- (step II) and ν^0 (step III) as illustrated in Figure 2. During this process, the "step-by-step" alterations of relative population numbers are given as the differences from the uniform distribution (one third of the nuclei in each energy level). Throughout the following is valid:

$$N_i^0 + N_i^- + N_i^+ = N \Rightarrow \Delta N_i^0 + \Delta N_i^- + \Delta N_i^+ = 0 \qquad (8)$$

The initial condition is:

$$\Delta N_0^0 - \Delta N_0^+ = Nh\nu^+/3k_BT = k\nu^+ \Delta N_0^0 - \Delta N_0^- = Nh\nu^-/3k_BT = k\nu^- \qquad (9)$$

N is the total number of nitrogen nuclei and $k = Nh/3k_BT$.

In the calculation also alterations in the proton reservoir after each level crossing are considered. The following results for α, β and γ, introduced in Figure 2 and describing the gradual population changes are obtained:

$$\begin{aligned}
\alpha &= (\nu_H - \nu^+)(1-\rho) \\
\beta &= \frac{1}{2}(\nu_H - \nu^+)(1-\rho)^2 + (1-\rho)\nu^0 \\
\gamma &= \frac{3}{4}(\nu_H - \nu^+)(1-\rho)^3 + (1-\rho)(\nu^- - \nu^0) + \frac{3}{2}(1-\rho)^2 \nu^0
\end{aligned} \tag{10}$$

The enhancement factors for ν^+, ν^- and ν^0 of ^{14}N NQR signal intensities are the ratios of the final and initial population differences:

$$EF^+ = \Delta N_3^+/\Delta N_0^+ = (\nu^+ + \alpha + \beta/2 + \gamma/2)/\nu^+$$

$$EF^- = \Delta N_3^-/\Delta N_0^- = (\nu^- + \alpha/2 + \beta - \gamma/2)/\nu^-$$

$$EF^0 = \Delta N_3^0/\Delta N_0^0 = (\nu^0 + \alpha/2 - \beta/2 + \gamma)/\nu^0$$

$$\begin{aligned}
EF^+ &= \frac{1+9\rho-5\rho^2+3\rho^3}{8} + \frac{13-21\rho+11\rho^2-3\rho^3}{8\nu^+}\nu_H + \frac{-2+8\rho-6\rho^2}{8\nu^+}\nu^- \\
EF^- &= \frac{-2+4\rho+6\rho^2}{8} + \frac{5-3\rho-5\rho^2+3\rho^3}{8\nu^-}\nu_H + \frac{1+3\rho-\rho^2-3\rho^3}{8\nu^-}\nu^+ \\
EF^0 &= \frac{-4\rho+12\rho^2}{8} + \frac{8-18\rho+16\rho^2-6\rho^3}{8\nu^0}\nu_H + \frac{10\rho-16\rho^2}{8\nu^0}\nu^+
\end{aligned} \tag{11}$$

In the calculation of enhancement factors an ideal adiabatic demagnetization and 100% cross-polarization transfer effectiveness were assumed. In each level crossing only two of the three levels are involved, hence the effective number of nitrogen nuclei is only two third of the total number and the value for ρ is:

$$\rho = \frac{\frac{2}{3}N_N}{N_H + \frac{2}{3}N_N} \tag{12}$$

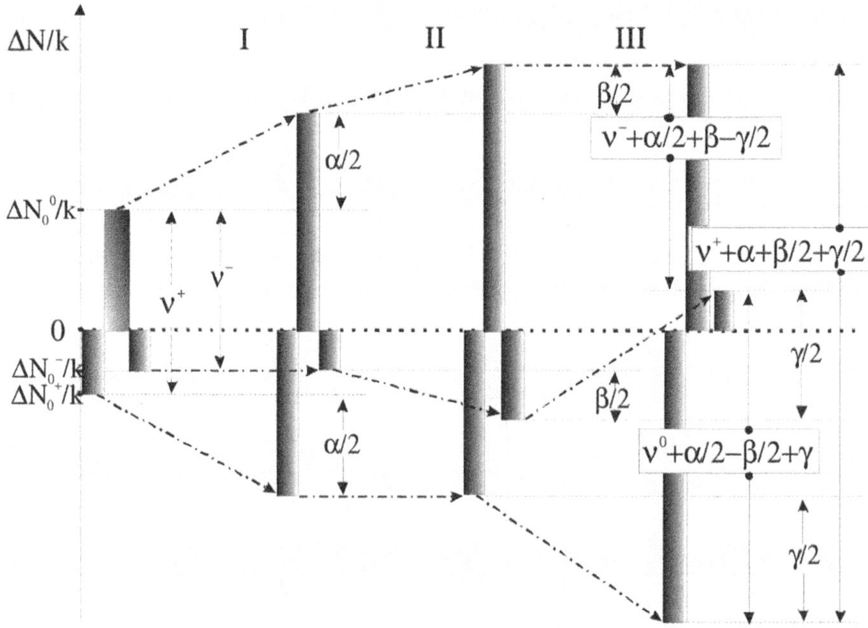

Figure 2. The occupation number alteration of different energy levels in a three step cross-polarization transfer. α, β and γ denote the increments of the population differences after each level crossing.

3. Experimental

In our experiments, protons in polycrystalline trinitrotoluene (TNT) sample were polarized by a block shaped NdFeB permanent magnet (70 mm × 65 mm × 65 mm). The polarizing device consisted of a computer controlled linear robotic arm, carrying the magnet towards and away from the sample. A cylindrical sample with diameter of 18 mm was at stable room temperature in modestly rf shielded space. In the nearest position the magnet was distant 10–12 mm from the sample surface. The vertical sample direction was perpendicular to the polarizing magnetic field. With this setup, it was possible to optimize the duration of level crossing. When the magnet is in the far position, its field value did not exceed the double value of the earth's magnetic field. The experimental procedure is shown schematically in Figure 3. At time t = 0, protons were polarized in a strong magnetic field. Then the magnet was quickly moved away (adiabatic demagnetization – fast compared to the proton T_1). During the demagnetization procedure when the proton NMR frequency became equal successively to each of the three ^{14}N NQR frequencies, the polarization was transferred from protons to nitrogen nuclei. Finally, in a zero magnetic field, standard pulsed spin-locking NQR detection was performed.

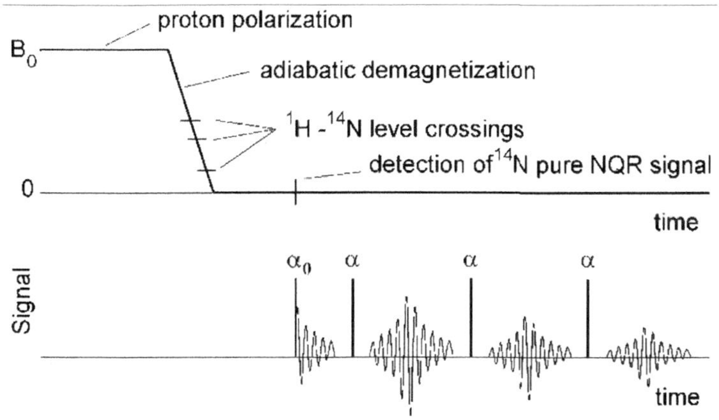

Figure 3. Proton–nitrogen level crossing polarization enhancement and pulsed spin-locking detection procedure (schematic).

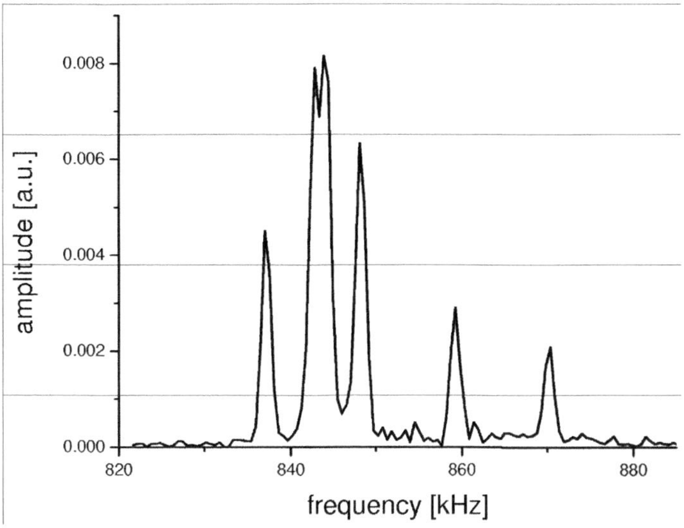

Figure 4. v^+ part of ^{14}N NQR spectrum of predominantly monoclinic TNT. Six lines for monoclinic TNT can be resolved: 837, 842, 844, 848, 859 and 870 kHz. This spectrum was obtained from nine averages in three sets where only the transmitter frequency was changed in three steps (837, 846 and 865 kHz). The reference frequency Rx was constant at 855 kHz. A polarizing field of about 250 ± 50 mT was used for this polarization enhancement experiment.

Polarization enhanced NQR was tested by measuring the ^{14}N NQR spectrum of TNT ($C_7H_5O_6N_3$). TNT exists in two crystal modifications, orthorhombic and monoclinic, with six non-equivalent nitrogen sites giving six different sets of ν^+, ν^- and ν^0 ^{14}N NQR lines (Figure 4) for each crystal modification. The monoclinic phase is considered to be the stable phase at room temperature.[24–25] TNT samples of different age and origin (Italian, German, Swiss, Czech and Yugoslav) were investigated. No significant differences were observed between them (Table 1). All samples were predominantly in the monoclinic phase, probably due to the remelting procedure which was generally carried out.

TABLE 1. ^{14}N NQR frequencies and asymmetry parameters for monoclinic TNT at room temperature.

No.	ν^+, kHz	ν^-, kHz	ν^0, kHz	η
1	870	714	156	0.295
2	859	751	108	0.201
3	848	739	109	0.206
4	844	714	130	0.250
5	842	769	73	0.136
6	837	743	94	0.178

4. Results and discussion

Having six different ν^+, ν^- and ν^0 ^{14}N NQR lines for this particular case means that each level crossing in the three-step polarization transfer is itself a six step process and exact calculation of the enhancement factor is much more complicated. However, if we taking into account the Zeeman broadening of ^{14}N NQR lines in polycrystalline or powder samples in the magnetic fields where the cross polarization transfer occurs, the transfer can be considered as a simultaneous process for all six lines (12 lines for mixed samples), because the broadenings of the lines is much stronger than the distances between them.[28] Therefore, the three level model is very close to the real situation. For TNT we have $\rho = 2/7$ and the enhancement factors for ν^+, ν^- and ν^0 can be calculated (Figure 5):

$$\begin{aligned} EF^+ &= 0.413 + 0.978\frac{\nu_H}{\nu^+} + 0.026\frac{\nu^-}{\nu^+} \\ EF^- &= -0.046 + 0.476\frac{\nu_H}{\nu^-} + 0.213\frac{\nu^+}{\nu^-} \\ EF^0 &= -0.020 + 0.503\frac{\nu_H}{\nu^0} + 0.211\frac{\nu^+}{\nu^0} \end{aligned} \tag{13}$$

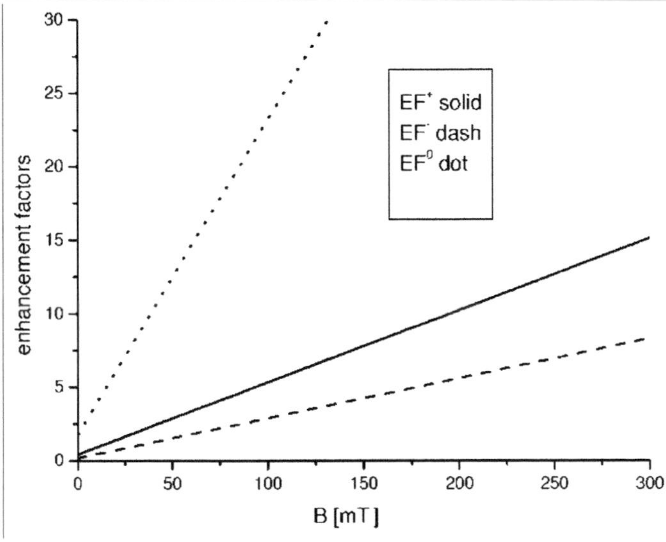

Figure 5. Calculated enhancement factor vs the polarizing field for ν^+, ν^- and ν^0 lines in TNT. Average enhancement factors for ν^+, ν^- and ν^0 are given assuming that ν^+ and ν^- transitions are grouped around 850 and 750 kHz, respectively.

In polarization enhanced NQR detection the initial polarization time must be long enough to allow complete proton polarization, i.e. the proton polarization time T_{pol} must be long compared to the proton spin-lattice relaxation time T_1 ($T_{pol} >> T_1$). The proton T_1 in TNT strongly depends on temperature and the Zeeman magnetic field.[11] From an asymptotic approach

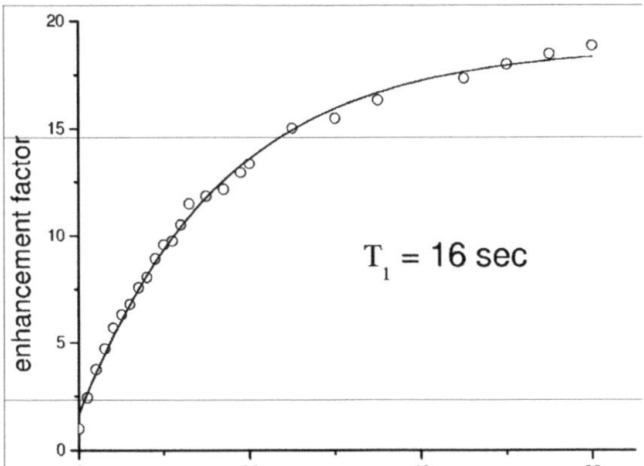

Figure 6. Average signal intensity vs polarization time for the triplet (842, 844 and 848 kHz) NQR line in the polarizing field 400 ± 50 mT. The characteristic time for an exponential fit to the measured data is 16 s. This time is approximately equal to the proton spin-lattice relaxation time T_1.

to the maximal signal intensity of the ^{14}N NQR signal (Figure 6) we estimated that T_1 of protons in a field of about 400 mT is about 16 s, which is in agreement with the reported measurements.[17] The required proton polarization time is thus around 30 s and this part of the measuring procedure is the most time consuming part of the whole experiment (obviously, for lower polarization fields T_1 of protons is shorter and correspondingly also the required polarization time).

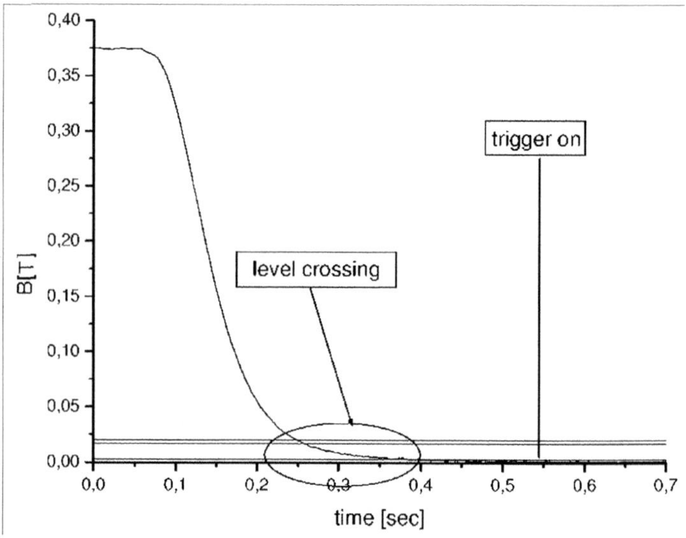

Figure 7. A typical time dependence of the magnetic field at the place of the sample.

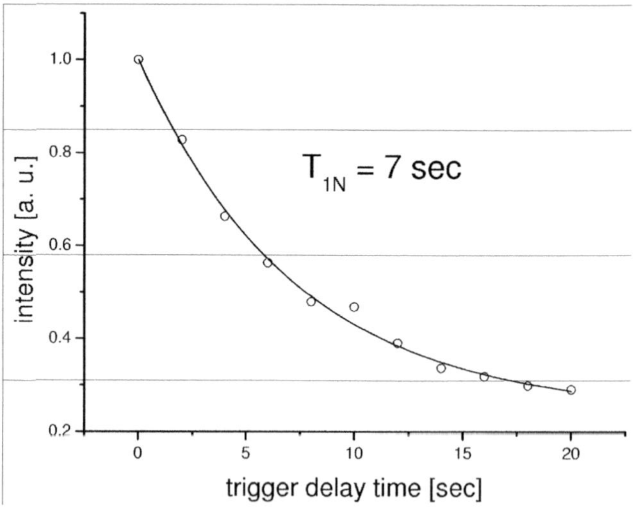

Figure 8. Average ^{14}N NQR signal intensity vs trigger delay time for the triplet line (842, 844 and 848 kHz) in TNT at room temperature.

Polarization of protons was followed by fast adiabatic demagnetization and the trigger onset for the beginning of the NQR pulse sequence was programmed at the point when zero magnetic field was reached at the end of adiabatic demagnetization. The demagnetization times in our experiments (Figure 7) were 300–500 ms and were short enough compared to the proton T_1. The time interval between the final level crossing and the trigger onset (at zero field) must be short compared to the nitrogen relaxation time T_{1N}. By altering the trigger delay we could estimate the nitrogen spin lattice relaxation times. For the ν^+ (average for triplet at 842, 844 and 848 kHz) transitions of the monoclinic TNT we obtained a T_{1N} of approximately 7 s at room temperature (Figure 8).

The efficiency of cross-polarization transfer depends on the strength of the proton-nitrogen spin coupling ν_{NH}. For TNT $\nu_{NH} \sim 500$ Hz.[26] For successful polarization transfer the cross-polarization time must be long compared to $1/\nu_{NH} \sim 2$ ms, giving the limit for the field sweep speed dB/dt at level crossings. Considering that the proton NMR line-width $\Delta\nu_H$ is around 20 kHz (0.5 mT), we can write the condition:

$$dB/dt \ll 0{,}5\ \text{mT}/2\ \text{ms} = 250\ \text{mT/s}$$

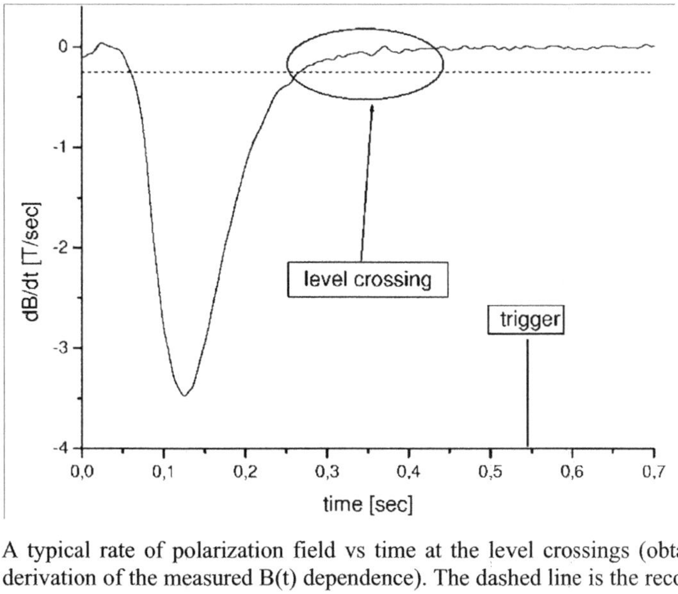

Figure 9. A typical rate of polarization field vs time at the level crossings (obtained with numerical derivation of the measured B(t) dependence). The dashed line is the recommended limit value.

In our experimental arrangement we were close to this limit (Figure 9) and the experimental enhancement factors for PE NQR in TNT are close to the calculated ones in the ideal case. To confirm this we measured the signal intensity enhancement factor versus polarizing magnetic field (Figure 10).

The measured enhancement factor was only slightly lower than the calculated one.

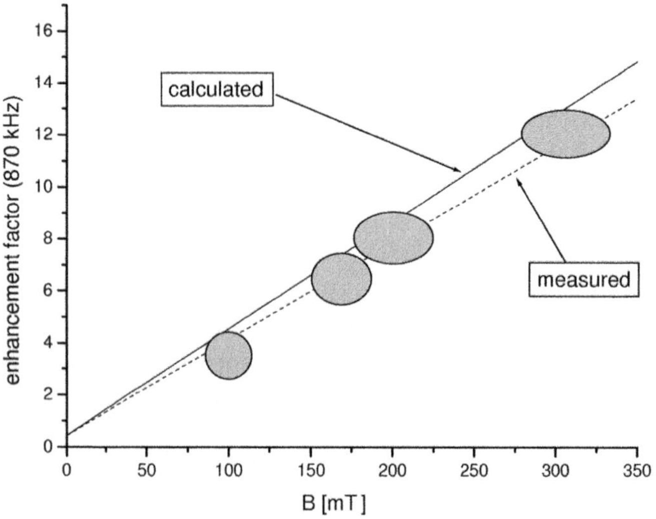

Figure 10. The signal intensity enhancement factor for the ν^{+} (870 kHz) line in TNT at room temperature. The measured uncertainties, denoted by the ellipsoidal surfaces, are mainly due to the dimensions of the sample and very inhomogeneous polarization magnetic field obtained with the NdFeB permanent magnet.

With polarization enhanced NQR and the proper selection of measuring pulse sequences, the s/n ratio of NQR signals can be increased up to three orders of magnitude. The required measuring time was thus adequately reduced.[27,28] The multi-pulse pulse spin-locking (PSL) sequence applied in our experiments was: $\alpha_0 - (\tau - \alpha_{90} - \tau -)_n$, where n refers to the number of pulse-train repetitions and the pulse width α was chosen to optimize the signal. The sequence allowed several echoes in a single shot and thus effectively enhanced the s/n ratio with averaging them within the single shot. A typical time response of the detector is given on Figure 11. For the resolution in the frequency space around 500 Hz usually 10–20 echoes could be averaged in one shot. The single shot ^{14}N signals at 842, 844 and 848 kHz in the central part of the NQR spectrum of 15 g of TNT at room temperature are shown in Figure 12. The total measuring time for proton polarization, demagnetization and PSL sequence was about 15 s. Fifteen echoes in the single pulse sequence (single shot) were averaged. A sharp low pass filter of 5 kHz, 12 dB/octave at the receiver output was used. The polarization field was ~250 mT.

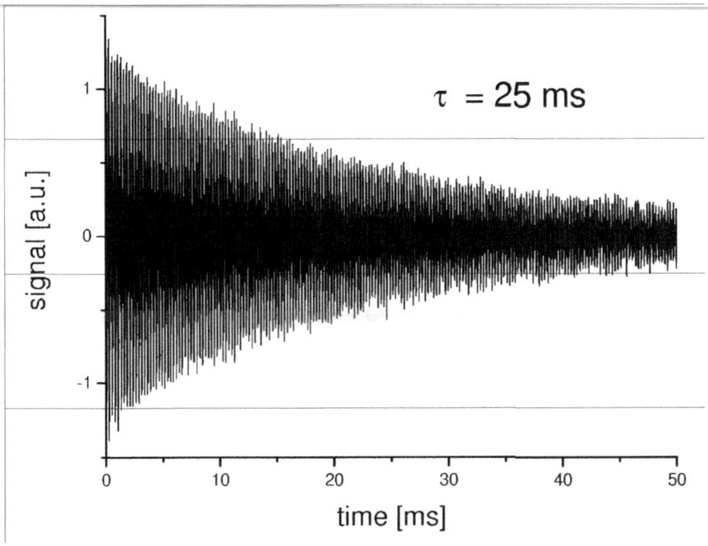

Figure 11. A train of 200 echoes showing the typical decay time approximately 25 ms (~300 averages).

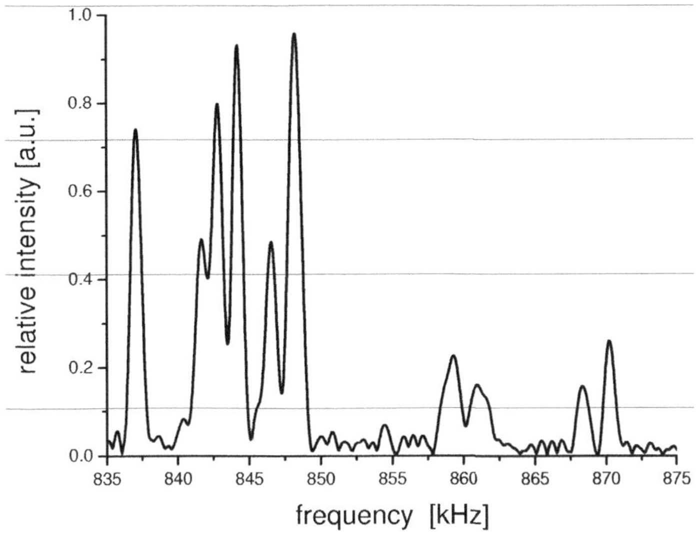

Figure 12. The central part of the ^{14}N NQR spectrum (signals at 842, 844 and 848 kHz) in pure monoclinicTNT with reference frequency 845 kHz filtered, with a sharp LP filter at the receiver output, obtained from a single shot experiment.

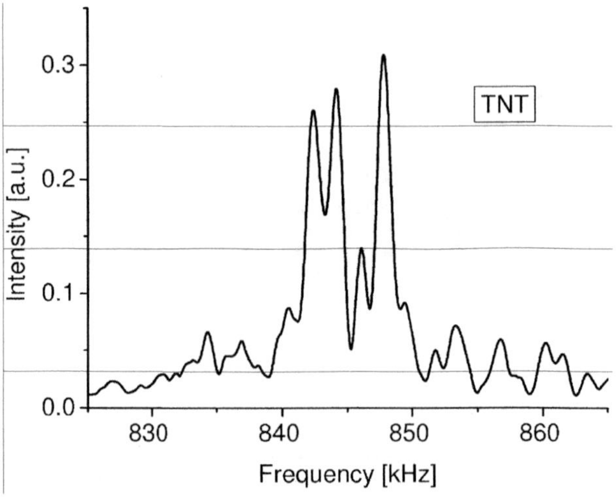

Figure 13. ν^+ part of ^{14}N NQR spectrum in an old sample of TNT where the mixture of two phases monoclinic/orthorhombic (~50%/50%) can clearly be resolved. With polarization enhancement such a spectrum can be obtained in 10–5 min (~30 averages).

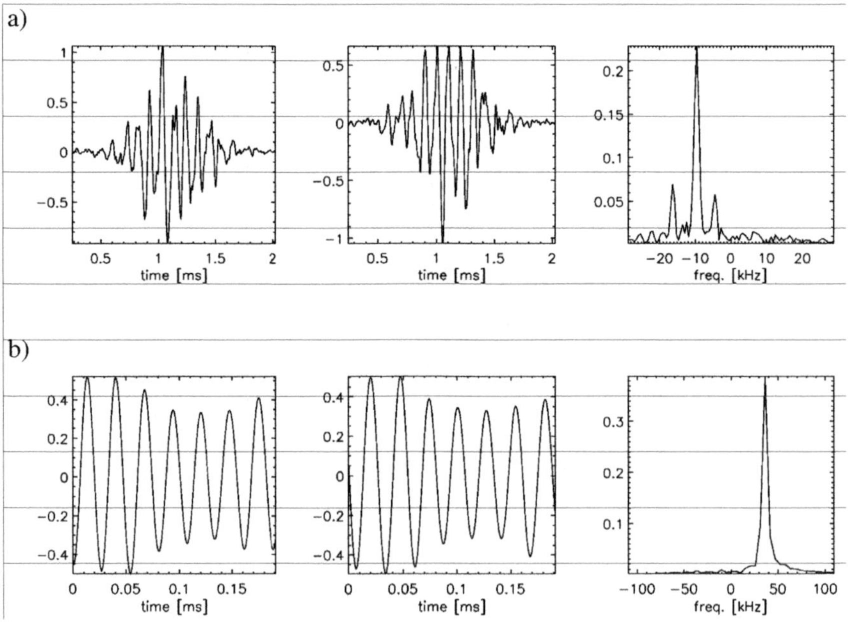

Figure 14. Two examples of single shot ^{14}N NQR signals in TNT obtained from averaged echoes: (a) the pulse separation time 2τ = 2,000 μs, average of 50 echoes and (b) the pulse separation time 2τ = 400 μs, average of 400 echoes Only the second half of echoes is averaged in (b), because of the after pulse detection dead time.). In (a) we can resolve 837 and 848 kHz lines while the 842 and 844 lines appear as a broader single line in the middle. In (b) the resolution in frequency space is so low that only one line can be observed, but with much higher signal/noise ratio. The receiver reference frequencies were: 855 kHz (a) and 820 kHz (b). The three panels from left to right: absorption, dispersion and power spectra.

In relatively short time the complete ^{14}N NQR TNT spectrum of sufficient quality can be obtained (Figure 13). With the reduced resolution we can apply pulse sequences with much shorter pulse distances and increase the number of echoes for averaging (depending on the echo effective decay time). Namely, for the detection of an illicit substance the sensitivity is more important than the resolution. Figure 14 shows two examples of single shot ^{14}N NQR signals in monoclinic TNT where pulse separations are 2,000 and 400 μs. Fifty and 400 echoes were averaged respectively.

Conclusions: Our results demonstrate that polarization enhanced single shot pulse spin-locking detection of ^{14}N NQR is fast and sensitive enough to be applied for routine detection of explosives in, e.g., mail services, customs control and in prevention of terrorist actions, as well as in the pharmaceutical industry for nondestructive chemical analysis of solid samples and for polymorph determination.

Acknowledgements

This research was supported by: (1) the Slovenian Research Agency of the Ministry of Higher Education, Science and Technology (Grant No. 3311-04-828010) and (2) NATO Science for Peace (Grant No. SfP-978007).

References

1. Grechishkin V.S., Sinjavsky N.J.: Phys. Usp. 40, 393 (1997)
2. Yesinowski J.P., Buess M.L., Garroway A.N.: Anal. Chem. 67, 2256 (1995)
3. Grechishkin V.S.: Appl. Phys. A: Solids Surf. 55, 505 (1992)
4. Garroway A.N., Buess M.L., Miller J.B., Suits B.H., Hibbs A.D., Barrall G.A., Matthews R., Burnett L.J.: IEEE Trans. Geosci. Remote Sens. 39, 1108 (2001)
5. Smith J.A.S., Deas R.M., Gaskell M.J.: International Conference on Requirements and Technologies for the Detection, Removal and Neutralization of Landmines and UXO, 15–18 September 2003. Vrije Universiteit Brussels, Belgium, Vol. 2, p. 715 (2003)
6. Smith J.A.S.: Chem. Soc. Rev. 15, 225 (1986)
7. Balchin E., Malcolme-Lawes D.J., Poplett I.J.F., Rowe M.D., Smith J.A.S., Pearce D.E.S., Wren S.A.C.: Anal. Chem. 77, 3925 (2005)
8. Blinc R., Seliger J., Zidansek A., Zagar V., Milia F., Robert H.: Solid State Nucl. Magn. Reson. 30, 61 (2006)
9. Seliger J., Blinc R., Mali M., Osredkar R., Prelesnik A.: Phys. Rev. B 11, 27 (1975)
10. Slusher R.E., Hahn E.L.: Phys. Rev. 166, 332 (1968)
11. Blinc R., Seliger J., Arcon D., Cevc P., Zagar V.: Phys. Stat. Sol. A 180, 541 (2000)
12. Nolte M., Privalov A., Altmann J., Anferov V., Fujara F.: J. Phys. D 35, 939 (2002)
13. Seliger J., Zagar V., Blinc R.: J. Magn. Reson., Ser. A 106, 214 (1994)
14. Kruk D., Altmann J., Fujara F., Gaedke A., Nolte M., Privalov A.F.: J. Phys. Condens. Matter 17, 519 (2005)
15. Thurber K.R., Sauer K.L., Buess M.L., Klug C.A., J.B. Miller J. Magn. Reson. 177, 118 (2005)

16. Luznik J. Pirnat J., Trontelj Z.: Solid State Commun. 121, 653 (2002)
17. Blinc R., Apih T., Seliger J.: Appl. Magn. Reson. 25, 523 (2004)
18. Grechishkin V.S., Anferov V.P., Sinjavsky N.J.: Adv. Nucl. Quadrupole Reson. 5, 13 (1983)
19. Mikhaltsevitch T., Beliakov A.V.: Solid State Commun. 138, 409 (2006)
20. Abragam A.: *The Principles of Nuclear Magnetism*, Clarendon Press, Oxford, 1961
21. Goldman M.: *Spin Temperature and Nuclear Magnetic Resonance in Solids*, Oxford University Press, Oxford, 1970
22. Edmonds D.T.: Phys. Rep. 29, 233 (1977)
23. Das T.P., Hahn E.L.: *Nuclear Quadrupole Resonance Spectroscopy*, Academic Press, New York/London, 1958
24. Marino R.A., Connors R.F.: J. Mol. Struct. 111, 323 (1983)
25. Deas R.M., Gaskell M.J., Long K., Pierson N.F., Rowe M.D., Smith J.A.S.: Proc. SPIE 5415, 510 (2004)
26. Naito A., Root A., McDowell C.A.: J. Phys. Chem. 95, 3578 (1991)
27. Luznik J., Pirnat J., Jazbinsek V., Apih T., Gregorovic A., Blinc R., Seliger J., Trontelj Z.: Appl. Phys. Lett. 89, 123509 (2006)
28. Luznik J., Pirnat J., Jazbinsek V., Apih T., Gregorovic A., Blinc R., Seliger J., Trontelj Z.: J. Appl. Phys. 102, 084903 (2007)

EFFICIENT EXCITATION AND RINGING SUPPRESSION IN NUCLEAR QUADRUPOLE RESONANCE

JOEL B. MILLER[1*], KAREN L. SAUER[1,2],
CHRISTOPHER A. KLUG[1], MICHAEL L. BUESS[3]
[1]*Naval Research Laboratory, Code 6120, Washington, DC 20375-5342 USA*
[2]*Department of Physics and Astronomy, George Mason University, 4400 University Drive MS 3F3, Fairfax, VA 22030 USA*
[3]*Advanced Technology Division, SFA, Lanham, MD 20785 USA*

Abstract Detection of materials using nuclear quadrupole resonance (NQR) is often hampered by interference from remote radio frequency sources such as radio stations, and from local sources such as piezoelectric and magneto-acoustic ringing. NQR pulse sequences have been devised that reduce or eliminate the ringing, but these pulse sequences also reduce the NQR signal. We show that the signal reduction is the result of destructive interference between spin echoes created by multiple echo formation pathways. The destructive interference can be greatly reduced using composite refocusing pulses in the NQR pulse sequences to force the echoes into selected echo formation pathways.

Keywords: NQR, SLSE, phase-alternated SLSE, piezoelectric and magnetoacoustic ringing

1. Introduction

The popularity and value of magnetic resonance techniques derives from their (unsurpassed) sensitivity to molecular structure and the molecule's local environment. The techniques have proven to be important tools in areas as diverse as protein structure determination, solid state physics, and applications such as magnetic resonance imaging (MRI) and nuclear quadrupole resonance (NQR) detection of illicit substances, primarily explosives.[1] The primary limitation of magnetic resonance techniques is signal sensitivity; the transition

[*] E-mail: joel.b.miller@nrl.navy.mil

J. Fraissard and O. Lapina (eds.), *Explosives Detection Using Magnetic and Nuclear Resonance Techniques*,

frequencies are in the radio frequency range, therefore the population differences are small, the signals weak, and the signal-to-noise ratio (SNR) poor.

Although the NQR technique suffers from poor sensitivity, this limitation is generally outweighed by its high specificity for the target substance. High specificity is important in reducing false alarms in heterogeneous environments such as the contents of suitcases or the ground in a mine field. Although NQR signals from benign material are not observed when searching for illicit substances due to the wide dispersion of NQR frequencies and the limited excitation bandwidth of NQR detectors, the RF pulses used to generate NQR signals *can* generate from benign material non-NQR signals which can be mistaken for, or mask, true NQR signals. Examples of such spurious signals include magnetoacoustic ringing[2,3] in spring mechanisms and piezoelectric ringing[4] in quartz crystals. Furthermore, in the unshielded environment of landmine detectors radio frequency interference (RFI) from radio stations (for example) can also hide or mimic a true NQR signal. Spurious ringing and RFI can be much stronger than an NQR signal and detecting explosives in the presence of such interference is a major challenge.

Much effort has gone into devising new RF excitation schemes to differentiate NQR from spurious ringing signals. One method that has been proposed to separate the NQR signal from the interference of ringing is circularly polarized excitation and detection. In NMR the excited nuclear magnetization precesses about the static magnetic field creating a circularly polarized signal in the plane perpendicular to the field. In most NMR spectrometers only one of the two linearly polarized components comprising the circularly polarized signal is detected, but in systems designed to detect the full circularly polarized signal, a SNR advantage is realized.[5,6]

In contrast, the NQR signal results from nuclear magnetization oscillating along the direction of the exciting RF field, resulting in a linearly polarized signal. In a powder sample only the crystallites whose quantization axis projects onto the RF field direction contribute to the signal. If, however, the NQR signal is excited with a circularly polarized RF pulse, a pulse whose direction rotates in a plane at the NQR resonance frequency, many more crystallites' quantization axes projects onto the RF field direction, at least part of the time.[7–9] All the crystallites oscillate at the same frequency, but the phase of the oscillation depends on their orientation in the plane of the RF excitation. The net result is a circularly polarized NQR signal with SNR enhancement of 21%.

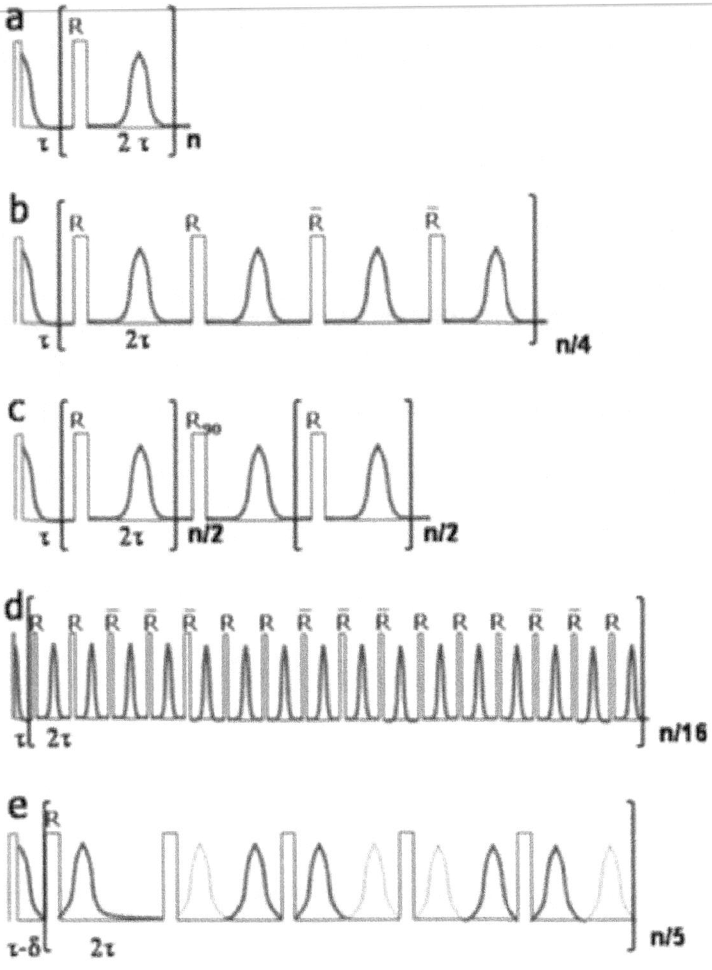

Figure 1. Pulse sequences discussed in the text: (a) the SLSE pulse sequence; (b) phase-alternated SLSE pulse sequence; (c) SLIME pulse sequence; (d) MLEV-16 phase-alternated SLSE pulse sequence; (e) timing for the time-shifted SLSE pulse sequences. The refocusing pulses, R, are either a single pulse of phase 0°, or a composite pulse with net phase of 0°; in all cases the excitation pulse is assumed to have a phase of ±90°. The bar over the R indicates a 180° phase shift. In the time-shifted pulse sequence the two echo pathways are indicated by black and gray echoes.

Like NQR signals, piezoelectric and magnetoacoustic ringing signals are linearly polarized. However, unlike NQR, there are rarely more than one or a few crystals or domains contributing to the ringing signal, therefore even under circularly polarized excitation one expects ringing signals to be substantially linearly polarized. Circularly polarized NQR can be used to differentiate NQR from ringing signals, at the expense of reducing the SNR by $\sqrt{2}$.[8,9]

Both the NQR and ringing signals respond coherently to the RF pulses. With the RF field strengths generally employed to excite NQR the ringing signal amplitude is proportional to the energy of the RF pulse whereas the NQR response is non-linear. It is possible, in principle, to design NQR pulse excitation schemes that exploit the nonlinear nature of the NQR response to separate the NQR signal from ringing signals. The non-linear NQR response can be exploited through spin echo pulse sequences such as the spin-locked spin-echo (SLSE),[10] the NQR variant of the Carr–Purcell–Meiboom–Gill (CPMG)[11] pulse sequence (Figure 1). The SLSE pulse sequence is well suited for detecting materials which have a long NQR spin-lattice relaxation time, T_1, which can be several seconds. The SNR obtained in each scan is significantly enhanced by summing the train of echoes generated by the pulse sequence. Unfortunately, spurious ringing is coherent with the pulses and is also enhanced by echo summation as a result. Spurious ringing can be canceled to some extent by repeating the SLSE sequence with an inverted preparatory pulse. That inverts the NQR signal but not the ringing and proper summation of the two scans will enhance the NQR signal and reduce the ringing. However, the scans must be separated by intervals on the order of T_1, which can be several seconds, and in those intervals the properties of the material that is ringing can change due to RF heating. This makes it impossible to cancel the ringing entirely. Because RFI is not coherent with the pulses a multiple scan detection is as likely to enhance RFI interference as to suppress it. A successful cancellation scheme must work on time scales for which ringing and RFI are essentially constant.

To cancel the ringing on a shorter time scale, numerous variants of SLSE have been proposed that modulate the phase of the spin echoes relative to the phase of the preceding pulses.[12–14] Periodically inverting the phase of the refocusing pulses leaves the phase of the spin echoes unchanged while inverting the ringing. The so-called spin-locked-inversion-mid-echo (SLIME)[15] sequence periodically inserts an inverting pulse at the echo maximum to invert the phase of the echoes while leaving the ringing phase unchanged (Figure 1). These schemes can produce good ringing cancellation, however, each inversion results in a significant loss of echo intensity. To maintain sensitivity only a few, widely spaced, inversions are used, resulting in reduced ringing cancellation.

The loss of echo intensity is a consequence of the powder distribution of quantization axes in the laboratory frame in NQR. The spatial distribution of quantization axis directions relative to the RF field direction leads to large variations in the effective RF field strength, and hence in the pulse flip angle, from crystallite to crystallite. To use an NMR analogy, in general then, the refocusing pulse splits the magnetization into longitudinal and transverse components, so that there are two echo formation pathways, analogous to spin echoes and stimulated echoes in NMR. In multiple pulse cycles the potential number of echo formation pathways grows with the number of applied pulses. The different echo formation pathways produce echoes whose phases depend in a complex way on the pulse phase cycles[12–14] and resonance offset,[16] an effect not generally observed in NMR. The result is interference of the echoes and loss of signal intensity.

We illustrate this effect by simulating time domain NQR data. Time domain data for variants of the SLSE pulse sequence, as defined in Figure 1, were simulated using the Bloch equations in a fictitious spin-1/2 system. Such simulations are useful for spin-1 NQR when only one transition is excited, that is when the asymmetry parameter, η, is not zero (or one). The simulations are done on resonance, with averaging over the powder distribution of quantization axes, using line width (1 kHz), T_1 (5 s)and T_2 (40 ms) relaxation times, and RF field strength (10 kHz) similar to those encountered in our laboratory for glycine hemihydrochloride (diglycine HCl) and 2,4,6-trinitrotoluene, TNT. The excitation pulses have a nutation angle of 119°. The optimal maximum set value of the refocusing pulse nutation varies somewhat with sample and pulse sequence parameters but is often found to be near 150°.[17] Here the refocusing pulses have a nutation angle of 150°. For all SLSE pulse sequences the inter-pulse spacing was 2 ms.

The interference effect is simulated in Figure 2a. The phase-alternated SLSE echoes decay much more rapidly than the SLSE echoes. We can get a qualitative view of the effects of the two echo formation pathways by adjusting the timing of the pulse sequence (Figure 1) to temporally separate the echoes of the two pathways. The results are shown in Figure 2b. In the SLSE pulse sequence the echoes from the two pathways are essentially equal amplitude. The echoes from the two pathways are also phase coherent whereas the phase-alternated SLSE results in phase-modulated echoes that eventually interfere destructively. Note that in the time-shifted pulse sequence the phase-alternated SLSE echoes persist over a much longer time than in the standard timing sequence. Time-shifting might be a partial solution to the interference of the different echo pathways, however we note that: (1) the full intensity of the echoes may not recovered by time-shifting; (2) spreading the echoes out temporally reduces the SNR by requiring signal integration over a broader range of times; (3) half the echoes are shifted closer to the refocusing pulses and in many cases will be lost to probe ringing.

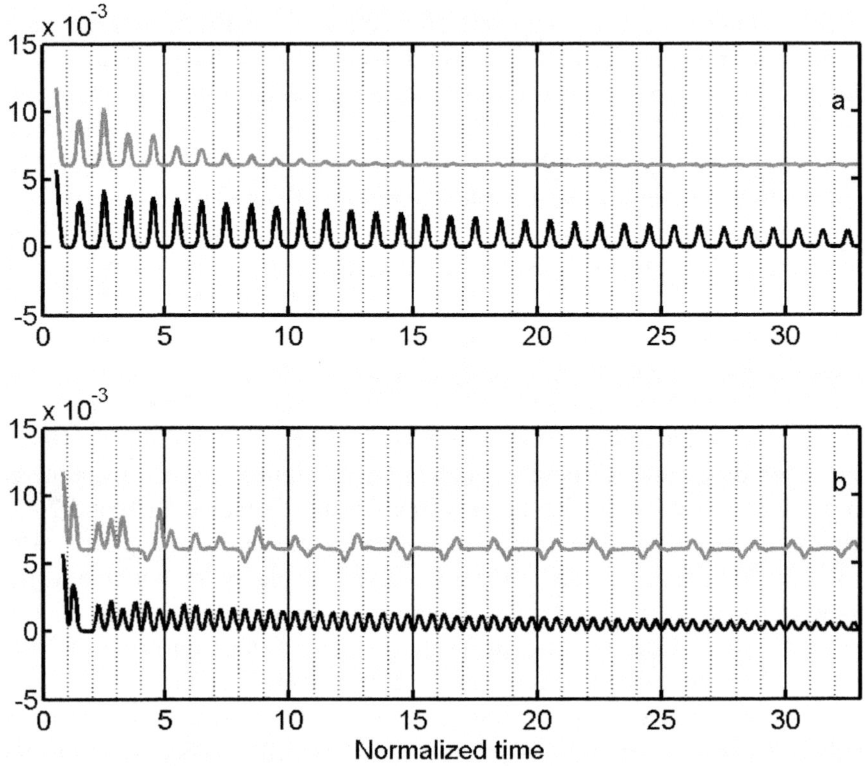

Figure 2. Simulations of the SLSE and phase-alternated SLSE pulse sequences: (a) comparison of the SLSE (lower, black trace) and phase-alternated SLSE (upper, gray trace) time domain data; (b) comparison of the time-shifted SLSE (lower, black trace) and time-shifted phase-alternated SLSE (upper, gray trace) time domain data. The offset of the phase-alternated data was added to separate the traces. The time axis is normalized to the refocusing pulse interval.

In principle, more complicated phase alternation patterns can alleviate, or at least reduce, the destructive interference of the echo pathways. A good case in point is SLSE combined with the MLEV-16 phase alternation scheme.[18] The MLEV-16 pulse sequence alternates the phase of the RF pulses (Figure 1) such that summing the echoes produces a net NQR signal but no net ringing. It does not cancel interference from RFI. Ringing cancels quite well in this pulse sequence because cancellation occurs within each 16-pulse cycle (milliseconds rather than seconds). Unfortunately, it is not so efficient at enhancing the NQR signal. The echo decay time for MLEV-16 phase alternation is similar to that obtained with SLSE alone, but the net NQR signal for the MLEV-16 pulse sequence is approximately 50% smaller than that produced by a SLSE pulse sequence of the same length (Figure 3a). Again, time-shifted pulse sequences can help shed some light on the echo behavior. We see in Figure 3b that the MLEV-16 phase alternation shifts intensity between the echo pathways, enhancing some echoes and reducing others. The enhanced echoes are phase coherent; the result is that the destructive interference of the echoes is reduced but not completely eliminated.

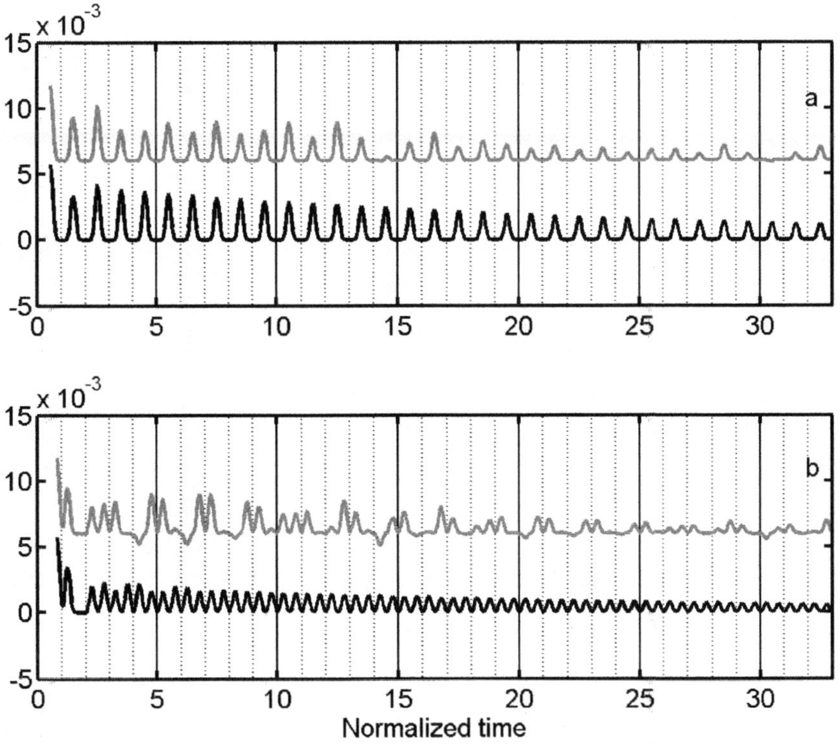

Figure 3. Simulations of the SLSE and MLEV-16 SLSE pulse sequences: (a) comparison of the SLSE (lower, black trace) and MLEV-16 SLSE (upper, gray trace) time domain data; (b) comparison of the time-shifted SLSE (lower, black trace) and time-shifted MLEV-16 SLSE (upper, gray trace) time domain data. The offset of the MLEV-16 data was added to separate the traces. The time axis is normalized to the refocusing pulse interval.

Although the MLEV-16 phase alternation provides a significant improvement in sensitivity over other phase alternation schemes, while maintaining the attributes of ringing cancellation, in most applications the signal loss is still unacceptably large. What we really need is a method to eliminate one of the main echo pathways. The effects of powder averaging on RF excitation in NQR are similar to those of inhomogeneous RF fields on RF excitation in NMR. The NMR problem has been addressed successfully using "shaped" pulses. Shaped pulses are well known in NMR spectroscopy for RF field inhomogeneity compensation; they can also be designed to enhance or restrict the pulse excitation bandwidth. In form, shaped pulses fall into two categories: composite pulses[19] composed of several back-to-back rectangular pulses of varying phase, amplitude, and/or duration; and pulses whose phase and/or amplitude vary continuously.[20]

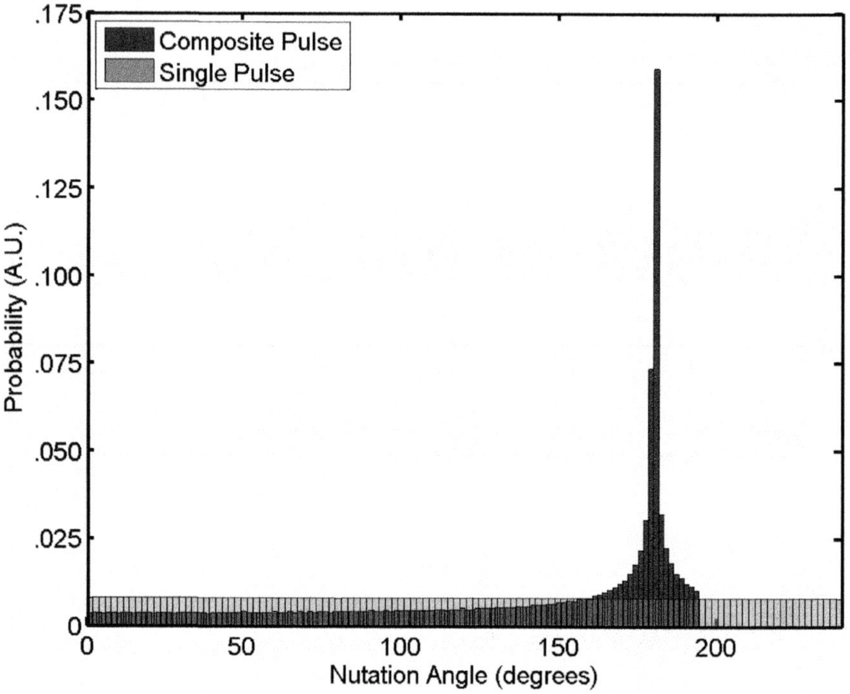

Figure 4. Plot of the probability that a nucleus in a crystallite in a powder sample will experience a given pulse nutation angle. For a single pulse, all nutation angles from the maximum set value to zero are equally probable. For the composite refocusing pulse consisting of three pulses with nutation angles of 240° and RF phases of 300°, 60° and 300°, there is a high probability that nuclei will experience a net 180° nutation angle, as designed. In the simulations and experiments presented here, single refocusing pulse nutation angles were 150°, however for comparison to the composite pulse, the single pulse maximum nutation angle in this figure is 240°.

In NQR, composite pulses have been proposed for compensation of resonance offset[21–25] and RF field inhomogeneity.[21,22,26–28] Several reports on continuously variable shaped pulses for use in NQR have also appeared.[29–31] Shaped pulses that compensate for RF field inhomogeneity also partially compensate for the powder average effect. Theoretically, shaped excitation pulses can produce uniform excitation of virtually all observable nuclei, resulting in a SNR enhancement of 15% in powder samples.[31]

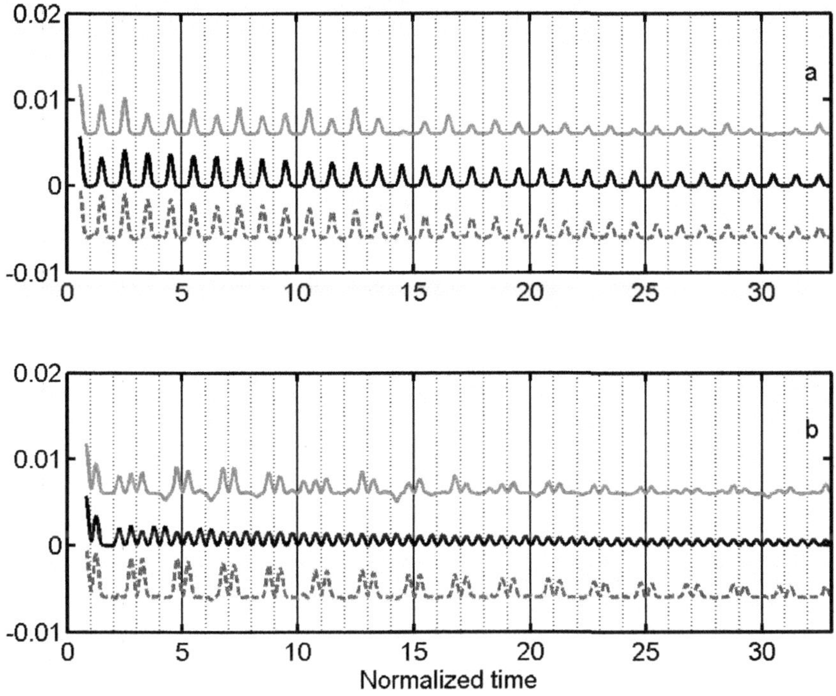

Figure 5. Simulations of the SLSE and MLEV-16 SLSE pulse sequences, with single and composite pulses: (a) comparison of the SLSE (middle, black trace), MLEV-16 SLSE (upper, gray trace), MLEV-16 SLSE with composite refocusing pulse (lower, dashed, gray trace) time domain data; (b) comparison of the time-shifted SLSE (middle, black trace), time-shifted MLEV-16 SLSE (upper, gray trace), and time-shifted MLEV-16 SLSE with composite refocusing pulse (lower, dashed, gray trace) time domain data. The offsets of the MLEV-16 and MLEV-16 with composite pulse data were added to separate the traces. The time axis is normalized to the refocusing pulse interval.

Most NQR applications of composite pulses have focused on their use as excitation pulses. A recent paper suggests combinations of shaped excitation and refocusing pulses may also provide better response from multi-echo pulse sequences such as SLSE.[17] To test this idea the NQR response to the MLEV-16 phase-alternated SLSE, using a composite refocusing pulse was simulated. A composite refocusing pulse consisting of three pulses with nutation angles of 240° and RF phases of 300°, 60° and 300°, with a net nutation angle of 180°, was chosen. This composite pulse preserves the overall pulse nutation angle over a wide range of effective RF field strengths, as illustrated in Figure 4. In Figure 5a we compare simulations of the NQR response to the MLEV-16 phase-alternated SLSE using a single pulse to the response with our composite pulse. It is clear that MLEV-16 with the composite pulse provides superior refocusing, with the echo decay rate nearly identical to the standard SLSE pulse sequence. Once again, the time-shifted pulse sequences provide an explanation of the results. As shown in Figure 3b, the standard SLSE pulse sequence generates a series of echoes, following both echo formation pathways, that are phase coherent whereas the MLEV-16 phase alternation results in some phase alternation of the echoes created by the two pathways. In contrast, MLEV-16 phase alternation with the composite results in nearly all the signal intensity appearing in echoes from only one of the two pathways, ensuring phase coherence of the signals (Figure 5b).

2. Discussion and results

NQR experiments were performed on a sample of glycine hemihydrochloride: the NQR frequencies, line widths and relaxation times of glycine hemihydrochloride are similar to those of TNT making it a reasonable compound on which to test pulse sequences designed for TNT detection. All experiments were carried out using Tecmag Apollo spectrometer (Tecmag, Houston, TX), and a homebuilt probe with a coil of volume of approximately 50 cc operated at room temperature. Pulse nutation angles were the same as those used in the simulations. The single refocusing pulse length was 60 μs with an inter-pulse delay of 930 μs; the total composite refocusing pulse length was 276 μs with an inter-pulse delay of 730 μs.

Figure 6 compares the NQR response of glycine hemihydrochloride to the SLSE pulse sequence with its response to an MLEV-16 phase-alternated pulse sequence comprised of either single pulses or the three-pulse composite pulse. The echo intensities are averaged over several consecutive phase cycles to improve SNR. The echo decay times are nearly identical for all three pulse sequences, but the echo intensities are reduced for the MLEV-16 phase-alternated sequences. The integrals of the echo trains in MLEV-16 sequences with single and composite pulses are 50% and 90% as large as that in the SLSE sequence demonstrating the improvement in signal intensity provided by this composite pulse.

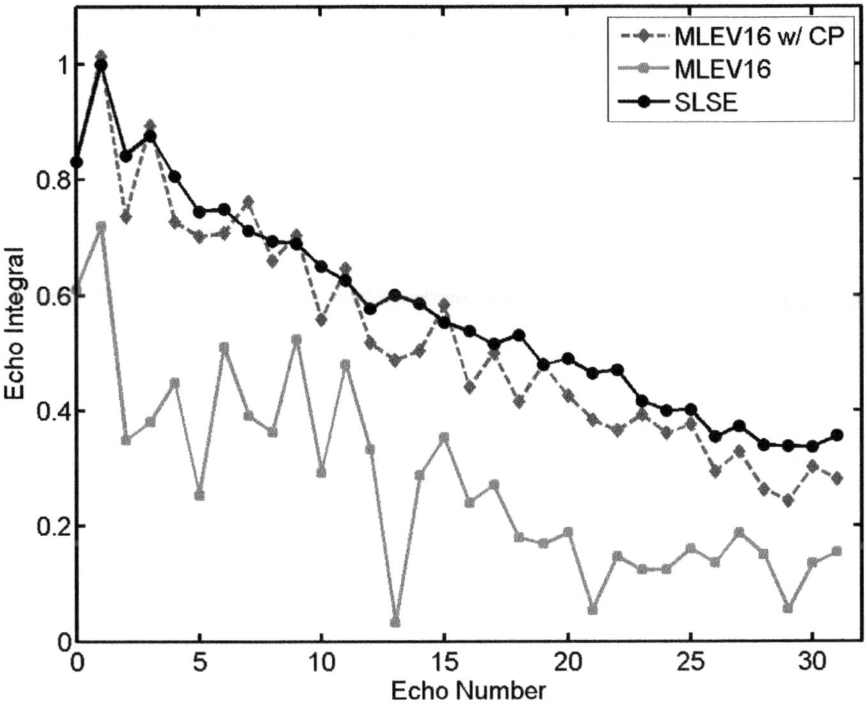

Figure 6. Plot of the integrated echo intensities as a function of echo number for the SLSE (black, circles), MLEV-16 SLSE (gray, squares), and MLEV-16 SLSE with composite refocusing pulse (gray, dashed, diamonds) pulse sequences.

Preserving signal intensity has been the main focus of the discussion presented to this point, but other operational factors are equally important to good detection efficiency. Resonance offset can also be an important factor for NQR detection outside the laboratory. NQR resonance frequencies are temperature dependent, and the temperature coefficients can easily exceed 1 kHz/°C. This can result in a significant uncertainty in the resonance frequency. For example, in land mine detection or luggage screening, the temperature of the ground surface or luggage surface may be known, but the temperature of the buried mine or the contents of the luggage will not be known precisely.

We measured the NQR response to SLSE and MLEV-16 phase alternated SLSE, with single pulses and composite pulses, as a function of resonance offset. SLSE with composite pulses provides slightly larger SNR on resonance at the expense of a small reduction in bandwidth (Figure 7). For standard refocusing pulses the bandwidth of MLEV-16 phase-alternated SLSE is approximately one-half of that for SLSE. The MLEV-16 bandwidth is the same for single and composite refocusing pulses, but the composite refocusing pulses nearly double the SNR. Overall, MLEV-16 phase alternation results in ringing cancellation with only a small lose in SNR and bandwidth.

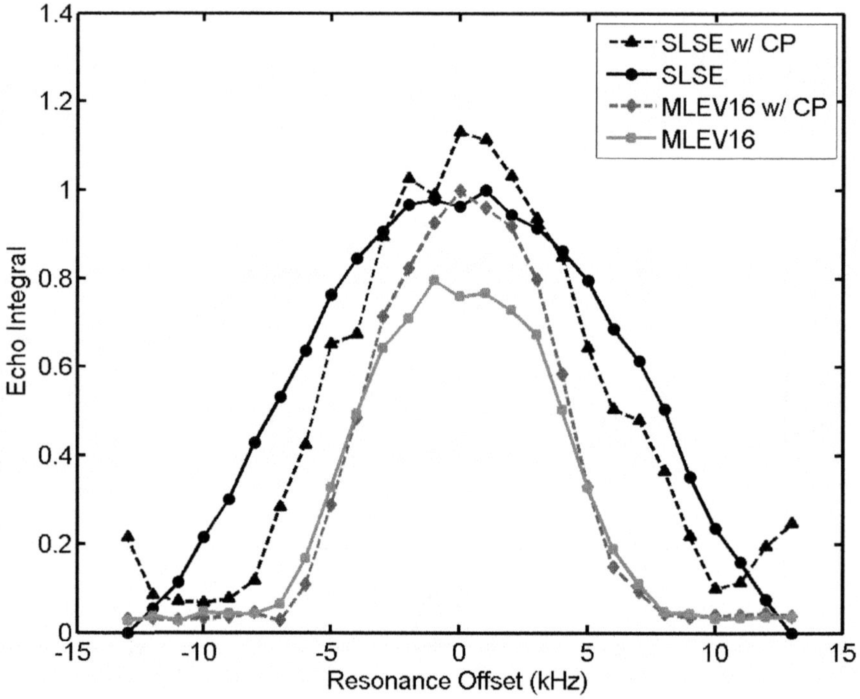

Figure 7. Plot of the integrated echo intensities as a function of resonance offset for the SLSE (black, circles), SLSE with composite refocusing pulse (black, dashed, triangles), MLEV-16 SLSE (gray, squares), and MLEV-16 SLSE with composite refocusing pulse (gray, dashed, diamonds) pulse sequences.

The pulse phase alternation of MLEV-16 does not cancel RFI because RFI is not coherent with the RF pulses. We have proposed a time-shifted phase-alternated SLSE pulse sequence to separate the two echo pathways and move the resulting echoes away from the center of the acquisition window.[32] When the resulting echoes can be shifted far enough apart to avoid destructive interference it is possible to invert the echoes after every other pulse by alternating the phase of the refocusing pulses. (This is generally possible only for lines that are wide enough that their echoes are narrower than the refocusing pulse interval.) Signal processing for a phase alternated pulse sequence involves digitally inverting every other echo prior to summation to obtain a net NQR signal. With the refocusing pulse intervals commonly used in explosives detection (<2 ms) RFI will likely be the same during adjacent pulse intervals and will cancel when processed to maximize the NQR signal. The practical downside to this scheme is that half the echoes are shifted close to the refocusing pulse and can be lost to the probe recovery time.

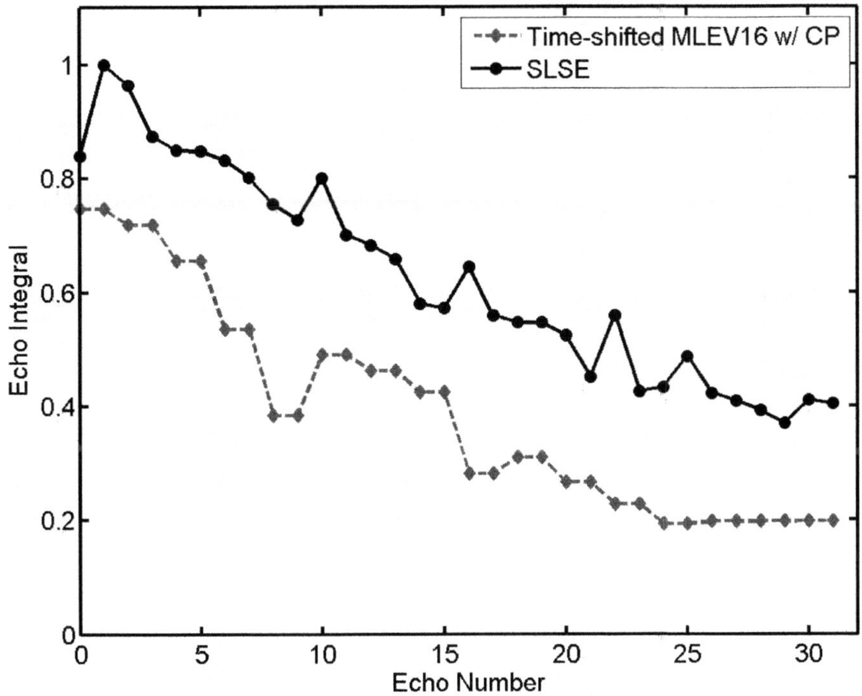

Figure 8. Plot of the integrated echo intensities as a function of echo number for the SLSE (black, circles) and MLEV-16 SLSE with composite refocusing pulse (gray, dashed, diamonds) time-shifted pulse sequences.

The time-shifted echoes in the MLEV-16 phase-alternated SLSE can be combined to suppress both ringing and RFI. An examination of the MLEV-16 pulse phases reveals that ringing cancels among the odd-numbered and even-numbered pulses separately. Therefore, an MLEV-16 phase-alternated sequence in which the signal from every other pulse is inverted will significantly reduce RFI and still suppress spurious ringing. For the MLEV-16 phase-alternated SLSE with composite pulses, every other detection period has no echo, therefore inverting the signal from every other detection period and adding the signals does not change the SNR compared to not inverting the signals before addition.

Figure 8 compares the NQR response of glycine hemihydrochoride to the SLSE pulse sequence with its NQR response to the time-shifted MLEV-16 phase alternated SLSE with composite refocusing pulses. The phase corrected NQR signal in the time-shifted MLEV-16 SLSE actually fluctuates a great deal from pulse to pulse and the echo data shown in Figure 8 are the averages of the signals in the echo pairs. These results indicate that MLEV-16 SLSE with composite pulses and time-shifted echoes will suppress both RFI and spurious ringing at the cost of a 40% reduction in the integrated NQR signal of glycine hemihydrochloride.

Composite pulses can be used to manipulate the echo formation pathways in multiple pulse sequences used for NQR detection. This opens the possibility of complex phase modulation schemes to cancel spurious ringing and RFI. Our results with composite refocusing pulses demonstrate significant SNR enhancement with SLSE when combined with MLEV-16 phase alternation for ringing cancellation compared to the same pulse sequence using single refocusing pulses. Additional SNR enhancement can be expected if shaped excitation pulses are used in combination with composite refocusing pulses.

Acknowledgement

This work was supported in part by the Office of Naval Research.

References

1. J. B. Miller, "Nuclear Quadrupole Resonance Detection of Explosives", in "Counterterrorist Detection Techniques of Explosives", J. Yinon ed., Elsevier, Amsterdam/London, 157 (2007).
2. M. L. Buess and G. L. Petersen, Rev. Sci. Instrum., 49, 1151 (1978).
3. E. Fukushima and S. B. W. Roeder, J. Magn. Reson., 33, 199 (1979).
4. K. Choi and I. Yu, Rev. Sci. Instrum., 60, 3249 (1989).
5. D. I. Hoult, C. -N. Chen, and V. J. Sank, Magn. Reson. Med., 1, 339 (1984).
6. C. -N. Chen and D. I. Hoult, "Biomedical Magnetic Resonance Technology" Adam Hilger, New York (1989).
7. Y. K. Lee, H. Robert, and D. K. Lathrop, J. Magn. Reson., 148, 355 (2001).
8. J. B. Miller, B. H. Suits, and A. N. Garroway, J. Magn. Reson., 151, 228 (2001).
9. J. B. Miller, A. N. Garroway, and B. H. Suits, Nuclear Quadrupole Resonance (NQR) Method and Probe for Generating RF Magnetic Fields in Different Directions to Distinquish NQR from Acoustic Ringing Induced in the Sample, US Patent No. 6 522 135 (2003).
10. R. A. Marino and S. M. Klainer, J. Chem. Phys., 67, 3388 (1977).
11. H. Y. Carr and E. M. Purcell, Phys. Rev., 94, 630(1954); S. Meiboom and D. Gill, Rev. Sci. Instrum., 29, 688 (1958).
12. D. Ya. Osokin, J. Mol. Struct., 83, 243 (1982); D. Ya. Osokin, Mol. Phys., 48, 283 (1983); D. Ya. Osokin, V. L. Ermakov, R. H. Kurbanov, and V. A. Shagalov, Z. Naturforsch., Teil A 47, 439 (1992).
13. A. K. Dubey and P. T. Narasimhan, J. Mol. Struct., 192, 321 (1989).
14. V. T. Mikhaltsevitch, T. N. Rudakov, J. H. Flexman, P. A. Hayes, and W. P. Chisholm, Appl. Magn. Reson., 25, 449 (2004).
15. G. A. Barrall, L. J. Burnett, and A. G. Shelton, Method and System for Cancellation of Extraneous Signals in Nuclear Quadrupole Resonance Spectroscopy, US Patent No. 6 392 408 (2002).

16. M. L. Buess, A. N. Garroway, and J. P. Yesinowski, Removing the Effects of Acoustic Ringing and Reducing Temperature Effects in the Detection of Explosives by NQR, US Patent No. 5 365 171 (1994).
17. K. L. Sauer and C. A. Klug, Phys. Rev. B, 74, 174410 (2006).
18. M. H. Levitt, R. Freeman, and T. A. Frenkiel, J. Magn. Reson., 47, 328 (1982).
19. M. H. Levitt, Prog. NMR Spectroscop., 18, 61 (1986).
20. M. Garwood and L. DelaBarre, J. Magn. Reson., 153, 155 (2001).
21. A. Ramamoorthy and P. T. Narasimhan, J. Mol. Struct., 192, 333 (1989).
22. G. -Y. Li and X. -W. Wu, Chem. Phys. Lett., 204, 529 (1993).
23. A. Ramamoorthy, N. Chandrakumar, A. K. Dubey, and P. T. Narasimhan, J. Magn. Reson., 102, 274 (1993).
24. A. Ramamoorthy, Mol. Phys., 93, 757 (1997).
25. V. T. Mikhaltsevitch, T. N. Rudakov, J. H. Flexman, P. A. Hayes, and W. P. Chisholm, Sol. St. Nucl. Magn. Reson., 25, 61 (2004).
26. G. -Y. Li, Y. Jiang, and X. -W. Wu, Chem. Phys. Lett., 202, 82 (1993).
27. V. P. Anferov and G. V. Mozzhukhin, Russ. Phys. J., 42, 826 (1999).
28. K. L. Sauer, C. A. Klug, J. B. Miller, and A. N. Garroway, Appl. Magn. Reson., 25, 485 (2004).
29. S. Z. Ageev and B. C. Sanctuary, Chem Phys. Lett., 225, 499 (1994).
30. C. Schurrer and S. C. Perez, Appl. Magn. Reson., 16, 135 (1999).
31. J. B. Miller and A. N. Garroway, Appl. Magn. Reson., 25, 475 (2004).
32. Internal report, March 1995.

DETECTION OF CONCEALED LIQUID EXPLOSIVES AND ILLICIT DRUGS IN UNOPENED BOTTLES

SANKARAN KUMAR[1], PABLO J. PRADO[2*]
[1]*1009 Honeysuckle Dr., San Marcos, CA 92078, USA*
[2]*69 Paul Gore St., Boston, MA 02130, USA*

Abstract This article discusses the use of Magnetic Resonance relaxometry as a non-invasive method to detect concealed liquid explosives and drugs in unopened containers. The safe, non-ionizing detection is based on sensing liquid alteration without opening its container. The technique has proven to be effective with non-metallic bottles, independently of their shape and size. Other spectroscopic and bulk methods to identify illicit liquid substances as well as the potential of combining screening technologies are presented.

Keywords: explosives, drugs, bottles, security, airports, screening, magnetic resonance

1. Introduction

1.1. CONCEALED LIQUID EXPLOSIVES AND ILLICIT DRUGS

Security and trafficking concerns have motivated the development of new technologies to detect concealed explosives and drugs.

Following sabotage of Pan Am Flight 108 with explosives in 1988, the US Congress passed the Aviation Security Improvement Act, directing the Federal Aviation Administration to create standards for explosive detection systems. After attacks to civilian targets such as those in September 2001 in New York and in July 2005 in London, the population and government awareness for the need of effective threat detection systems increased significantly.

The security concerns extend beyond airports, including seaports, rail stations, points of entry, government and military facilities, and secured commercial buildings.

[*] E-mail: pablojprado@gmail.com

J. Fraissard and O. Lapina (eds.), *Explosives Detection Using Magnetic and Nuclear Resonance Techniques*,

Of great magnitude is also the problem of illicit drug trafficking. There is evidence[1] of a recent surge in the supply of illicit drugs. During 2007 there were record opium harvests, and cocaine cultivation increased in Bolivia, Peru and Colombia. Opium production almost doubled between 2005 and 2007. Seizures of opium and morphine grew 10% and 31% respectively in 2006. The amount of cocaine seized augmented by about 80% during the early 2000s, to a present levels of 700 t per year.

The current expansion in supply together with the development of new trafficking routes could strengthen demand and create new markets for deadly substances. Decreasing drug trafficking is a method to control availability.

Effective methods exist for detection of hazardous and illicit substances, for example vapor and trace detectors, but they require direct contact with the compounds. As a result, a practical solution implemented today to avoid transporting undetectable substances in unopened containers with liquids is simply to confiscate the containers. However, there are obvious limitations in confiscating bottles that purport to contain legitimate beverages. Since transporting contraband in beverages is a security and customs concern, a method to effectively inspect unopened containers could improve control point protocols, while increasing security and reducing illicit substance traffic.

This article discusses the use of Magnetic Resonance (MR) as a method to noninvasively inspect unopened liquid containers. We show the strengths of this bulk measurement technique, list other emerging methods, and mention the possibility of combining methods as a high performance solution.

1.2. THE NEED FOR NEW DETECTION METHODS

Often high performance liquid explosives contain nitrogen in nitro-, nitrato-, or nitramino-groups, like nitroglycerine, nitroglycol, nitromethane, nitroethane and plasticizers. Also peroxide-containing compounds like triacetontriperoxide (TATP) are used. Liquid explosives are achieved in emulsion, slurries or solutions with combinations of fuels and oxygen like in the case of nitric acid/nitromethane or solutions with hydrogen peroxide. The high vapor pressure of most these nitrogen containing liquid explosives makes them easier to detect than their solid counterparts, but only if the vapor is in an open environment.

Even though explosives in liquid form are highly sensitive to shocks, they may be transported in sealed containers by preparing solutions that keep the compound more stable[2] or by transporting the components of the solution in separate containers. In general, they are chemically easier to combine and simpler to ignite than solid explosives.

Explosives and drugs can be concealed in sealed environments, keeping them out of reach of commonly used substance detecting methods, such as vapor and trace based technologies.[3–6] It is also noted that emulsion and slurry based compounds like ammonium-nitrate have low vapor pressure, making them more difficult to detect.

Bulk measurement detection methods are therefore necessary to screen unopened containers.

2. Detection by magnetic resonance relaxometry

Conventional MR, routinely used for chemical analysis is not a practical configuration for non-invasive security screening methods. This is because the sample size for chemical analysis is typically of the order of milliliters and the equipment requires the use of cryogenic materials and dedicated rooms. Potentially effective substance discrimination methods under development include MR spectroscopy with portable sensors[7] and dielectrometry.[8]

As an alternative to spectroscopic methods, MR relaxometry has shown to discriminate between the expected parameters of a known and an altered liquid in an unopened container. This makes MR relaxometry an effective screening method. Statistical tests showing the efficiency of this low resolution MR approach have been carried out under FAA programs at Quantum Magnetics (Bottle Scanner).[9]

As a method of screening, an alert signal is issued if there is a discrepancy between the measured parameters and those expected from the claimed contents and/or if the contents of the bottle are deemed to be hazardous. MR relaxation times, T_2 and T_1, have been used as discriminators.[9,10] It is noted that proton amplitude evaluated during the MR measurement may improve classification.[11] Further parameters such as diffusion constants and dedicated pulse sequences need to be evaluated as a means to improve detection performance.

Low resolution scanners can be deployed in compact configurations utilizing electromagnets[9] or permanent magnets.[12] Open magnet configurations are a solution to scan a variety of container sizes.[13–15]

3. Results with concealed explosives and drugs

3.1. DETECTION OF LIQUID EXPLOSIVES AND FLAMMABLES

MR has demonstrated high potential to detect altered liquids from unopened bottles.[9] Due to their greater volatility at room temperature compared to beverages, there is a larger motional averaging of the local dipolar magnetic field in flammable and explosive liquids which, in general, helps to increase relaxation times (Figure 1). A number of FAA sponsored tests with a dedicated bottle scanner have indicated that a wide variety of liquids can be classified correctly by the MR technique.[11]

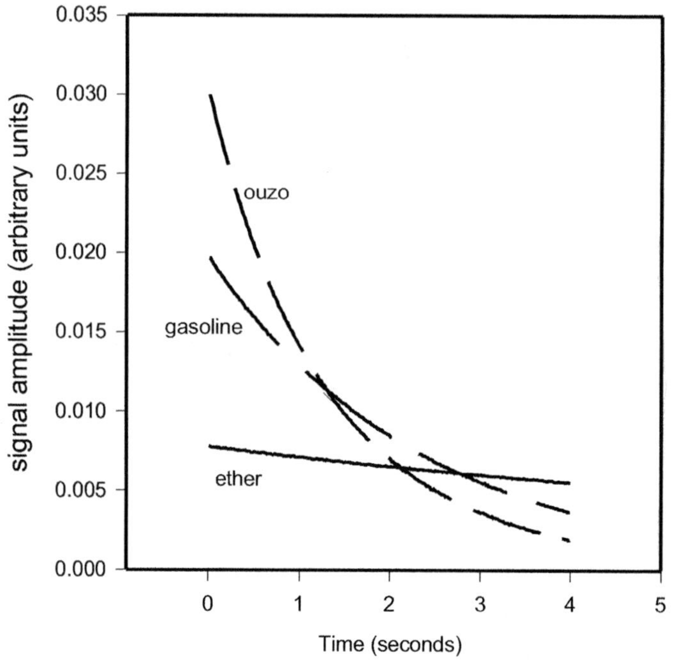

Figure 1. T_2 relaxation of Ouzo (a hard liquor), gasoline (a flammable liquid), and diethyl ether (a flammable and a toxic liquid). The differences in the relaxation times allow liquid classification. FAA sponsored tests with Quantum Magnetics' Bottle Scanner.[9]

However some liquid explosives, such as nitroglycerin are relatively viscous and have short relaxation times. Since both liquid hazards and benign liquids can have a range of relaxation times that could overlap, this can pose some challenges to the classification.

Combination of detection methods is likely the solution for achieving high detection performance (high probability of detection with low false positive levels). For example, it has been suggested that MR with microwave techniques may improve performance.[10]

3.2. DETECTION OF DISSOLVED DRUGS

A dedicated scanner has been used to measure the change in MR relaxation times due to dissolved cocaine hydrochloride (CocHCl).[16] Since drugs are often smuggled by dissolving them in common beverages, this ability can potentially be put to use at checkpoints to detect dissolved illicit drugs.

Dissolving cocaine hydrochloride in various beverages including water, wine, beer and whiskey significantly lowered the relaxation times.[16] The change in the relaxation times increased with increasing amounts of dissolved CocHCl.

Figure 2 displays the decrease in T_2 in whiskey as a function of increasing CocHCl concentration. The highest concentration on the plot corresponds to 6 g of CoCHCl dissolved in 30 ml of whiskey.

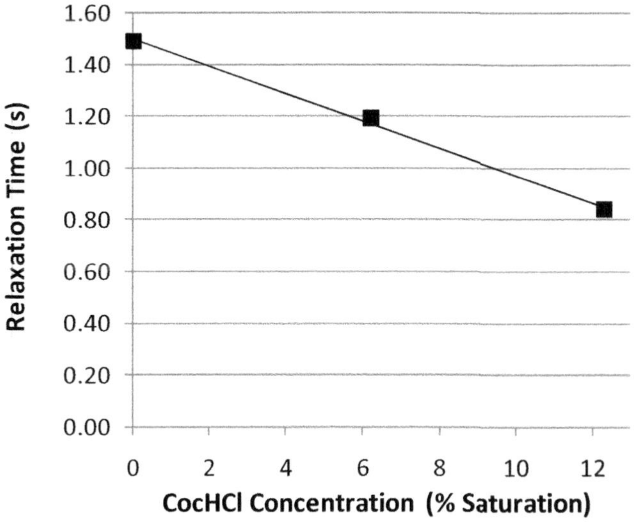

Figure 2. Relaxation time T_2 versus dissolved CocHCl concentration in whiskey.[16]

4. Discussion and conclusions

MR Relaxometry is an effective non-invasive, non-contact method to detect concealed liquid materials. Relaxometry results allow discriminating alteration of liquids without opening their containers. MR relaxometry is robust in the sense that it is independent of bottle shape and size and can be performed through the non-metallic packaging. MR is also a safe, non-ionizing technology that can be implemented at public checkpoints. Other bulk methods have been used to identify liquids and inspect bottles noninvasively. Raman spectroscopy has been used to identify liquid explosives in plastic and glass bottles.[17] Spatially offset Raman spectroscopy (SORS)[18] has been used to detect aqueous solutions of hydrogen peroxide, a key component of explosives such as TATP, inside transparent, colored and diffusely scattering plastic and glass containers. SORS permits the elimination of the fluorescence and Raman signals from the container itself that can be a significant source of interference in conventional Raman spectroscopy. FT-IR has also been used to identify explosives. For instance, Schulte–Ladbeck et al.[19] have measured the FT-IR spectrum of TATP in conjunction with liquid chromatography. Brown and Gozani[20] have measured the 10.8 MeV neutron capture gamma ray from the nitrogen nucleus to inspect bottles for liquid explosives. Harding and Harding[21] report the use of x-ray diffraction imaging to inspect bottles. Dielectrometry has also been used to classify liquids in bottles based on the dielectric response to microwave radiation.[6,22] A combination of bottle screening techniques may allow for a robust bottles inspection system. MR with its flexibility, robustness to bottles shape and size, and safety can play a central role in bottle screening.

References

1. United Nations' World Drug Report, United Nations Office of Drugs and Crime, 2008.
2. Schubert H., Kuznetsov A., Detection of Liquid Explosives and Flammable Agents in Connection with Terrorism, Springer, NATO Science for Peace and Security Series B: Physics and Biophysics (2008).
3. Krausa M., and Reznev A.A., Vapor and Trace Detection of Explosives for Anti-Terrorism Purposes: Proceedings of the NATO Advanced Research Workshop, Moscow, Russia, 19–20 March (2004).
4. Yinon J., Forensic and Environmental Detection of Explosives, Wiley, New York (1999).
5. Woodfin R.L., Trace Chemical Sensing of Explosives, Wiley, New York (2007).
6. Theisen L., Hannum D.W., Murray, D.W., Parmeter, J.E., Survey of Commercially Available Explosives Detection Technologies and Equipment. Document 208861, Sandia National Laboratories. Published by the U.S. Department of Justice (2005).
7. Perlo J., Demas V., Casanova F., Meriles C.A., Reimer J., Pines A., and Blümich B., High-Resolution NMR Spectroscopy with a Portable Single-Sided Sensor, Science 308, 1279 (2005).
8. Sears J., Interdigital dielectrometry based detection and identification of dangerous materials for security applications. Massachusetts Institute of Technology, Master Thesis (2003).
9. Kumar S., McMichael W.C., Kim Y.-W., Sheldon A., Magnuson E.E., Ficke L., Chhoa T.K.-L., Moeller C.R., Barrall G.A., Burnett L.J., Czipott P.V., Pence J.S. and Skvoretz D.C., Screening sealed bottles for liquid explosives, Proceedings of the SPIE Conference on Security Systems and Nonlethal Technologies for Law Enforcement, Boston, MA, Vol. 2934, pp. 126–137 (1997).
10. Mauler J. and Blümich B., Identification of liquids encountered in carry-on luggage by mobile NMR. NATO Workshop. St. Petersburg, Russia (2008).
11. Kumar S., Liquid-Contents Verification for Explosives, Other Hazards and Contraband by Magnetic Resonance, App. Magn. Reson. 25: 585–597 (2004).
12. Demas V., and Prado P.J., Compact Magnets for Magnetic Resonance. Concepts Magn. Reson. A, 34A, 48–59 (2009).
13. Blümich B., Blümler P., Eidmann G., Guthausen A., Haken R., Schmitz U., Saito K., and Zimmer G., The NMR-MOUSE: Construction, Excitation, and Applications, Magn. Reson. Imaging 16: 479–484 (1998).

14. Prado P.J., NMR Hand-Held Moisture Sensor. Magn. Reson. Imaging, 19: 506–508 (2001).
15. Prado P.J., Single-Sided Imaging Sensor. Magn. Reson. Imaging, 21: 397–400 (2003).
16. Kumar S., McMichael W.C., Magnuson E.E., Lee Y.K., Moeller C.R., Czipott P.V., Rayner T.J., Newman D.E., and Wroblewski D., Liquid contents verification for explosives, chemical agents, and dissolved narcotics, Proceedings of the SPIE Conference on Enabling Technologies for Law Enforcement and Security, Boston, MA, USA, Vol. 4232, pp. 206–216 (2001).
17. Lewis I.R., Daniel Jr. N.W., Chaffin N.C., Griffiths P.R., and Tungol M.W., Raman Spectroscopic Studies of Explosive Materials: Towards a Fieldable Explosives Detector, Spectrochimica Part A: Molec. Biomol. Spectrosc., 51(12): 1985–2000 (1995).
18. Eliasson C., Macleod N. A., and Matousek P., Noninvasive Detection of Concealed Liquid Explosives Using Raman Spectroscopy, Anal. Chem. 79: 8185–8189 (2007).
19. Schulte-Ladbeck R., Edelmann A., Quints G., Lendl B. and Karst U., Determination of Peroxide-Based Explosives, Using Liquid Chromatography with On-Line Infrared Detection, Anal. Chem, 78(23): 8150–8155 (2006).
20. Brown D.R. and Gozani T., Thermal Neutron Analysis Technology, Proc. SPIE 2936: 85–94 (1997).
21. Harding G. and Harding A., X-ray diffraction imaging for explosives detection, Counterterrorist Detection Techniques of Explosives, Ed. J. Yinon, Elsevier, London, pp. 199–235 (2007).
22. Seward IV D.C., LaChapelle J. and Seward III, D.C., Multi-field Microwave Reflective Sensors, Published United States Patent Application: US2002/0050828A1 (2002).

PROSPECTIVES AND LIMITATIONS OF NQR SIGNAL ENHANCEMENT BY POLARISATION TRANSFER

ALEXEI F. PRIVALOV[1*], ACHIM GÄDKE[1],
HOLGER STORK[1], DANUTA KRUK[2]
[1]*Inst. f. Festkörperphysik, TU Darmstadt, Hochschulstr. 6, 64289 Darmstadt, Germany*
[2]*Institute of Physics, Jagiellonian University, Reymonta 4, 30-059 Krakow, Poland*

Abstract In this contribution we discuss the prospectives and limitations of enhanced ^{14}N NQR of nitrogen containing compounds. Our considerations are exemplified by Nuclear Magnetic Resonance (NMR) and NQR field cycling experiments on two systems, namely TNT and urea. It turns out that the optimum signal-to-noise ratio depends in a subtle way on the spin-lattice relaxation rates of the involved spins, $T_1^{-1}(^1H)$ and $T_1^{-1}(^{14}N)$, as well as on the polarization transfer rate $R(^1H, ^{14}N)$ between the proton and the nitrogen nuclear spin reservoirs.

Keywords: NQR, NMR, TNT detection, enhanced ^{14}N NQR

1. Introduction

Research work aiming at the enhancement of ^{14}N Nuclear Quadrupole Resonance (NQR) signals of nitrogen containing compounds has largely been motivated by the humanitarian necessity to look for detection methods for anti-personal land mines with a metal content too low to be detectable by common types of metal detectors. Among such mines TNT-containing and even TNT-only mines are known to be of significant relevance.[1] Here, a specific challenge arises which is due to the fact that for TNT, in contrast to most other nitrogen containing explosives, the ^{14}N quadrupole frequencies are below 1 MHz.[2] These low frequencies render conventional zero field NQR detection of ^{14}N in TNT, under realistic and practical circumstances,

* Tel.: +49-6151-16-5122, Fax.: +49-6151-16-2833, E-mail: alexei.privalov@physik.tu-darmstadt.de

almost impossible. Although research has been conducted in this field for several decades – for two selected review papers we refer to references[3,4] – a substantial progress in the field is still missing. Partially, this suboptimal situation has been caused by private or military interests/considerations prohibiting early and free publications in an outspoken manner. The scene seems to have changed in the last years, and a number of detection schemes enhancing the TNT ^{14}N signal have been thought of and readily published, a representative selection of which is the following: advanced antenna technologies thus improving the S/R of pure ^{14}N NQR[5]; double resonance techniques[6]; two frequency irradiation techniques[7]; multipulse techniques in combination with two-frequency irradiation[8,9]; multipulse techniques in combination with polarisation transfer enhancement.[10,11] Most of these contributions deal with volume geometries; only a minority of groups have gone a step further in applying surface coils. In this respect, to our mind, the Ljubljana group[10,11] seems to be leading.

Our own approach to this field, so far also being restricted to volume geometries, has been triggered by having at hand an electronic field cycling NMR spectrometer.[12] Using this spectrometer, we have been looking at the TNT detection problem in a twofold manner, namely comparing a "direct" ^{14}N detection technique and an "indirect" ^{1}H detection technique with each other. Basically, the experiment works as follows: first, we polarise the TNT protons in a magnetic field as high as possible; second, we transfer this polarisation to the ^{14}N spins and, third, we detect either the enhanced ^{14}N NQR signal at zero field (direct method) or we detect the reduced ^{1}H polarisation at high field (indirect method). Applying the latter method in a volume coil geometry, an amount of as little as 0.5 g TNT could be detected in a single shot experiment.[13] Despite this encouraging preliminary result it has always been obvious that this success still leaves the major problems unsolved: (a) an indirect detection scheme imposes higher requirements upon the homogeneity of the applied magnetic field, and mainly (b) the development of a coil system which allows for performing such experiments in a surface geometry. We are well aware of the fact that the latter problem represents the huge challenge which has to be solved if ever hybrid NQR/NMR techniques will be applied to TNT landmine detection. For the present purpose, let us restrict ourselves to problem (a), thus critically comparing the feasibility of direct and indirect detection schemes. For many applications a restriction to volume geometries is still relevant enough, for example, the detection of flat slices of explosives or narcotics such as mail or bag sheets. In the following (Section 2) we present basics of polarisation transfer and the techniques we are dealing with. In Section 3 we recall experimental results on two materials which will serve to demonstrate two alternative situations which one may encounter when doing such experiments.

For all technical and instrumental details we will refer to previous publications. These representative sets of results are the basis for the discussion (Section 4) which also contains some information on basic requirements on surface geometry experiments. Finally, we collect the most important conclusions (Section 5).

2. Polarization transfer spectrum

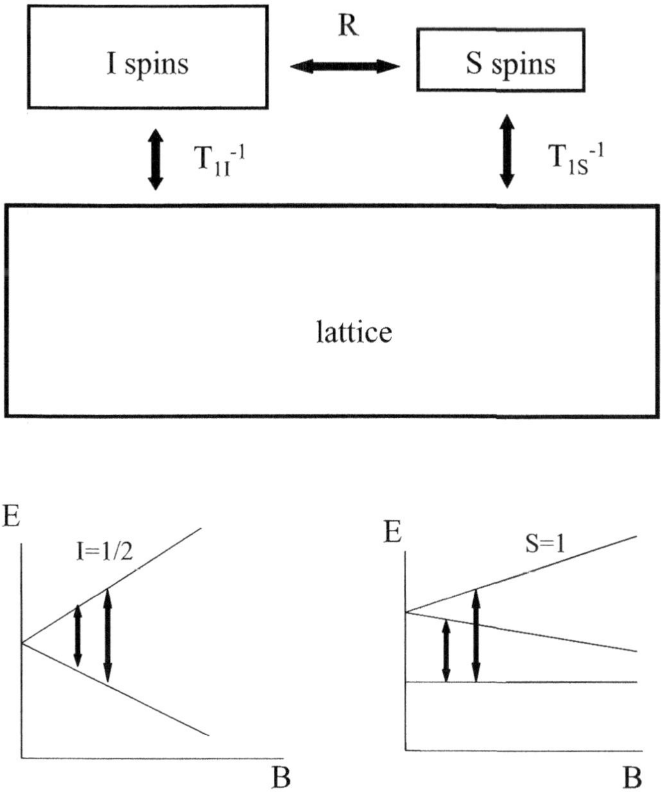

Figure 1. Top: Energy reservoirs of two spin systems (I-spins and S-spins) and the lattice. The leak rates (spin-lattice relaxation rates, T_{1I}^{-1} and T_{1S}^{-1}, and the polarization transfer rate R among the two spin systems) are indicated. Bottom: schematic energy level diagrams for the case I = 1/2 (Zeeman interaction only) and S = 1 (with nonzero electric quadrupole coupling constant); the arrows indicate magnetic field positions where energy conserving flip-flop processes among I- and S-spins are possible.

Given a system of spins I (e.g. ^{1}H-spins, I = 1/2) and a system of spins S (e.g. ^{14}N-spins, S = 1) in an external magnetic field B. An instructive thermodynamic description involves three subsystems, namely the I-spin system, the S-spin system and the lattice (Figure 1, top). As indicated in the figure, the "leaks" among the subsystems are described by corresponding spin-lattice relaxation rates and a polarisation transfer rate R, occasionally also called cross-relaxation rate.[14] It is well known that the polarization transfer rate R is nonzero only if energy conserving flip-flops between I-spins and S-spins are possible. If the S-spins are subjected to electrical quadrupolar interactions, the situation is as represented in Figure 1, bottom. Here, polarisation transfer occurs only at certain values of the external magnetic field characteristic for the quadrupole energy splitting of the S-spins. In this sense the polarisation transfer spectrum contains information on the S-spin quadrupole spectrum.

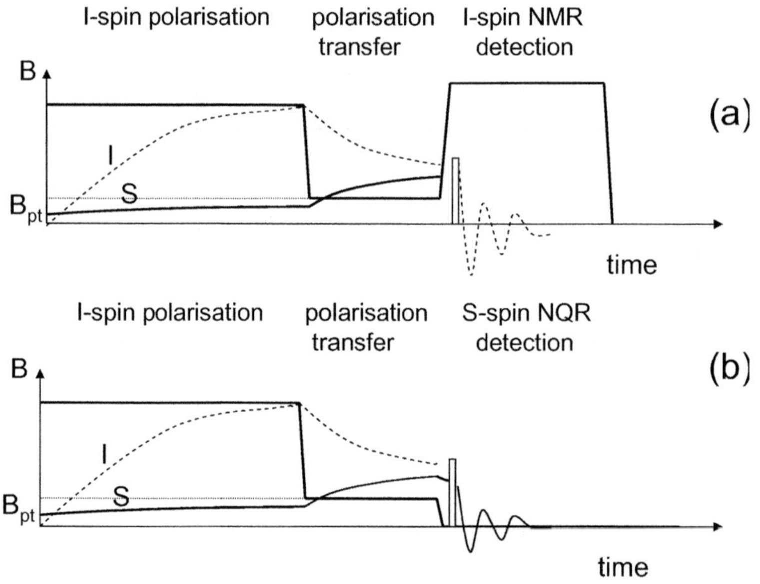

Figure 2. Timing diagrams of (a) the indirect and (b) the direct detection scheme. In (a) the I-spin signal (^{1}H-NMR) is measured in high field after a polarisation transfer period; in (b) the S-spin signal (^{14}N) is measured in zero field.

There are two ways to measure the polarisation spectrum (and thereby the S-spin quadrupole spectrum), either by I-spin NMR (indirect method) or by S-spin NQR (direct method) relaxometry. Both methods can provide a strong signal enhancement. This is depicted in Figure 2. In a polarisation period a high magnetic field is applied and a I-spin polarisation is built up, stronger than that of the S-spins. If subsequently the magnetic field is switched to a value such that polarisation transfer occurs (arrows in Figure 1, bottom) energy will flow among the two spin systems and a common thermal equilibrium will be built up with a rate R. The new equilibrium polarisations of either spin system can now be detected by I-spin NMR (indirect method, Figure 2a) or by S-spin NQR (direct method; Figure 2b).

For instance, Figure 2b qualitatively demonstrates the enhancement of the S-spin NQR signal as compared to conventional pure NQR. In practical cases an enhancement factor of about one order of magnitude may be reached, see below.

3. Experimental results

3.1. INDIRECT DETECTION SCHEME: TNT

TNT, the most relevant molecule, has been detected using the indirect detection scheme.[13] The experimental results, together with the corresponding transition energy diagram, are plotted in Figure 3. Note that in contrast to Figure 1, Figure 3a does not show the energies but the energy splittings.

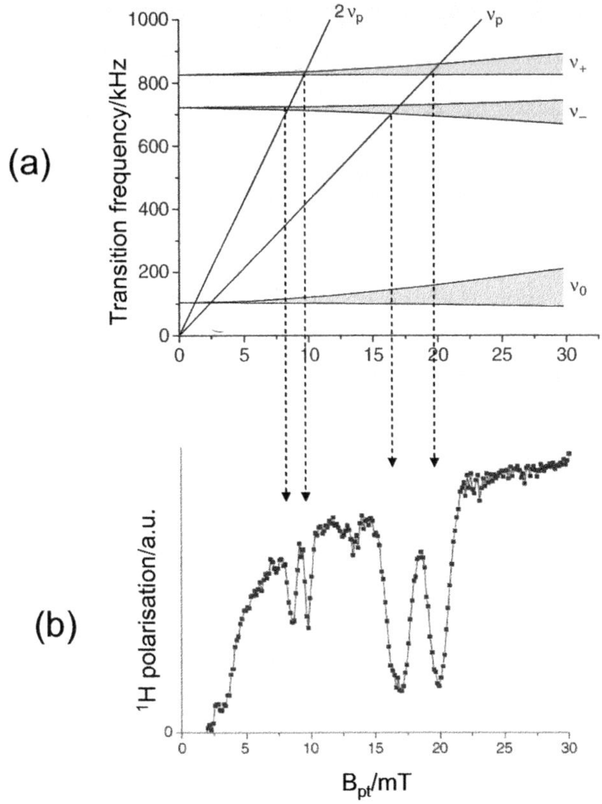

Figure 3. (a) Energy level splitting (given in frequency units) of ^{14}N, denoted by ν_+, ν_0 and ν_-, and those of ^1H (denoted as ν_p) and its double ($2\nu_p$) in TNT according to the quadrupole coupling parameters $e^2qQ/h = 1.037$ MHz, $\eta = 0.21$.[17] (b) Measured ^1H NMR signal intensity in TNT as a function of the polarisation transfer field (indirect detection scheme).

Such a plot shows directly the positions of the polarisation transfer fields. Whenever energy splittings of the I-system (^{1}H) equal those of the S-system (^{14}N), polarisation transfer occurs which is reflected as a dip in the ^{1}H NMR signal. From inspection of Figure 3 we learn that polarisation transfer processes do not only occur when a transition energy in the ^{14}N system, $\Delta E(^{14}N)$, equals to that in the ^{1}H system, $\Delta E(^1H)$, but also when $\Delta E(^{14}N) = 2 \cdot \Delta E(^1H)$.

For a quantitative understanding of the polarization transfer spectra the Liouville–von Neumann equation $\frac{d}{dt}\rho(t) = -\frac{i}{\hbar}[H, \rho(t)]$ (where $\rho(t)$ is the density operator of the considered system) has to be solved.[15,16]

To describe polarization transfer occurring within nuclei of TNT molecule one should consider at least the five spin system N–H–N–H–N[17] accounting for dipole-dipole interactions among these spins.

This approximation is justified by the strong distance dependence of the dipole–ipole interactions and the fact that the dipolar coupling between the methyl group protons and the adjacent nitrogens is partially averaged out because of the fast methyl group rotation. The full spin Hamiltonian

$$H = H_I + H_S + H_{DD} = \sum_{J=1}^{2} H_{I,Z}^{(j)} + \sum_{j=1}^{3}\left[H_{S,Z}^{(j)} + H_{S,Q}^{(j)}\right] + \sum_{\substack{j,k=1 \\ j<k}}^{5} H_{DD}^{(j,k)}$$

consists of the sum of those of all five particles, the two I-spins being subjected only to the Zeeman interaction ($H_{I,Z}$), the three S-spins subjected to the Zeeman ($H_{S,Z}$) and the quadrupolar ($H_{S,Q}$) interactions, and the mutual dipole–dipole interactions among any spin pair (j, k). Since H is time independent the formal solution of the Liouville–von Neumann equation assumes the form:

$$\rho(t) = e^{-\frac{i}{\hbar}Ht}\rho(0)e^{\frac{i}{\hbar}Ht} = \sum_{\alpha,\beta=1}^{N}\langle\alpha|\rho(0)|\beta\rangle\exp\{-i\omega_{\alpha\beta}t\},$$

where $|\alpha\rangle$ and $|\beta\rangle$ are eigenstates of the system obtained by diagonalizing the Hamiltonian, while $\omega_{\alpha\beta} = \omega_\alpha - \omega_\beta$ is equal to the difference in the eigenvalues between the states. In the present case the number of states is equal to $N = (2S+1)^3(2I+1)^2 = 108$. The proton polarisation is proportional to the expectation value $\langle I_z\rangle$ of the operator $I_z = I_{1z} + I_{2z}$, that is given as:

$$\langle I_z \rangle = Tr\{\rho(t) I_z\} = \sum_{\alpha,\beta=1}^{N} \langle \alpha | \rho(0) | \beta \rangle \langle \beta | I_z | \alpha \rangle \exp\{-i\omega_{\alpha\beta} t\}$$

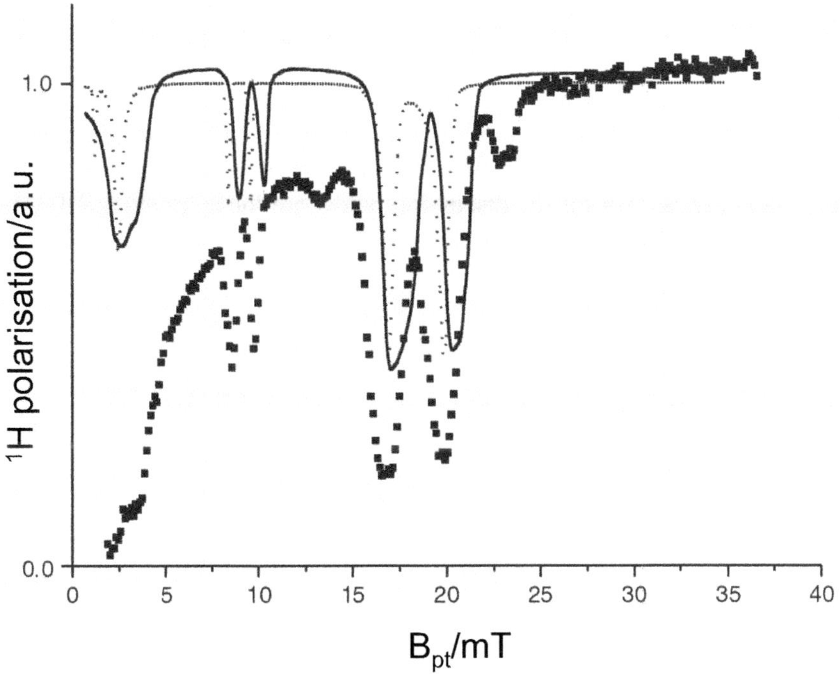

Figure 4. Experimental ^{1}H signal intensity in TNT as a function of B_{pt} and comparison to theoretical curves as described in the text. Dotted curve: only one site ($e^2qQ/h = 1.037$ MHz, $\eta = 0.21$); full curve: six sites.

Figure 4 shows the comparison of experiment and this kind of theoretical approach. The figure indicates that the polarisation transfer dips are correctly described only if more than one nitrogen site, corresponding to more than one quadrupolar coupling constant, is assumed.[17] The experimentally observed gradual decrease of the I-spin polarisation with lower magnetic field is not reproduced by the theoretical curve since spin-lattice relaxation effects are neglected in the calculation, that is, it is assumed that $T_{1I}^{-1} = T_{1S}^{-1} = 0$. Experimentally[18, 20] however, it is found that T_{1I}^{-1} (40 MHz) ≈ 0.02 s^{-1}, R ≈ 100 s^{-1} and T_{1S}^{-1}(B = 0) ≈ 0.2 s^{-1}. These remarks will be of importance for the discussion, below.

($\left[\frac{e^2qQ}{h}\Big/\text{MHz}, \eta\right]$ = [1.056, 0.303], [1.087, 0.170], [1.064, 0.208], [1.045, 0.253], [1.068, 0.172], [1.052, 0.179]), taken from reference[20].

3.2. DIRECT AND INDIRECT DETECTION SCHEMES: UREA

The direct detection scheme (Figure 2b) has successfully been applied to enhance the ^{14}N polarisation of urea (CON_2H_4) and to detect it in zero field (Figure 5).[19] The parameters are given in the figure caption. The enhancement factor, being ≈ 6 in the actual experiment, is determined by $t_{pol} = 300$ s ≈ $T_1(^1H)$, where the proton system had not yet reached its full polarisation. For an asymptotically long polarisation time ($t_{pol} = 900$ s) an enhancement factor of ≈ 8 is reached (not shown). For all experimental details we refer to reference[19]. Important for the discussion below are the spin-lattice relaxation rates and the polarisation transfer rate:

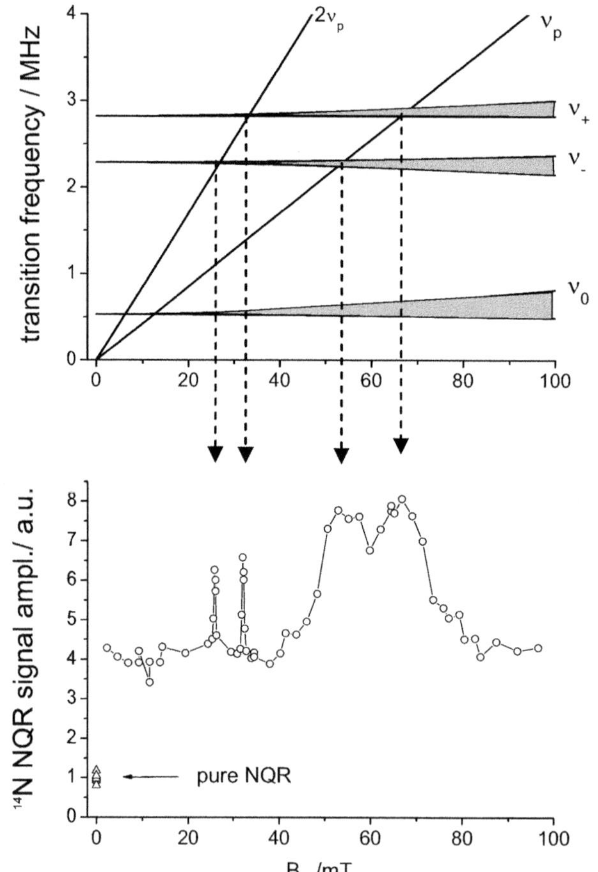

Figure 5. (a) Energy level splitting (given in frequency units) of ^{14}N, denoted by ν_+, ν_0 and ν_-, and those of ^1H (denoted as ν_p) and its double ($2\nu_p$) in urea according to the quadrupole coupling parameters $e^2qQ = 3.46$ MHz, $\eta = 0.29$.[21] (b) Measured ^{14}N NQR signal intensity in urea as a function of the polarisation transfer field (direct detection scheme).

$T_1^{-1}(^1H) \approx 0.003\ s^{-1}$ at 0.7 T; $T_1^{-1}(^{14}N) \approx 0.5\ s^{-1}$ at zero field; $R \approx 1{,}000\ s^{-1}$. As can immediately been seen from Figure 5, ^{14}N polarisation enhancement peaks also appear at those values of B_{pt} where double proton flips occur. And there is a B_{pt} independent ^{14}N polarisation enhancement by about a factor of 3 due to a mere fact that at B_{pol} also the ^{14}N spins reach a higher polarisation as compared to zero field and to slight contributions stemming from sweeping the magnet down thus crossing the polarisation transfer fields.

Urea has also successfully been observed by the indirect method detection scheme (Figure 2a),[17] plotted in Figure 6. The procedure very much resembles that of the measurement of TNT and needs no further explanation.

4. Discussion

4.1. COMPARISON OF ENHANCED ^{14}N NMR/NQR DETECTION SCHEMES

For discussing the feasibility of polarisation transfer enhanced NQR we must have in mind the quantitative values of the "leak" rates between the three energy reservoirs shown in Figure 1a. Let us recall these rates for TNT and urea.

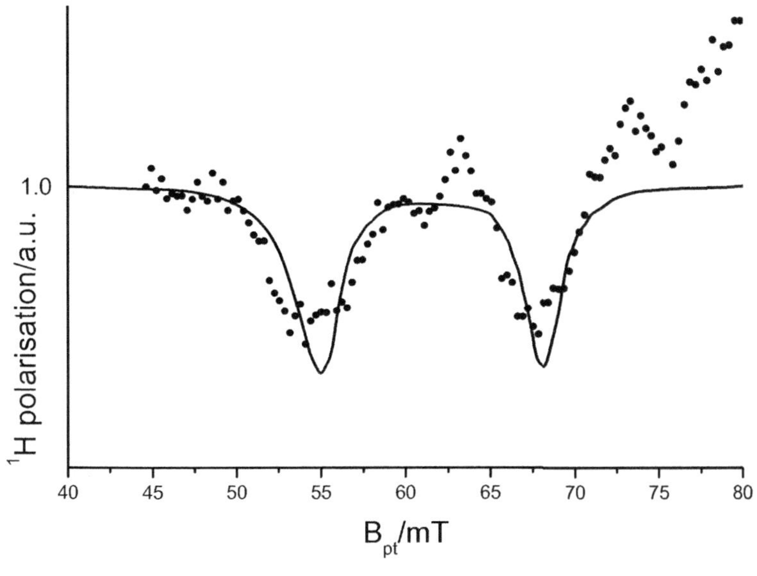

Figure 6. ^1H NMR signal (indirect detection scheme) as a function of polarisation transfer field B_{pt} for urea. The ^1H polarisation time has been set to 100 s and the polarisation transfer time to 20 ms.

In the case of TNT we observe polarization transfer dips applying the indirect method. But we are not able to detect directly an enhanced NQR spectrum. This is surprising since the ^{14}N spin-lattice relaxation rate[20] of 0.2 s^{-1} at zero field is significantly below the polarisation transfer rate R, which is of the order of 100 s^{-1}.[18] This implies that there should be some ^{14}N signal enhancement. However, the small proton polarisation rate in the polarisation field (about 0.02 s^{-1}) does not allow to transfer polarisation as fast as desired, thus representing a bottleneck. By the way, the difficulties of NQR detection might have been due to other technical problems involving piezoelectric vibrations being stronger than the desired NQR signal.

In the case of urea we observe an enhanced ^{14}N NQR spectrum by both, the indirect and the direct method. Here, again, the ^{14}N spin-lattice relaxation rate at zero field turns out to be much smaller than the polarisation transfer rate R. Thereby, polarisation can effectively be transferred to the ^{14}N system leading to an enhancement of the ^{14}N NQR signal of about 6 when the polarisation time is about equal to the proton spin-lattice relaxation time. Nevertheless, considering urea as exemplifying case, the usefulness of this method is still questionable. This is because at the polarisation field $T_1^{-1}(^1H)$ is of the order of 0.003 s^{-1} such that after having done a transport step once one has to wait very long time before the next one can be effected. Since $T_1^{-1}(^{14}N) \approx 0.5\ s^{-1}$ one can accumulate the conventional ^{14}N signal about (or even more than) 100 times yielding an enhancement given by the signal-to-noise ratio of 10 which significantly exceeds the enhancement factor of 6 by polarisation transfer enhanced NQR.

In conclusion, we do not see the usefulness of the direct detection method of polarisation transfer enhanced ^{14}N NQR. Except for possible special cases, which we do not have at hand, we should weigh the enhancement factors with those obtained from accumulating the standard NQR signal.

4.2. BASIC REQUIREMENTS FOR INDIRECT SURFACE DETECTION

As far as the practical use of the indirect method, which indeed works formidably well for TNT in a volume geometry, is concerned we have to be aware of the enormous problems when turning to a surface geometry. To our knowledge, an electronically or mechanically "switchable" magnetic field set up in surface geometry yielding a sufficiently large B_0 at a few cm depth which is sufficiently homogeneous over some relevant volume has never been demonstrated. In principle, such a magnet, which should either be able to be moved up and down fast enough ("mechanical field cycling") or the current of which should be able to be switched as required ("electronical field cycling"), can be constructed. First attempts are being done in our laboratory.[22] However, even disregarding the question of adequateness of such a technique for outdoor applications, its construction is an enormous technical task, and the success is not guaranteed. Here, we limit ourselves in discussing only some basic requirements for such a surface magnet system.

4.2.1. *Field homogeneity*

At room temperature the spin-spin relaxation time T_2 of TNT is of the order of 25 μs. Therefore it is of great advantage to detect the NMR signal directly from FID instead of using echos. This leads to some requirements concerning the field homogeneity during the detection phase. Since the length of the FID is given by $\frac{1}{T_2^*} = \frac{1}{T_2} + \gamma\,\Delta B$, the FID decays within the dead time of the spectrometer if the field inhomogeneity ΔB is too large. Since the dead time decreases linearly with frequency but, for a given magnet geometry, ΔB increases linearly, an optimum for the field strength has to been found. Also note that the lower the frequency is the higher the inductivity and the more sensitive the rf coils are.

4.2.2. *Field switching rate*

There are also some requirements for the field switching rate. In order to identify a polarization transfer dip in Figure 3 at least two NMR experiments at an evolution field within a dip and beyond it have to be performed. When the magnetization of the latter experiment is significantly higher, this is a clear evidence for the presence of explosives. If the switching rate is too slow, significant amounts of the magnetization will decay during passing the field range where cross-polarisation takes place. Especially at measurements with low signal-to-noise ratio this can reduce the contrast between the two magnetizations in a way that a clear detection of explosives is not longer possible. In the case of TNT the influence of the switching rate on that contrast has been studied (Figure 7). The contrast gets significantly reduced for switching rates below 2 T/s. This result is of some relevance since for a switching rate of this order of magnitude mechanical field cycling, that is, adequately moving a permanent magnet, still appears to be feasible.

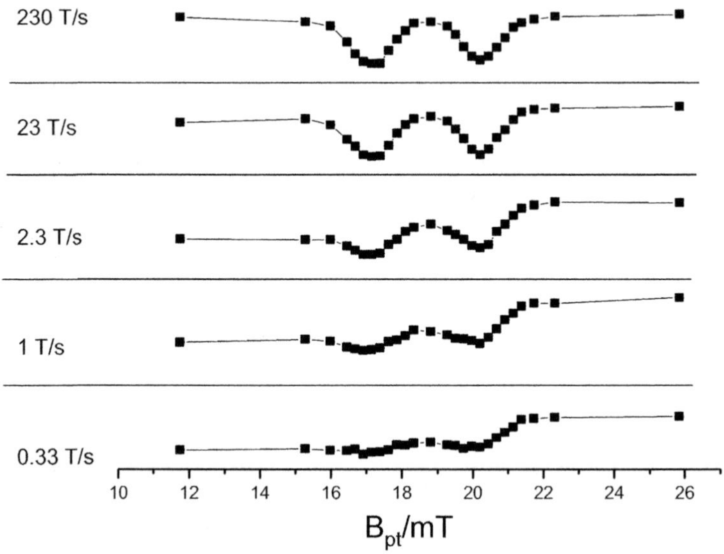

Figure 7. Evolution field dependent ^{1}H polarisation in TNT for various field switching rates as noted in the plot. The sample has been polarised at 0.7 T for 60 s. Then it has been exposed for a period of 4 s to the field as noted on the x-axis. The NMR signal has been detected at 0.94 T.

5. Conclusions

A critical discussion of potential polarisation transfer enhancement methods of ^{14}N NQR arrives at several conclusions:

1. The direct method with prepolarisation is not efficiently applicable to the detection of TNT and is not likely to be applicable to other substances as well.
2. If polarisation transfer enhancement of TNT is thought to be necessary, we are bound to indirect detection methods. In a volume coil, indirect detection works well.
3. Despite this success, so far the construction of a suitable surface magnetic field coil remains an unsolved, but not impossible, task. Estimates for the field homogeneity are straightforward, and the required field switching rates are low enough to permit mechanical field cycling.
4. Nevertheless, in contrast to our previously stated belief that pure ^{14}N NQR detection of TNT is impossible we now tend to propose to invest more efforts in trying to improve the direct (zero field) ^{14}N NQR signal by better rf-antennas, multi-pulse sequences, filtering methods, data processing procedures and so forth.

5. However, if volume geometry applications are thought of indirect detection of polarisation transfer enhanced ^{14}N NQR might still be a method the potential of which should be kept in mind.

Acknowledgments The work has strongly profited from significant contributions of our electronic and mechanical institute workshops. In its early state, the work has been funded by the Deutsche Forschungsgemeinschaft (FU 308/7). We also like to acknowledge contributions of Dr. J. Altmann (Dortmund), Dr. V. Anferov (Kaliningrad) and Y. Tchernychev (St. Petersburg).

References

1. Land Mines and Demining – A Global Problem, CD-ROM, National Ground Intelligence Center, Charlottesville VA
2. Miller JB, Barall GA, Am. Sci. 93, 50 (2005)
3. Grechishkin VS, Sinyavskii NY, Physics – Uspekhi 40, 393 (1997)
4. King JD, De Los Santos A, Appl. Magn. Res. 25, 535 (2004)
5. Pannetier M, Science 304, 1648 (2004)
6. Prescott DW, Olmedo O, Soon S, Sauer KL, J Chem. Phys. 126, 204504 (2007)
7. Seliger J, Zagar V, Blinc R, J. Magn. Res. 106, 214 (1994)
8. Mozjoukhine GV, Appl. Magn. Res. 22, 31 (2002)
9. Sauer KL, Suits BH, Garroway AN, Miller JB, J. Chem. Phys. 118, 5071 (2003)
10. Luznik J, Pirnat J, Trontelj Z, Solid State Commun. 121, 653 (2002)
11. Luznik J, Pirnat J, Jazbinsek V, Apih T, Gregorovic A, Blinc R, Seliger J, Trontelj Z, Appl. Phys. Lett. 89, 123509 (2006)
12. Lips O, Privalov AF, Dvinskikh SV, Fujara F, J. Magn. Res. 149, 22 (2001)
13. Nolte M, Privalov A, Altmann J, Anferov V and Fujara F, J. Phys. D 35, 939–942 (2002)
14. Fujara F, Stöckmann HJ, Ackermann H, Buttler W, Dörr K, Grupp H, Heitjans P, Kiese G, Körblein A, Z. Phys. B 37, 151 (1980)
15. Kruk D, Theory of Evolution and Relaxation of Multi-Spin Systems – Application to Nuclear Magnetic Resonance (NMR) and Electron Spin Resonance (ESR), Abramis Academic, Arima Publishing, UK (2007)
16. Kruk D, Lips O, Solid State Nucl. Magn. Reson. 28, 180 (2005)
17. Kruk D, Altmann J, Fujara F, Gädke A, Nolte M and Privalov AF, J. Phys. Condense. Matter 17, 519 (2005)
18. Nolte M, Grundlagen zur TNT-Detektion mittels magnetischer Resonaz, PD Thesis, TU Darmstadt (2005)
19. Privalov AF, Kruk D, Gädke A, Stork H, Fujara F, J. Phys. D: Appl. Phys. 40, 7555 (2007)
20. Blinc R, Seliger J, Arcon D, Cevc P, Zagar V, Phys. Statur Solidi A 180 (2000)
21. Landolt-Börnstein, Numerical Data and Functional Relationships in Science and Technology vol. 3.3.20a, Springer, Heidelberg, p. 160, 244 (1988)
22. Stork H, Privalov AF, Fujara F, work in progress

MODELING OF QR SENSORS FOR OPTIMIZED EXPLOSIVES DETECTION

HECTOR ROBERT*, ALEJANDRO BUSSANDRI, KEVIN DERBY
Quantum Magnetics, Inc. 15175 Innovation Dr. San Diego, CA 92128

Abstract Quadrupole Resonance (QR) sensors have the unique capability of detecting explosives with remarkably high detection rates and low number of false alarms. The sensitivity of a QR-based sensor in inductive detection can be assessed in terms of the signal-to-noise ratio (SNR), which determines the Receiver Operating Characteristics (ROC) curves of the detector and provides a fundamental limitation to the performance of the QR explosive detection system. The main goal of the QR sensor design is, therefore, to maximize the SNR to achieve the highest possible detection performance with the lowest number of nuisances.

This paper describes part of the work performed at Quantum Magnetics Inc. to model the characteristics of the QR signal detection process, which includes the receiver's response, data processing, and signal detection algorithm. These theoretical and numerical models allow us to predict ROC curves for the QR detector and evaluate trade-offs between experimental parameters, sample characteristics, data processing, and hardware features of the detector. Numerical and experimental results are presented to validate our models, and demonstrate their usefulness for QR sensor design and optimisation.

Keywords: QR sensor, explosive detection, SLSE pulse sequence, TNT

1. Introduction

QR is a radio-frequency spectroscopic technique where the signal frequency is related to the characteristics of the nucleus and the crystalline structure of

* Tel: 858.605.5500, Fax: 858.605.5504, www.gesecurity.com, hector.g.robert@ge.com

J. Fraissard and O. Lapina (eds.), *Explosives Detection Using Magnetic and Nuclear Resonance Techniques*,

the compound.[1,2] Therefore, QR has the capability of detecting and identifying numerous compounds, including explosives such as TNT, Tetryl, RDX, and PETN, and components of commercial explosives like ammonium nitrate (AN), potassium nitrate, and others. The chemical specificity of the QR response make this technique an excellent candidate to detect explosives concealed in non-metallic containers or buried in the ground. In a fully automated system configuration, a simple "alarm" or "clear" indicator informs the operator whether a threat is present or not in the inspected volume. Hand-portable and vehicle-mounted systems have been demonstrated to detect bulk explosives in antipersonnel and antitank landmines; closed-volume inspection system; such as, baggage, packages, and mail scanners, have also been developed and tested.[3,4]

QR detection of explosives predominantly focuses on materials containing ^{14}N nuclei[5] and it works by sending radio wave pulses with coils surrounding or nearby the volume of interest and detecting an induced voltage when an explosive is present. The nuclei of the explosive's molecules absorb energy from the specially tuned radio frequency (RF) pulses as they align themselves in the electromagnetic field produced by the sensor's resonant coil. When the RF pulse stops, the nuclei precess during a transient period, inducing a measurable voltage in the detection coil.

It is relatively straightforward to observe QR signals for a wide variety of explosive materials in laboratory conditions. There are, however, several difficulties to overcome with the QR techniques to detect the presence of these materials in real operating environments with a low probability of false alarms (PFA) and a high probability of detection (PD). One of the main limiting factors is the intrinsic low sensitivity of the QR technique and it is, therefore, imperative to extract the maximum possible sensitivity from the sensor to achieve optimum performance with the shortest possible scan time.

The purpose of this paper is to describe the modeling of a complete QR sensor for optimizing performance in the presence of thermal noise. Signal-to-Noise Ratio (SNR) per unit time; i.e., sensitivity, is the main parameter to be evaluated for QR explosive detection when the objective is to determine the presence or absence of a resonance line, while no other spectroscopic information (line shape, relaxation parameters) needs to be retrieved. The SNR determines the Receiver Operating Characteristic (ROC) curve of the detector, and provides a fundamental limitation to QR signal detection. Another important application of the numerical model for the QR sensor is in the design for quality and reliability in the production process; the model allows quick evaluation of the tradeoffs between design parameters and system's performance.

Suits et al.[6] have examined the SNR with increasing detection coil quality factor Q for the detection of known QR signals in noise. They found that in the absence of amplifier noise, the maximum obtainable SNR is achieved in the "super-Q" limit, when the bandwidth of the tuned sensor is smaller than that of the signal to be detected. Their analysis is limited to the case of a simplified echo signal. The sensor's analysis presented here considers the QR responses detected with multipulse spin-lock sequences[7,8] that are commonly used for enhanced explosives detection.

The purpose of a QR sensor is to discriminate the thermal noise from the QR responses evoked in the material by the excitation pulse sequences.

Figure 1 shows a block diagram of a QR sensor, which comprises three main components:

- Receiver subsystem: includes the QR coil and a low-noise preamplifier. In the absence of an explosive, the receiver picks up only the thermal noise voltage that determines the PFA. In the presence of a threat, the receiver picks up the voltage induced by the QR response plus the thermal noise, which determines the PD.
- Signal processing unit: aims at maximizing the signal-to-noise ratio. After digitalization of the voltages acquired and amplified by the receiver, the data is filtered using the known echo shape and the echo time decay constant, T_{2E}, of the QR response. Fourier transformation and proper phasing of the complex frequency spectrum yields the input to the detection algorithm.
- Signal detector: makes a decision on the presence of a threat. The simplest detection algorithm is a peak picking that extract the largest peak in the frequency domain within the expected frequency range of the QR signals and compares its intensity to a predefined threshold to determine whether an explosive is present or not.

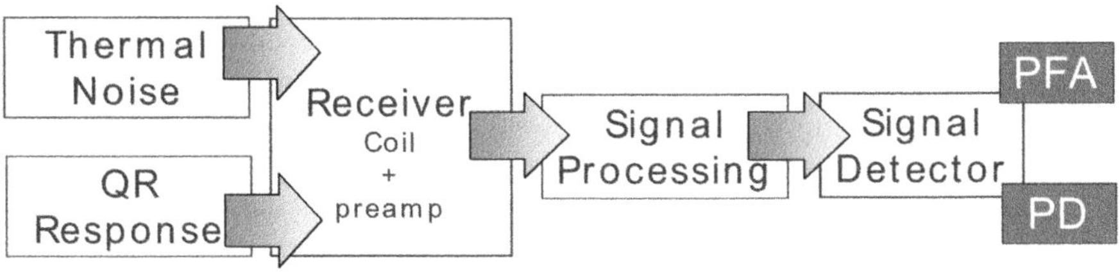

Figure 1. Block diagram of a QR sensor showing the main components for the model: receiver subsystem, signal processing unit, and signal detector.

2. QR detection performance modeling

The SNR of the detector is defined as the ratio of peak signal voltage to root-mean-square (rms) value of the noise voltage,

$$SNR = \frac{\text{peak signal voltage}}{\text{rms noise voltage}} = \frac{F_S}{F_N},$$

where $F_S = \max\{FFT(S_{Total}(t))\}$ is the peak height in the QR spectrum, resulting from the Fourier transform (FFT) of the total time-domain QR response. A number of techniques are utilized for efficient excitation and signal detection. For instance, the spin-lock spin-echo (SLSE) pulse sequence[9] allows significant SNR enhancement by prolonging the lifetime of the QR echo signals and enabling co-addition of multiple echoes for increased sensitivity. The time-domain QR signal is filtered and averaged and the total signal $S_{Total}(t)$ is then transformed to the frequency domain, resulting in total signal power F_S.

The noise voltage F_N is the mean noise amplitude in the frequency domain that originates in the total resistance R_c of the circuit at the resonance frequency, together with the resistance of any conductors inductively coupled to the coil. To simplify the present analysis, we assume only random thermal noise and disregard the presence of external interferences.

The numerical model has to address the calculation of QR responses and thermal noise levels as a function of pulse sequence parameters. In the following sections, we described the calculations of QR response; we present our model to calculate the noise power spectra, and we compute the SNR as a function of pulse sequence parameters and the determination of optimum values.

2.1. QR RESPONSE MODELING

Figure 2 shows the timing of a typical SLSE pulse sequence[7–9] and the definition of several parameters involved in our calculations. This pulse sequence is typically used for detection of materials with strong spin-lock response; such as PETN, AN and TNT. The sensitivity of the SLSE pulse sequence is evaluated numerically as a function of length of the acquisition window (Acqw), number of echoes, pw90, pw180, and dead time (Dt).

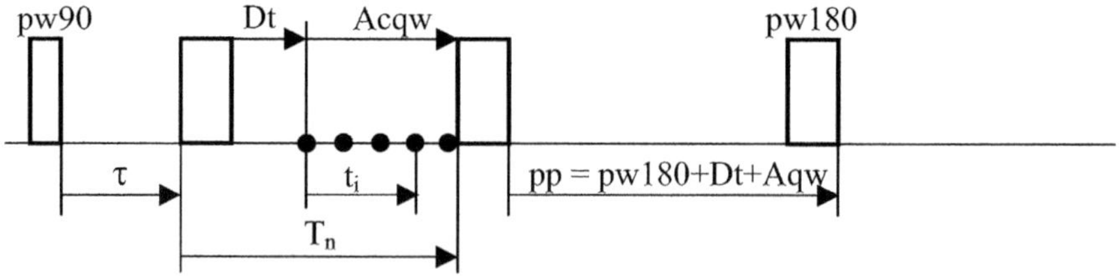

Figure 2. Schematic of the spin-lock spin-echo (SLSE) pulse sequence showing the definition of relevant parameters.

The total time-domain QR signal for a single SLSE scan is

$$S_{Total} = \sum_{n=1}^{N_E} F_n \sum_{i=1}^{Np} S_{i,n} E_i \, .$$

The function $S_{i,n} = S(t_i, T_n)$ is the time-domain QR echo signal (t_i and T_n are defined in Figure 2). For sake of simplicity, only on-resonance experiments are presented here, modifications for off-resonance signals are straightforward and will be address in future work. The QR echo shapes $S_{i,n}$ evoked by the excitation pulse sequence can be calculated from first principles by using quantum mechanical calculations of the spin evolution or simpler Bloch equations model. Detailed calculations of the echo shapes were published elsewhere[7,8,10] and they are not reproduced here. The calculations shall include the QR parameters (resonance frequency, lineshape, T_{2E}, etc.) and pulse sequence timing described above.

The echo signals are filtered with matched filter E_i of length N_p, $i = 1 \ldots N_p$, to maximize SNR, and the coefficients of the exponential filtering for the time-domain echo train are

$$F_n = \exp\left(-\frac{(n-1)\,dw}{T_{2E}} \right), \quad \mathrm{n} = 1 \ldots \mathrm{Ne},$$

where N_E is the number of echoes in the pulse train.

With the SLSE sequence, the on-resonance QR parameter T_{2E} is a function of pulse widths, pw180, and the pulse-to-pulse period T_n. The dependence of T_{2E} with the timing of the pulse sequence is determined experimentally, and not calculated from first principles. As an example, Figure 3 shows the dependence of the echo train time decay constant for a TNT sample at 842 kHz measured with RF field strengths of 20 and 40 G. For this sample, experimental results show a nearly linear dependence of T_{2E} with the pulse period of the SLSE pulse sequence over the interval of interest.

The data shown in Figure 3 can be fitted with a linear expression $T_{2E}[ms] = b_0 + b_1 T_n$, and the calculated coefficients of the linear fit are

$$b_0 = 197.7\ ms, \quad b_1 = -0.174 \frac{\mathrm{ms}}{\mu s} \text{ for } \mathrm{B_1} = 20 \text{ G}$$

$$b_0 = 202.4\ ms, \quad b_1 = -0.199 \frac{\mathrm{ms}}{\mu s} \text{ for } \mathrm{B_1} = 40 \text{ G}$$

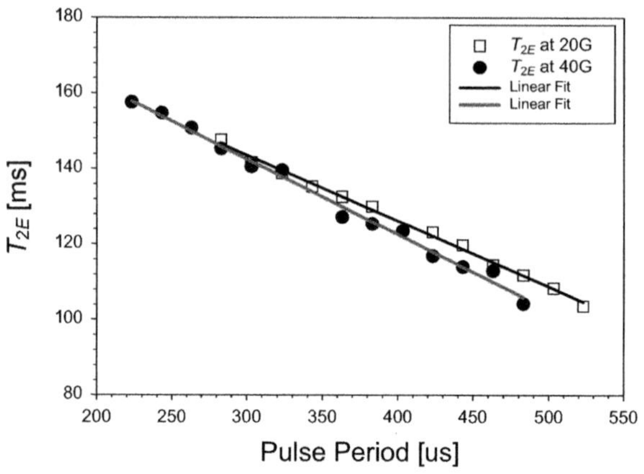

Figure 3. Spin-lock relaxation parameter with SLSE pulse sequence as a function of pulse period. The dependence of T_{2E} was measured with RF field strengths of 20 and 40 G.

To take into account the receiver's response during signal acquisition, the QR echo $S(t_i, T_n)$ is filtered using a normalized impulse response (IR) function of the receiver, described by a single decaying exponential, as

$$IR(t) = \frac{1}{T_c}\exp(-t/T_c), \text{ where } T_c = \frac{Q}{\pi \nu_0}$$

T_c is the filter time constant determined by the quality factor Q of the probe and the resonance frequency ν_0 of the probe.

2.2. NOISE MODEL FOR COIL AND PREAMP WITH INTEGRAL BROADBAND PROTECTION

In this section, we consider the noise model of a preamplifier with reactive protection impedance, finite input impedance, and feedback impedance connected to a high-Q QR coil.

Figure 4 shows the simplified model of the preamplifier with feedback and protection impedance.

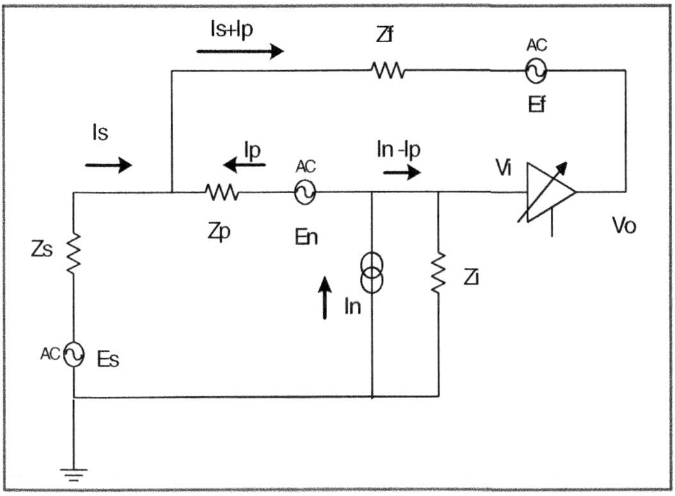

Figure 4. Simplified model of a preamplifier with feedback and protection impedance used for calculating the QR receiver's response.

In this model, the QR coil is represented by source impedance Z_s, with an associated source voltage ε_s, both of which are frequency dependent. The series protection element is represented by the impedance Z_p, and does not have a noise generator since it is assumed purely reactive. The preamp itself comprises noise generators ε_n and I_n, with a shunt input impedance of Z_i and an open loop gain – A. The feedback impedance is Z_f which will generally be resistive and have a voltage noise ε_f.

With this model the output voltage of the preamplifier is given by

$$V_0 = \frac{\left(1+\frac{Z_f}{Z_s}\right)\left(I_n Z_p + E_n\right) + \frac{Z_f}{Z_s} E_s + E_f + I_n Z_f}{1+\frac{1}{A}\left(1+\frac{Z_f}{Z_s}+\frac{Z_f}{Z_i}+\frac{Z_p}{Zi}\left(1+\frac{Z_f}{Z_s}\right)\right)}.$$

To calculate the output noise power we sum the noise voltages in quadrature (assuming no correlation between the different noise sources)

$$V_0^{\,2} = \frac{\left|1+\frac{Z_f}{Z_s}\right|^2 \left(I_n^2 \left|Z_p\right|^2 + E_n^2\right) + \left|\frac{Z_f}{Z_s}\right|^2 E_s^2 + E_f^2 + I_n^2 \left|Z_f\right|^2}{\left|1+\frac{1}{A}\left(1+\frac{Z_f}{Z_s}+\frac{Z_f}{Z_i}+\frac{Z_p}{Zi}\left(1+\frac{Z_f}{Z_s}\right)\right)\right|^2}$$

The impedance of the coil Z_s near resonance is approximately given as

$$Z_s = R_{\max} \frac{1 + i\left(\dfrac{2Q\Delta\nu}{\nu_0}\right)}{1 + \left(\dfrac{2Q\Delta\nu}{\nu_0}\right)^2},$$

where Q is the quality factor of the probe's coil and the frequency offset is $\Delta\nu = \nu - \nu_0$. The thermal noise voltages E_f and E_s are calculated as $E_{s,f} = \sqrt{4k_B T \,\mathrm{Re}(Z_{s,f})}$, where T is the temperature of the coil in kelvin.

The equation for the output voltage can then be used to calculate the noise power spectrum near the resonance frequency ν_0. The feedback impedance also has an effect on the gain of the preamplifier; therefore, the output voltages should be normalized by the factor

$$V_G^{\,2} = \frac{\left|\dfrac{Z_f}{R_{\max}}\right|^2}{\left|1 + \dfrac{1}{A}\left(1 + \dfrac{Z_f}{R_{\max}} + \dfrac{Z_f}{Z_i} + \dfrac{Z_p}{Zi}\left(1 + \dfrac{Z_f}{R_{\max}}\right)\right)\right|^2}.$$

Figure 5 shows the output voltage for an over-coupled preamp as a function of frequency offset, the broad frequency response results with a low Q of about 10. The response for a weakly coupled preamp is also displayed in Figure 5; the frequency response is limited by the unloaded coil quality factor Q. This behavior of the output voltage suggest the possibility of increasing the effective frequency bandwidth of the receiver by operating the preamplifier in overcoupled mode, which can be advantageous for detection of broad QR lineshapes; such as TNT resonances around 842 kHz, or increased temperature bandwidth. However, increasing the probe's bandwidth by decreasing the feedback impedance also results in higher off-resonance noise as discussed below. The final SNR depends on the bandwidth of the QR response (lineshape) and the RF excitation field levels that determine the excitation bandwidth of the pulse sequence.

A number of tests were performed to check that the equations for the noise power spectrum (NPS) reproduce the main features of the preamp response. Analysis of the equation for V_0 shows that there is a quadratic increase in the noise power for the over-coupled preamp (low Q), and a Lorentzian decrease in noise power near resonance with the preamp in weakly coupled mode (high Q). The experimental noise power spectra measured on a benchtop spectrometer with a low and a high Q are displayed in Figure 6.

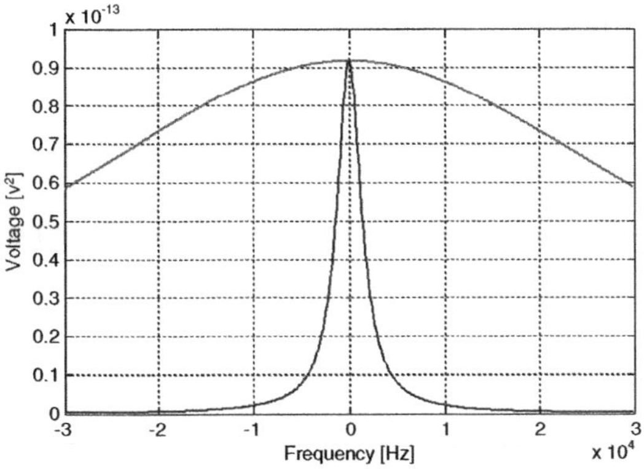

Figure 5. Calculated output voltage from the preamplifier in overcoupled (low-Q) mode and weakly coupled (high-Q) mode.

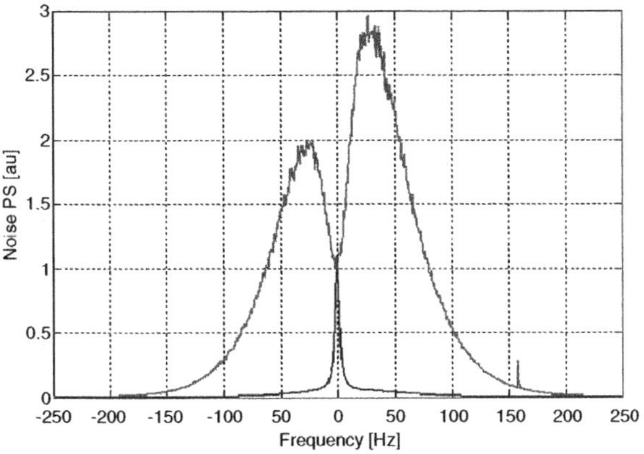

Figure 6. Noise power spectra measured on a benchtop QR spectrometer with and without feedback. The on-resonance spectral intensity is normalized and corrected by the difference in voltage gain. Note that the on-resonance noise power is identical for the preamp with low and high Q; however, there is a quadratic increase in noise power for over-coupled mode (low Q) and a Lorentzian decrease in noise power near zero frequency for the weakly coupled mode (high Q).

Figure 7 show the calculated and measured noise power spectra with an acquisition window of 1 ms in the frequency range of ±10 kHz. To account for the effects of a band-pass filter in the receiver chain of the spectrometer, the calculated noise power spectra were filtered in the frequency domain with a Gaussian function having a width of about 89 kHz. Excellent agreement was found between the experimental and calculated noise power spectra that validates the approach developed for modeling the QR receiver, as shown in Figure 7.

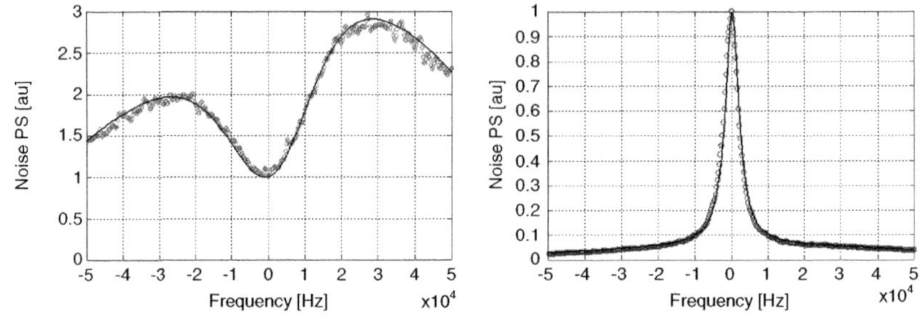

Figure 7. Experimental values (circles) and calculated (solid lines) noise PS with over-coupled (on the left) and weakly-coupled (on the right) preamp resulting in a low-Q and high-Q probes, respectively.

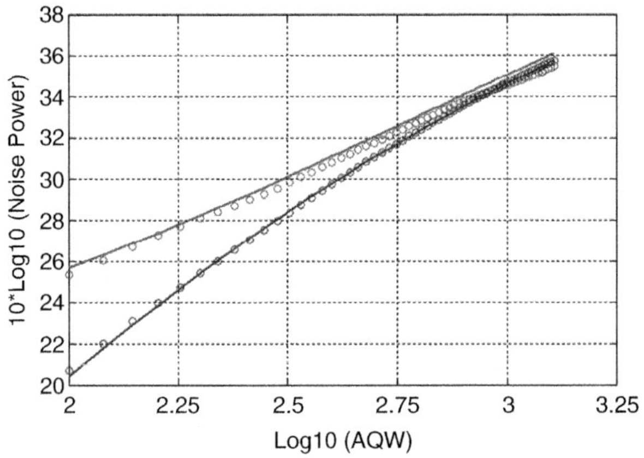

Figure 8. Calculated Noise power levels (solid lines) and experimental values (circles) at the central bin (zero frequency) as a function of the length of acquisition window. The upper curve shows the values for the over-coupled preamp and the lower curve for the weakly-coupled preamp.

The NPS calculated above corresponds to a measurement with infinite acquisition window; i.e. continuous wave condition. To calculate the noise power levels as a function of acquisition windows for different feedback impedances we shall assume the noise samples are uncorrelated in the frequency domain. Then the noise power level for various acquisition windows can be calculated by performing a convolution between the power spectrum of the window filter function and the NPS with infinite acquisition window.

Figure 9 shows the calculated and experimental noise power levels with over-coupled and weakly-coupled preamplifier with a Hanning filter function for the echo shape. There is a good agreement between the model and measured data.

Using this model, we can calculate the noise power level for arbitrary durations of the acquisition window, filter functions, and receiver's quality factor Q. The dependence of the on-resonance noise power level with acquisition window calculated for different Q values is shown in Figure 9.

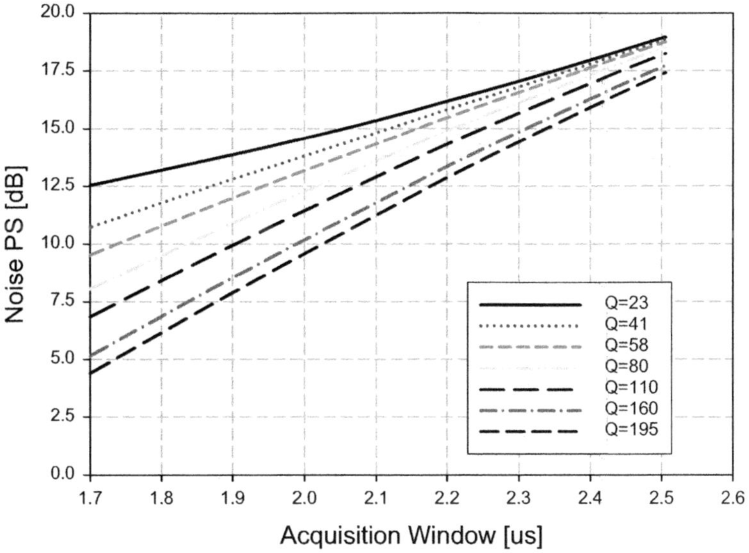

Figure 9. Calculated Noise power levels at the central bin (on resonance) as a function of the length of acquisition window. Note that the noise power level approaches the same value for long acquisition window.

To include the effects of data filtering on the noise power levels, we calculate the total thermal noise level as

$$F_N = \sigma \sqrt{\sum_{i=1}^{Np} E_i^2} \sqrt{\sum_{n=1}^{Ne} F_n^2},$$

where σ is the thermal noise level at zero frequency and it is a related to the NPS by the expression $\sigma = \sqrt{2/\pi}\, NPS$.

3. QR SNR calculations

The results in these sections correspond to the experimental conditions found for TNT detection, which is one of the most challenging materials to detect by QR, using the resonances around 842 kHz. Calculations of SNR levels, using the equations presented above, were carried out over a two-dimensional grid of acquisition windows and dead times values.

Figure 10 shows the results for low-Q and high-Q receivers. These numerical simulations performed for an overcoupled and weakly-coupled preamplifier (low and high Q) shows that the optimum acquisition window is about the same, 150 μs, for different levels of feedback impedance. The optimal dead time is shorter for the low-Q mode. These calculations enable us to estimate the optimal parameters for the SLSE pulse sequence to achieve the maximum SNR with this sensor.

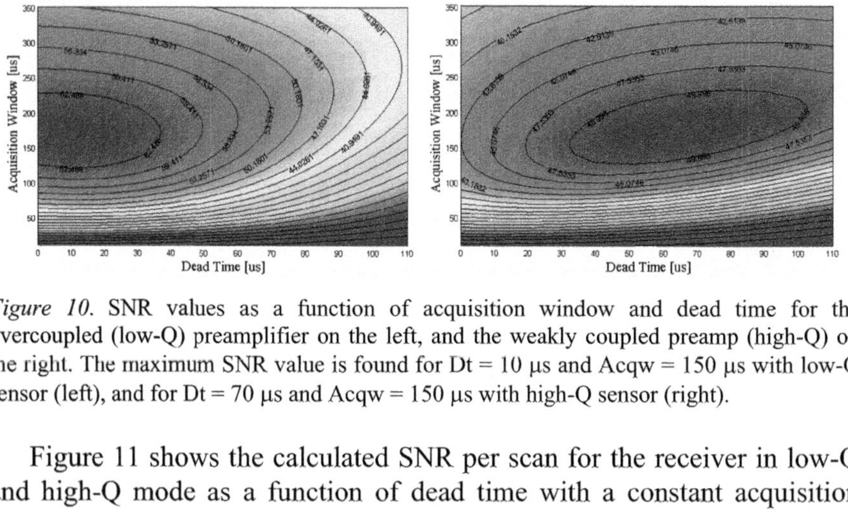

Figure 10. SNR values as a function of acquisition window and dead time for the overcoupled (low-Q) preamplifier on the left, and the weakly coupled preamp (high-Q) on the right. The maximum SNR value is found for Dt = 10 μs and Acqw = 150 μs with low-Q sensor (left), and for Dt = 70 μs and Acqw = 150 μs with high-Q sensor (right).

Figure 11 shows the calculated SNR per scan for the receiver in low-Q and high-Q mode as a function of dead time with a constant acquisition window of 150 μs. These results show that sensitivity can be improved by reducing dead time only when operating the receiver in low-Q mode. Reducing the dead time with a high Q receiver does not yield an increase in sensitivity over the investigated time interval due to the time lag in the QR signal induced by the IR function of the receiver. The higher the Q, the larger the time delay in the QR response after opening the receiver and reducing the dead time does not improve sensitivity. Figure 12 shows the calculated SNR levels as a function of acquisition windows and Q for a constant dead time of 70 μs, which is typical in our TNT detection systems. A slight improvement in sensitivity is observed with an optimum Q of about 100 relative to the SNR for a Q of 190. Through these simulations, it was found that the optimal receiver's Q depends on the bandwidth of the QR lineshape, the excitation bandwidth of the RF pulses, and the duration of the dead time.

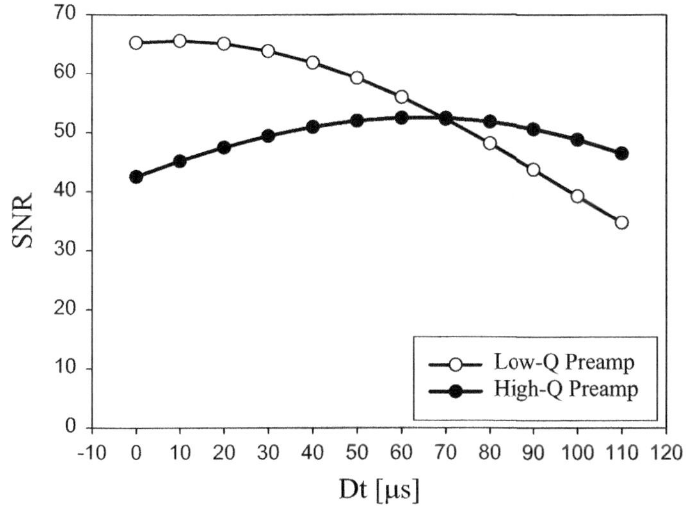

Figure 11. Sensitivity for the low-Q (open circles) and high-Q (solid circles) preamplifiers as a function of dead time for the optimum acquisition window of 150 μs.

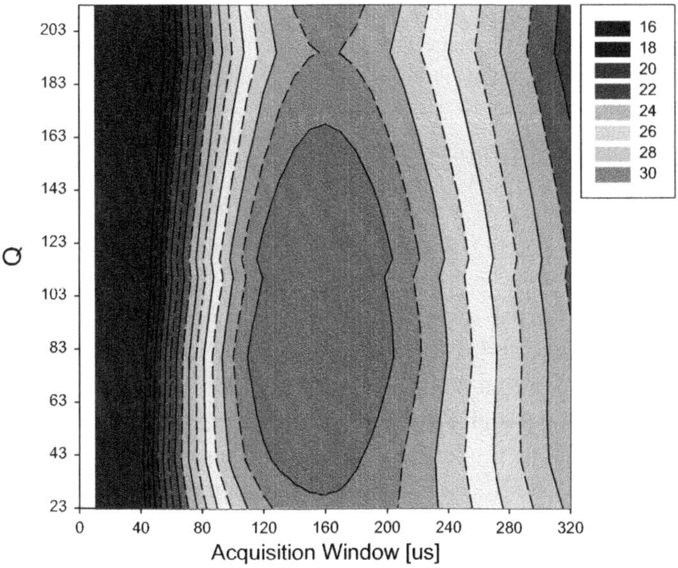

Figure 12. Sensitivity as a function of acquisition window for various Q values in the range 23–213. The dead time is constant at 70 μs. The simulations show that there is a maximum in SNR for a Q of about 100 for the experimental parameters used in these calculations. The optimum Q is a function of the QR lineshape and pulse bandwidth.

Experiments performed on a benchtop QR spectrometer with a TNT sample at room temperature to validate our numerical simulations. The experimental SNR curves in Figure 13 (on the left) show that the maximum SNR values with low-Q and high-Q receivers are found for the same acquisition window of about 150 μs, in agreement with the numerical results from the model. QR sensitivity as a function of quality factor Q of the receiver was measured by varying the feedback impedance of the receiver and keeping constant all parameters in the SLSE pulse sequence. The experimental data shown in Figure 13 (on the right) indicate a slight improvement in SNR with a Q of about 90, although variations are within the experimental error.

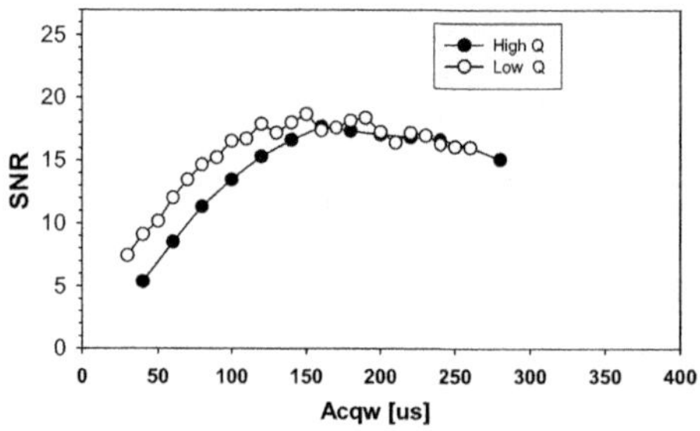

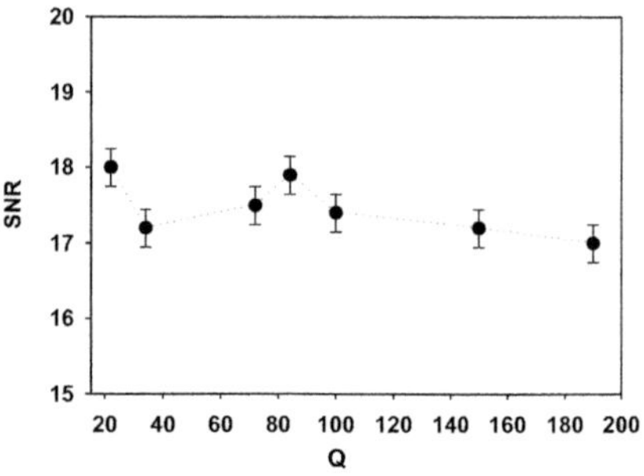

Figure 13. On the left: experimental SNR values for TNT as a function of dead time for low-Q (open circles) and high-Q (solid circles) receivers. The dead time is constant at 70 μs. On the right: Experimental SNR values for TNT as a function of Q of the receiver with a constant dead time of 70 μs, and constant acquisition window of 150 μs.

3.1. DETECTION ALGORITHM

In QR explosive detection, decisions about the presence of a threat are based on the *peak-picking* algorithm.[11,12] Its purpose is to determine whether the observation vector or spectrum belongs to class 1 (alarm) or class 2 (non-alarm). The peak-picking algorithm searches for the highest peak in the detection window and compares its amplitude with a predefined threshold value to assign the output of the measurement to one of the two classes. The decision rule is simply a one-dimensional problem, the maximum peak height $F_{\max}$ is compared to a predefined threshold Γ and the output is an alarm if $F_{\max} > \Gamma$ or clear if $F_{\max} \leq \Gamma$. The threshold level is set according to the desired false alarm rate *PFA* for the sensor.

The PFA is calculated for a given threshold from the cumulative probability distribution function, $PFA(\Gamma) = (1 - P^N(\Gamma))$. The function

$$P^N(\Gamma) = \prod_{i=1}^{N} P(\Gamma)$$

is the probability distribution function for detection over a window of N uncorrelated bins in the frequency domain, and

$$P(\Gamma) = \int_{-\infty}^{\Gamma} p(x)dx$$

is the probability distribution function (or cumulative probability distribution). The density probability function $p(x)$ is a normal distribution

$$p(x) = G(x \mid \mu, \sigma) = \frac{1}{\sigma\sqrt{2\pi}} \exp(\frac{-(x-\mu)^2}{2\sigma^2}),$$

where $G(x|\mu,\sigma)$ is a shorthand notation for a normal distribution with the expected or mean value μ and covariance or standard deviation σ. The background noise distribution; i.e. with an empty coil, is described with $\mu = 0$.

The QR signal only occurs at one bin in the frequency domain, therefore the probability distribution for the signal is $G(x|\mu_{QR},\sigma)$ where the mean value of the QR signal is $\mu_{QR} = SNR \times \sigma$. The probability of detection for a threshold value Γ is

$$PD(\Gamma) = \int_{\Gamma}^{\infty} G(x \mid \mu_{QR}, \sigma)dx .$$

These equations allow us to calculate the expected PFA and PD from the SNR and the noise values calculated using the model described in previous sections for arbitrary detector bandwidth. Calculations of PD and PFA as a function of threshold values Γ yields the ROC curve for the QR sensor.

4. Conclusions

The QR sensor model presented in this paper enables us to study the effect of hardware parameters (dead time, quality factor Q, etc.), pulse sequence parameters (acquisition window, number of echoes, pulse widths, RF field strength) on the final signal-to-noise ratio that determines the performance, PD and PFA, of the sensor. We presented explicit equations for calculating the noise power spectrum and the probabilities, PD and PFA, which determine the performance of the QR explosives detector.

The QR sensitivities for the SLSE pulse sequence as a function of the probe's Q or the preamplifier's feedback calculated with our models are in agreement with experimental results, within the experimental errors associated with our measurements. A preamplifier with variable feedback impedance allows further optimization of the detector by tailoring the probe's Q to the QR response to maximize sensitivity. Our results show that, for a detector using a spin-lock pulse sequence, there is an optimum Q for the system for a given dead time, QR lineshape, and excitation bandwidth (determined by RF field strength and pulse widths).

The model described here is also being used to determine engineering requirements from operational requirements (scan time, PD and PFA, temperature bandwidth, etc.). It allows us to perform quick evaluation of the effects on QR sensor's performance resulting from variations in system design and components specifications.

Acknowledgement This work was partially supported by the US Office of Navy Research (ONR) under contract N00014-03-C-0265.

References

1. Abragam, Principles of Nuclear Magnetism, Clarendon Press, Oxford (1961).
2. T.P. Das and E.L. Hahn, Nuclear Quadrupole Resonance Spectroscopy, in Solid State Physics, Supplement 1, Academic Press, New York (1958).
3. A.N. Garroway, M.L. Buess, J.B. Miller, B.H. Suits, A.D. Hibbs, G.A. Barrall, R. Matthews, and L.J. Burnett, Remote Sensing by Nuclear Quadrupole Resonance, IEEE Transactions on Geoscience and Remote Sensing, 39, 1108–1118 (2001).
4. J.B.L. Miller and G.A. Barrall, Explosive Detection with Nuclear Quadrupole Resonance, American Scientist, 93, 50–57 (2005).
5. R.A. Marino, R.F. Connors, and L. Leonard, Nitrogen-14 QR Study of Energetic Materials, U.S. Army Research Office 15556.4-PH (1982).
6. H. Suits, A.N. Garroway, and J.B. Miller, Super-Q Detection of Transient Magnetic Resonance Signals. Journal of Magnetic Resonance, 132(1), 54–64 (1998).
7. E.D. Ostroff and J.S. Waugh, Multiple Spin Echoes and Spin Locking in Solids, Physical Review Letters, 16, 1097–1098 (1966).
8. R.S. Cantor and J.S. Waugh, Pulsed spin locking in pure nuclear quadrupole resonance, Journal of Chemical Physics, 73, 1054–1063 (1980).
9. R.A. Marino and S.M. Klainer, Multiple Spin Echoes in Pure Quadrupole Resonance, Journal of Chemical Physics, 67, 3388–3389 (1977).
10. Y.K. Lee, Spin-1 Nuclear Quadrupole Resonance Theory with Comparisons to Nuclear Magnetic Resonance, Concepts in Magnetic Resonance, 14(3), 155–171 (2002).
11. M.K. Steven, Statistical Signal Processing, Volume II: Detection Theory, Prentice Hall, NJ (1998).
12. Tuzlukov Vyacheslav, Signal Processing Noise, CRC Press (2002).

DETECTION OF EXPLOSIVES BY NQR METHOD: MAIN ASPECTS FOR TRANSPORT SECURITY

TARAS N. RUDAKOV*
QSpin Technologies Ltd., 20 Bracadale Way, Canning Vale, WA 6155, Australia

Abstract The main aspects specific to the Nuclear Quadrupole Resonance (NQR) application for the transport security are analyzed and discussed. The continuing efforts to improve the function of the NQR based Explosive Detection Systems (EDS) through a more sensitive detection are described. The methods of reducing false alarm events generated from the RF excitation and interference are presented and discussed. The operational features of the NQR scanners are also presented and discussed.

Keywords: NQR, explosive detection systems, false alarm, RF excitation

1. Introduction

The detection and identification of explosives and other illicit materials is a very important part of transport security. The most common screening technique used at checkpoints employs x-ray method. This method offers excellent image quality, and with a skilled operator, a high probability of detection for a range of both explosive and non-explosive threats. However the detection technique on the basis of x-rays can not provide the required probability of detection of certain configurations of explosives.

The nuclear quadrupole resonance (NQR) method, often referred to as QR, is currently being evaluated for integration into security equipment for the detection of concealed explosives. This method provides automated detection of explosives types in configurations missed by the imaging x-ray method.[1]

* E-mail: trudakov@iinet.net.au

J. Fraissard and O. Lapina (eds.), *Explosives Detection Using Magnetic and Nuclear Resonance Techniques*,

The NQR method is a radio-frequency (RF) spectroscopic technique, defined as a phenomenon of resonance RF absorption or emission of electromagnetic energy. In NQR atomic nuclei with a non-spherical charge distribution (a nuclear spin $I \geq 1$) which possess nonzero electric quadrupole moment, interact with the local inhomogeneous electric field E or the electric field gradient (EFG) E produced by nearby charges. This interaction removes the degeneracy of the nuclear ground state and establishes the nuclear quadrupole levels with different energy. The resonance occurs when the transitions between the nuclear quadrupole energy levels are exited by an RF field. The transition frequencies between the energy levels depend both on the nuclear quadrupole moment which is a property of the nucleus, and the EFG tensor which is determined by the electric charge distribution around the nucleus. The resonance frequencies are specific to each compound and they can be used for their practically unique detection and identification. Therefore NQR method is of great interest for a number of applications in particular in detecting illicit materials such as explosives.[1–3] At the moment the NQR method is able to detect plastic explosives containing RDX, PETN, HMX, tetryl, TNT, ammonium nitrate (AN), ammonium perchlorate, peroxides (such as HMTD) and others. Mixed explosives such various grades of SEMTEX are readily identifiable from their chemical mixture also.

A key advantage of NQR is that a magnetic field is not required to create magnetization because the electric quadrupole moments of the nuclei interact with the internal electric field gradients of the surrounding charges in molecules. That makes NQR technology to be a simpler, less costly and more efficient platform technology. Besides the NQR technology is quite safe and successfully trialled by authorities worldwide. The detection process is automated so no reliance on operator vigilance or skill is required. The NQR technology is non-disruptive to normal screening operations. NQR is complementary and adds important functionality to all currently deployed x-ray systems. Explosives can be automatically detected regardless of size, shape, weight or bag clutter.

The NQR method is very similar to the well-known Nuclear Magnetic Resonance (NMR) method and is also termed as nuclear magnetic resonance (NMR) in a zero magnetic field.

In the present paper we offer an analysis of main issues specific to the NQR application for the transport security. The NQR detection technique for the different substances have been developed and discussed. This paper also describes the continuing efforts to improve the function of the NQR based Explosive Detection Systems (EDS) through a more sensitive detection, and methods of reducing false alarm events generated from the RF excitation and interference. The operational features of the NQR scanners are also presented and discussed.

2. NQR methods for transport security

In contrast to NMR methods, NQR can operate without a strong external dc magnetic field. This technique is known as "pure NQR", or direct NQR detection, and has a lot of advantages for many applications, such as detection and identification of explosives and remote NQR. Direct NQR methods can be subdivided into: oscillator-detectors methods (continuous-wave and super-regenerative techniques) and pulsed methods. At present the pulse method is the most widely used direct NQR method. It should be noted that only direct NQR methods are currently used for transport security.

Two different pulsed NQR techniques are being used in the transport security to detect explosives, namely detection using large ordinary coils and one side (or remote) detection using special coils (antennas). The first technique is used to scan baggage. It is a development of ordinary pulsed NQR techniques for use with large specimens and cannot be utilized to detect samples beyond certain distance from the spectrometer coil. At the same time, the remote NQR technique is more universal. However besides extremely complicated highly sensitive receiving equipment, it requires the use of specific signal processing methods for eliminating external interference and the establishing of the required RF field at a definite distance from the coil. In remote NQR special flat surface coils of various designs are used, with irradiating and receiving coils (antennas) sometimes set separately.

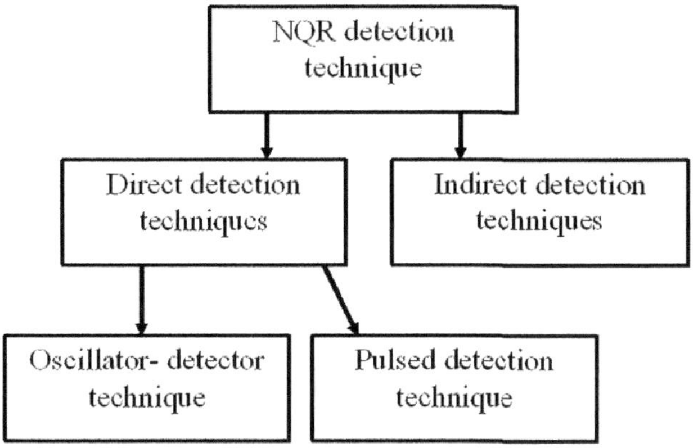

Figure 1. NQR methods for explosives detection.

Besides direct methods of detection, indirect NQR detection methods have also been developed. Indirect NQR detection methods are used when the sensitivity of direct methods is not sufficient. These methods can be conventionally divided into double resonance techniques and cross-relaxation spectroscopy method. To apply indirect methods, it is necessary for the sample to have two spin systems connected by dipole–dipole interactions. Similar to NMR, for most indirect experiments a constant magnetic field is needed. Indirect NQR technique is not currently used in transport security. However, it can be considered as a potentially important explosives detection technique. Block-diagram demonstrating the main NQR methods for explosives detection is given in Figure 1.

3. Main issues and solutions

When using NQR method for explosive detection a relatively poor signal-to-noise ratio (SNR) is often an issue.[1] Most explosives contain quadrupolar nuclei, for example the nitrogen (^{14}N), the spectral lines of which are usually located at low RF frequencies (0.4–6 MHz). The signal intensity in this frequency range is quite low because the low energy difference between excitation levels. Another specific issue for NQR baggage screening is a small filling factor: the detection of relatively small amounts of explosives in large volumes which often exceeding 150 l. This also makes it necessary to detect weak NQR signals. In order to improve sensitivity it is common to employ multi-pulse techniques, special equipment and signal processing.[2]

In practical situations the investigated baggage can contain objects which, when irradiated with strong RF pulses, can become sources of coherent spurious signals (coherent noise): magneto-acoustic and/or piezoelectric ringing. These spurious signals could be very strong and are the main reason for the false alarm rate (FAR) for explosive detection systems. Therefore the third specific issue for NQR baggage screening is the coherent noise signals. This problem is also can be solved by using the special multi-pulse technique.[4]

The NQR method for explosive detection is sensitive to the temperature and temperature gradient in the sample. This is also an issue for the transport security. The problem can be solved by choosing the appropriate NQR line for detection, improving the receiver and transmit circuits, measuring the sample temperature before screening and modifying the detection technique. These modifications of the detection technique may include multi-pulse sequences, composite and shaped pulses.

Another issue for NQR baggage and people screening is an external interference which could arise from AM radio stations and other sources. The problem is that strong external RF intrusion into the machine would lead to a higher FAR. The method of data accumulation, coupled with good shielding of the device, significantly reduces the consequence of external RF signals. However in such cases as shoe screening it may not be enough for the required performance of the detection systems. In this case the special receiving equipment including additional antennas and advanced signal processing methods can be used to solve the problem.

In order to obtain high sensitivity it is necessary for the receiving system of the machine to have a low noise factor and to ensure matching of its input impedance with the circuit. Without this matching the noise factor can increase considerably. The signal to noise ratio (SNR) can also be increased by increasing the Q-factor of the circuit. The SNR is proportional to $\sqrt{Q}$, because with the increase in quality the amplitude of the signal voltage increases Q times, while the effective noise voltage increases by $\sqrt{Q}$ times only. However, with the increase of the Q-factor of the circuit, transient processes (ringing) after the irradiating RF pulses become longer. At low frequency the length of transient processes can reach hundreds of micro-seconds, which creates considerable difficulties for detecting inductance signal and even spin-echo in samples with broad line and short relaxation times. The problem of ringing can be solved by several methods which include active Q-damping techniques and special multi-pulse sequences.

The RF field applied to excite an NQR response can potentially induce significant circulating currents and voltages in conductive sections of electronic devices that may be immersed in this field. The value of induced current and voltage depends on many factors such as the magnitude of the applied field, effective area of the circuit, its layout on the board, its orientation relative to the applied field, electromagnetic shielding, the electronic components that may be used, applied frequency and duration etc. The alteration or failure in operation of an electronic item due to the induced current and voltage depends specifically on its design and the sensitivity of its components to these induced currents and voltages. In order to avoid a damage of electronic items the RF field intensity should be minimized. However, this approach may cause a reduction in the overall performance of the NQR system. A field magnitude is therefore sought with sufficient SNR and the likelihood of damage to electronic goods is made as low as possible.

Table 1 demonstrates results of testing different electronic items applying the pulsed RF field. The total number of items tested near 890 kHz, 3.4 and 5.3 MHz was 144, of which 32 were mobile phones, 22 laptop computers, ten digital cameras, ten film cameras, seven organisers, four CD players, one DVD player, eight calculators, three pocket PCs, four USB keys, ten electronic games, two watches, three clocks/calendar, ten portable mini TVs/ radios/walkman, five items of medical equipment, an I-pod, an Mp3 player, a liquid crystal monitor and various other electronic items. Due to the continuously changing population of electronic goods Table 1 only attempts to indicate the current state.

TABLE 1. Magnetic component of RF field for which 97% of tested electronic items survive.

Frequency	Any effect (G)	Effects requiring intervention (G)	Permanent damage (G)
890 kHz	<1.8	<1.8	<2.7
3.4 MHz	<0.8	<0.8	<2.3
5.2 MHz	<0.5	<0.5	<1.7

The NQR method does not allow the detection of explosives enclosed in an effective RF shield. Therefore the discovery of potential RF shields in a piece of baggage would benefit the effectiveness of NQR technology. This discovery could be used to alert a user of the NQR machine that a more thorough search is required. The problem can be solved by using the metal detection technique, but without increasing the FAR to an unacceptably high level. Any metal detection technique would require some differentiation of items containing metal from shielded explosives to be effective.

4. Current developments

4.1. MULTI-PULSE NQR TECHNIQUE FOR EXPLOSIVE DETECTION

The multi-pulse NQR techniques can give a significant improvement in sensitivity of explosive detection equipment.[5–14] A choice of the optimal multi-pulse sequence is to a great extent determined by fixed parameters related to the NQR "nature" of the explosive such as NQR frequencies and relaxation times. These parameters vary considerably for different substances, and the most effective multi-pulse sequence that maximizes the SNR per unit time is selected for each substance. The effectiveness of a pulse technique strongly depends on parameters of the sequence, for example the pulse duration, pulse spacing and pulse phase. Therefore in order to achieve high sensitivity it is necessary to find the optimal value of these parameters for specific detection conditions.

In general we can divide all pulse sequences used for explosive detection into two main groups: the Steady-State Free Precession (SSFP) type multi-pulse sequences and Spin-Locking Multi-Pulse (SLMP) sequences. The first group of sequences is very effective for detection of explosives with short spin lattice relaxation time T_1, such as RDX; this however does not exclude its use for detection of substances with long T_1. The main element of this method is the creation of the continuous steady-state (SS) signals which recur for as long as the pulse sequence is applied.[7] This is very convenient for the fast coherent accumulation of the NQR signals. In order to achieve a strong SS signal the pulse repetition time in the sequence should be less or close to the time T_2^* (line shape parameter) which is inversely proportional to a line width. The simplest version of the SSFP pulse sequence is a long train of phase coherent RF pulses and can be represented as

$$(\tau - \theta^1_{0^\circ} - \tau -)_N, \quad (1)$$

where θ^1 is the flipping angle of the other pulses in the sequence, and τ is the time interval between the pulses.

A typical NQR signal train obtained with SSFP multi-pulse sequence is presented in Figure 2a.

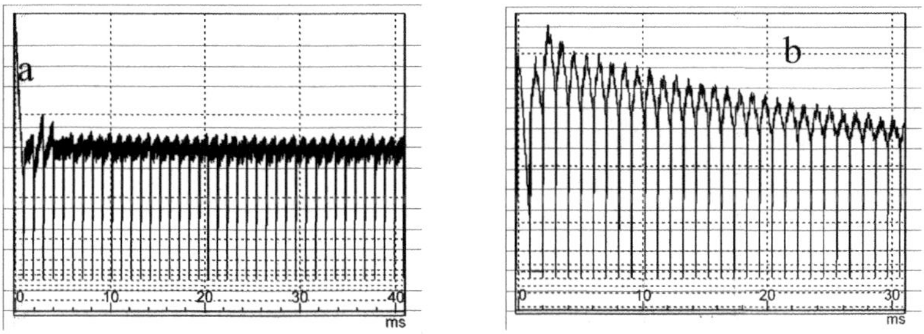

Figure 2. Typical NQR signals obtained in a powdered sample of sodium nitrite (Na NO_2) with: (a) SSFP pulse sequence, (b) SLMP sequence.

The second group of sequences is based on using the spin-echo effect and normally used for detection of explosives with long T_1, for example PETN and AN. The basic spin-locking multi-pulse (SLMP) sequence which was proposed in NMR by Ostroff and Waugh[6] can be represented as

$$\theta^0_{90^\circ} - (\tau - \theta^1_{0^\circ} - \tau -)_N, \quad (2)$$

where θ^0 is the flipping angle of the preparatory pulse and θ^1 is the flipping angle of the other pulses in the sequence and is phase shifted by 90° in relation to the preparatory pulse. The first experimental investigation of this type of sequence in pure NQR was made by Marino and Klainer.[5] Note that the results they obtained are very similar to those obtained earlier in the NMR experiments.[6]

In order to achieve high sensitivity it is necessary to ensure that the value of the decay constant of NQR signals (effective time T_{2e}) for the sequence is as close to the spin-spin relaxation time T_1 as possible. It is also desirable that the detection time or pulse sequence duration does not exceed excessively the T_{2e} time because the signal to noise ratio will generally degrades. Figure 2b demonstrates typical NQR signal train obtained with the SLMP sequence.

For the SSFP pulse sequences the amplitudes of SS signals increase with the decrease of the pulse repetition time 2τ. This fact is demonstrated in Figure 3a and b where NQR signals obtained with sequence (1) at different τ are presented. Therefore the optimal value of 2τ should be quite short and normally less than the line shape parameter T_2^*. It should be taken into account that the pulse repetition time should include the acquisition time and the duration of "ringing" after the RF pulse.

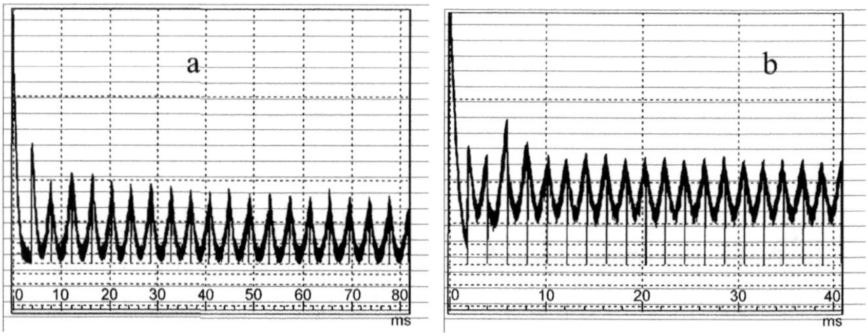

Figure 3. The NQR signal (magnitude) train from ^{14}N in sodium nitrite ($NaNO_2$) obtained with a 20 pulse SSFP sequence at resonance frequency offset $\Delta f = -0.4$ kHz: (a) $\tau = 2.1$ ms, (b) $\tau = 0.6$ ms.

In the case of SLMP sequences the amplitudes of NQR signals also increase with the decrease of the pulse repetition time 2τ. This effect is shown in Figure 4a and b.

As is well known for the SSFP pulse sequences when the steady-state condition is established the signal amplitude varies in a cyclic manner as a function of the frequency offset from resonance and pulse spacing.[8] This effect is known as intensity variations and takes place both in NMR and NQR spectroscopy.[9,10] It should be noted that in the case of SLMP sequences that intensity variations are also observed. However the frequency dependence of the signal amplitude is changed in this case. The dependences of the NQR signal intensity on frequency offset obtained using the conventional SSFP pulse sequence (1) and pulse sequence (2) are presented in Figure 5. The duration of all pulses equals the "90°-pulse". The A curve describes the frequency dependence for the conventional SSFP sequence (1) and the B curve describes the dependence for the sequence (2). Primarily note that the strong intensity variations are observed in both cases and the signal amplitude varies periodically. However the period is different for each sequence. One can see that the introduction of a preparatory pulse in the SSFP pulse sequence results in twice as long duration of maximum and minimum periods of the signal amplitude. This fact is quite interesting and obviously should be taken into account for practical applications.

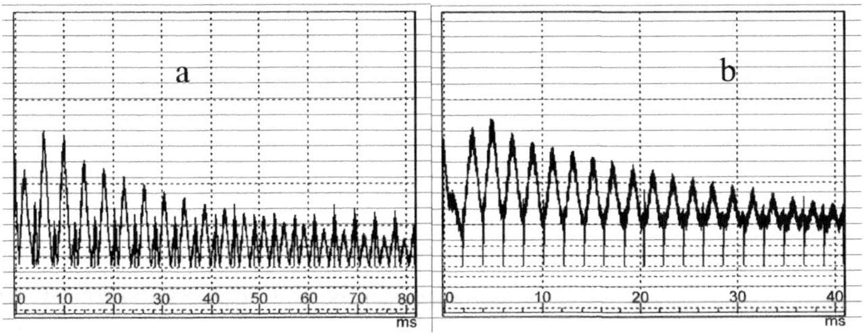

Figure 4. The NQR signal (magnitude) train from ^{14}N in sodium nitrite ($NaNO_2$) obtained with a 20 pulse SLMP sequence at resonance frequency offset $\Delta f = -0.4$ kHz: (a) $\tau = 2.1$ ms, (b) $\tau = 0.6$ ms.

Besides SSFP and SLMP sequences, a multi-pulse technique combining the properties of both of these groups of sequences can also be used. It is important that in many cases it is this technique that gives better results. It was shown that at certain conditions the total intensity of accumulated NQR signal can be increased by introducing the preparatory pulse in the SSFP sequence. This happens due to the SE signals which in all appearance are added to the level of the steady-state signals and decay with the effective T_{2e} time. The observed effect looks like a result of the joint impact of spin-locking and the SSFP phenomena. Therefore the NQR signal in the "observation windows" between RF pulses of the sequence has a complex structure and apparently contains the steady-state and the SE components.

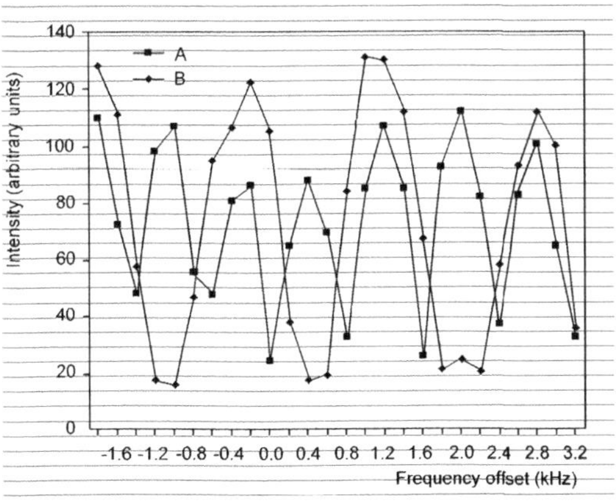

Figure 5. The intensity of the NQR signal from ^{14}N in sodium nitrite ($NaNO_2$) as a function of the resonance offset obtained with: SSFP pulse sequence (curve A) and the same pulse sequence with a preparatory "90°-pulse" (curve B). Pulse spacing $\tau = 0.6$ ms.

Note that this representation seems to be very approximate and can only be used for a simple explanation of the improvement in effectiveness of the multi-pulse technique obtained by introducing a preparatory pulse.

4.2. COHERENT NOISE ELIMINATION

In practical situations the investigated volume can contain objects which, when irradiated with strong RF pulses, can become sources of coherent noise (or spurious signals): magneto-acoustic and/or piezoelectric ringing. These "ringing" signals have a lifetime longer than the exciting RF pulse and correspond to the phase of this pulse. They are therefore picked up the sensitive receiver electronics. The spurious signals could be very strong and are the main reason for the FAR for explosive detection systems. As these signals are coherent with RF pulses, they can be considerably reduced by using a special multi-pulse technique.[4] It should be noted that the transient signals (also called "ringing") emerging in a resonant tank circuit after the RF pulses have been activated, refer to coherent noise, as their phase corresponds to the phase of the RF pulse. This coherent noise can be eliminated using the same methods that are used for magneto-acoustic or piezoelectric signals.

A general approach to rejection of spurious coherent RF signals includes accumulation of the total signal and the signal without a contribution from the nuclei and the subtraction of the latter from the former. Therefore a multi-pulse method is comprised of applying at least two excitation sequences. The first sequence is intended for the excitation of the nuclear resonance signal and coherent noise together, and the second one is used for the excitation of coherent noise only. In order to avoid detecting a nuclear resonance signal in the second block, a certain delay is set between the two blocks and/or the spin-saturation technique or a phase flip is applied.[15,16] If the structures of the first and the second blocks are identical, this method ensures very efficient coherent noise elimination even in the case of low frequencies.

A simple version of the SLMP type sequence for coherent noise elimination is

$$\theta^0_{90^\circ} - (\tau - \theta^1_{0^\circ} - \tau -)_N - \tau - \theta_\varphi - \tau - (\tau - \theta^1_{0^\circ} - \tau -)_N \tag{3}$$

where θ_φ is the flipping angle of the intermediary pulse, φ is the phase of the intermediary pulse.

The refocusing effect caused by the intermediary pulse takes place when the duration of this pulse is equal to or about an "180°-pulse", which is optimal for the refocusing of the transverse magnetization.[17] The typical

NQR signals obtained with pulse sequence (3) at on resonance frequency are presented in Figure 6. In this figure one can see the magnitude (Figure 6a) and the real component (Figure 6b) of the total NQR signal. The duration of the intermediary pulse was 180μs and the phase $\varphi = 90°$. Note that in this case the NQR signal is detected in the second block and the intermediary pulse inverts the phase of this signal. A similar result is obtained if the phase $\varphi = 270°$. The refocusing effect is also observed when $\varphi = 0°$ or 180°. However in this case the intensity of the signal is less and the intermediary pulse does not invert the phase of the signal.

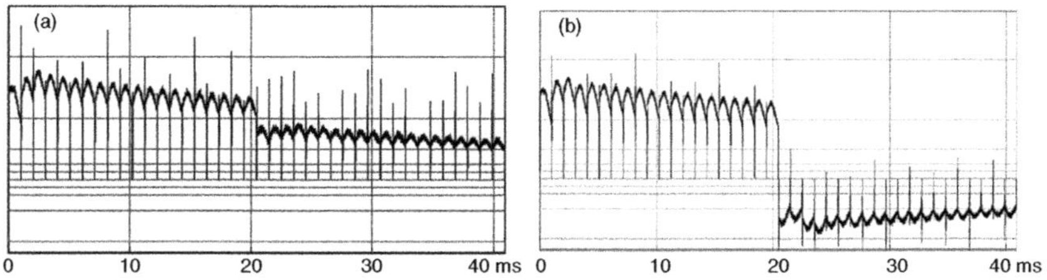

Figure 6. The NQR signal trains in a powdered sample of sodium nitrite ($NaNO_2$) obtained with the pulse sequence $\theta^0_{90°} - (\tau - \theta^1_{0°} - \tau -)_{N_1} - \tau - \theta_\varphi - \tau - (\tau - \theta^1_{0°} - \tau -)_{N_2}$ exactly at resonance frequency: (a) The total NQR signal (magnitude), (b) The total NQR signal (real component). The duration of all pulses equals the "90°-pulse", $\tau = 0.6$ ms and $\varphi = 90°$.

The main disadvantage of sequence (3) is quite strong dependence of the phase of the NQR signal in the second block on the frequency offset. It means that at several offsets the second block may significantly reduce the total NQR signal after subtraction.

The second version of the SLMP type sequence can be written as

$$\theta^0_{90°} - (\tau - \theta^1_{0°} - \tau -)_N - (\tau_1 - \theta_\varphi - \tau_1)_n - (\tau - \theta^1_{0°} - \tau -)_N \qquad (4)$$

In this sequence the additional multi-pulse excitation is applied between the two main blocks of the sequence and $2\tau_1$ is the pulse repetition time of the additional intermediary pulse sequence. It was found that when the parameters of the base pulse sequence and intermediary pulse sequence differ significantly, the intermediary pulse sequence causes a considerable loss of the NQR signal in the second block. For instance the signal is practically eliminated in all cases when $\varphi = 90°$ or 270°. This fact is illustrated in Figure 4 for a pulse sequence (4), with $\theta^1_{0°} = 180°$, $\varphi = 90°$, $\tau_1 = 0.3$ ms,$\tau = 0.6$ ms and $n = 10$. Therefore the second block of sequence (4) does not reduce the total NQR signal after subtraction.

The SSFP pulse sequences are a very efficient technique for the detection of substances with relatively short spin-lattice relaxation time T_1.

In this case the suppression of the coherent noise can also be achieved by using at least two blocks of pulse sequences. However, due to the fast signal recovering the NQR signal canceling technique is not effective. The blocks are not designed to be fully identical and all of them can detect NQR signals. The required effect is achieved by the appropriate phase flipping (Figure 7).[17,18]

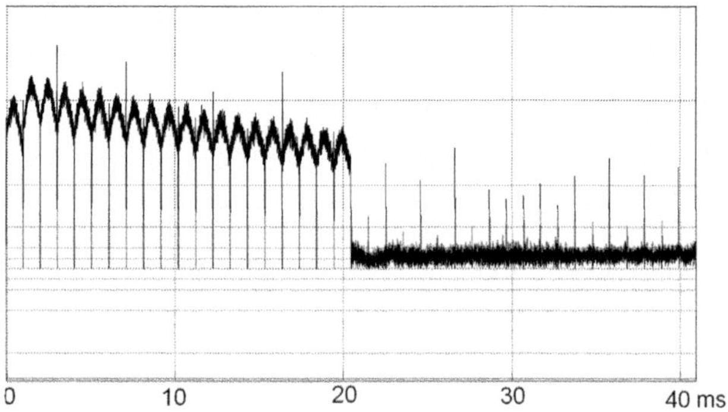

Figure 7. The total NQR signal (magnitude) in a powdered sample of sodium nitrite ($NaNO_2$) obtained exactly at the resonance frequency with the pulse sequence $\theta^0_{90^\circ} - (\tau - \theta^1_{0^\circ} - \tau -)_{N_1} - (\tau_1 - \theta_\varphi - \tau_1)_n - (\tau - \theta^1_{0^\circ} - \tau -)_{N_2}$. The duration of all pulses equals the "180°-pulse", $\tau = 0.6$ ms $\tau_1 = 0.3$ ms, n = 10 and $\varphi = 90°$.

As one of the commonly used version let us consider the pulse sequence which contains two blocks of RF pulses and can be written as

$$(\theta_{0^\circ} - \tau - \theta_{0^\circ} - \tau -)_n - \tau_1 - (\theta_{0^\circ} - \tau - \theta_{180^\circ} - \tau -)_n \tag{5}$$

where θ is the flipping angle of the pulses, τ and τ_1 are the time intervals between the pulses and blocks accordingly. The first block of the sequence is the simplest version of the SSFP sequence containing a train of the pulses with the same phases. The second block is the sequence containing a train of the pulses with alternating phases and can be referred as a SSFP phase alternated pulse sequence (PAPS). Therefore the first block may be called as a SSFP non-phase alternated pulse sequence (NPAPS).

In order to subtract the coherent noise the phases of RF pulses and reference phases of the receiver should be the same in the first block and they should be opposite in the second one. Sequence (5) is used for the canceling the coherent noise caused by the immediately preceding RF pulse. However if the coherent noise time constant is long the noise carries across observation window and can not be completely distinguished from NQR signals.

Sequence (5) ensures the canceling of coherent noise of up to 20 dB that is not enough for the practice. The reason for this is that each RF pulse of the sequence excites an independent coherent noise signal, the phase of which is determined by the phase of the respective pulse,[4] whereas sequence (5) contains an unequal number of 0° and 180° phases of RF pulses. The problem is solved by adding the third block containing the pulses with 180° phases.[16] This sequence can be written in the form

$$(\theta_{0^\circ} - \tau - \theta_{0^\circ} - \tau -)_{2n} - \tau_1 - (\theta_{180^\circ} - \tau - \theta_{180^\circ} - \tau -)_{2n} - \tau_1 - (\theta_{0^\circ} - \tau - \theta_{180^\circ} - \tau -)_{4n} \quad (6)$$

Sequence (6) contains an equal number of 0°and 180° phases of RF pulses, and therefore it is quite effective for the canceling the coherent noise caused by the immediately preceding RF pulse. However, if the duration of the coherent noise signals exceeds the duration of the time interval between the pulses (i.e. does not fit into the observation window) the level of the canceling decreases significantly. The situation with the long coherent noise time constant is clearly shown in Figure 8. We can see that in this case the part of the noise signal goes into the next pulse interval and is added to the coherent noise, created by the following RF pulse and having a different phase. In order to cancel this part of the noise signal we need to modify sequence (6). The simple and quite effective way is to use an additional block of RF pulses designed to be able to subtract the exceeding noise signals.

The further improvement can be achieved by adding the pair-wise alternated sequence (PWAS)[18] which can be written as

$$(\theta_{0^\circ} - \tau - \theta_{0^\circ} - \tau - \theta_{180^\circ} - \tau - \theta_{180^\circ} - \tau -)_m \quad (7)$$

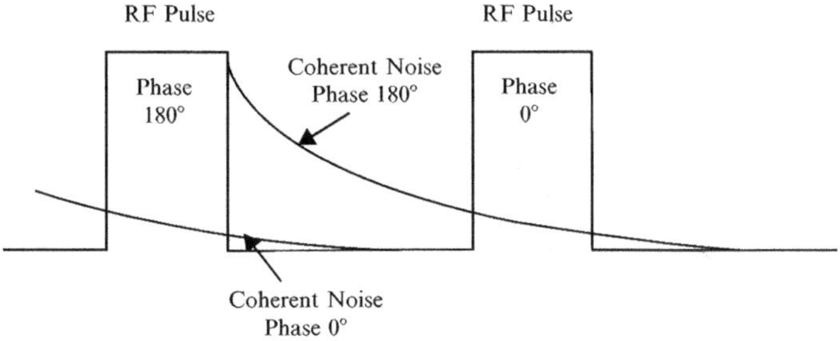

Figure 8. Long time constant coherent noise obtained with SSFP PAPS.

One can see that in this sequence the phase is alternated in each pair of RF pulses. In order to subtract the exceeding noise signals the reference phase of the receiver is set to be equal to the phase of the RF pulse in the first observation window of each pair and opposite to the phase of the RF pulse in the second window. It should be noted that PWAS is well known in NMR[19] and its properties have been also investigated in the case of NQR.[20]

For convenience in further discussion, we can divide the proposed modified pulse sequence into four sequences and write it in the form

First $$(\theta_{0^\circ} - \tau - \theta_{0^\circ} - \tau - \theta_{0^\circ} - \tau - \theta_{0^\circ} - \tau -)_n - \tau_1 \quad (8.1)$$

Second $$(\theta_{180^\circ} - \tau - \theta_{180^\circ} - \tau - \theta_{180^\circ} - \tau - \theta_{180^\circ} - \tau -)_n - \tau_1 \quad (8.2)$$

Third $$(\theta_{0^\circ} - \tau - \theta_{180^\circ} - \tau - \theta_{0^\circ} - \tau - \theta_{180^\circ} - \tau -)_{2n} - \tau_1 \quad (8.3)$$

Fourth $$(\theta_{0^\circ} - \tau - \theta_{0^\circ} - \tau - \theta_{180^\circ} - \tau - \theta_{180^\circ} - \tau -)_{4n} - \tau_1 \quad (8.4)$$

We will call the part of the coherent noise which does not exceed the duration of the observation window as the coherent noise I. Accordingly the exceeding part of the noise will be called as the coherent noise II.

Phases and polarities of the coherent noise signal observed in the observation windows of the pulse sequences are presented in Table 2. Note that the polarities of the noise signal are determined by the reference phase of the receiver. One can see that combination of pulse sequences (8.1)–(8.4) is able to cancel both the coherent noise I and II.

It should be noted that additional sequence (8.4) gives substantial contribution into the total NQR signal. Thus the proposed technique increases detection time insignificantly.

The proposed modified technique base on using four blocks of sequences (8.1)–(8.4) is quite effective for canceling the coherent noise with a long time constant and distinguishing NQR signals. This fact is demonstrated in Figure 9 where one can see the signals obtained from both samples RDX and sand, placed together in the coil. Figure 9a shows the signal obtained with sequence (6). In this case the NQR signal is covered by the strong PE signal and can not be detected. Figure 9b demonstrates that using the proposed technique we can distinguish the pure NQR signal. Note that very similar results were obtained for canceling MA signals from ferrites and nickel plated items.

Therefore the modification based on the use of additional blocks of the pulses and phase cycling technique enables the suppression of the coherent noise signals such as the magneto-acoustic and piezoelectric signals or the ringing signals from the NQR probe. The obtained results demonstrate effectiveness of the proposed technique for cancelling the coherent noise with a long time constant and distinguishing NQR signals.

TABLE 2. Phases and polarities of the coherent noise signal observed in the observation windows of the pulse sequences.

	First window	Second window	Third window	Fourth window	Sum	Total
First sequence						
Coherent noise I	+(0°)	+(0°)	+(0°)	+(0°)	+4(0°)	+4n(0°)
Coherent noise II	+(0°)	+(0°)	+(0°)	+(0°)	+4(0°)	+4n(0°)
Second sequence						
Coherent noise I	−(180°)	−(180°)	−(180°)	−(180°)	−4(180°)	−4n(180°)
Coherent noise II	−(180°)	−(180°)	−(180°)	−(180°)	−4(180°)	−4n(180°)
Third sequence						
Coherent noise I	−(0°)		−(0°)		−2(0°)	−4n(0°)
		+(180°)		+(180°)	+2(180°)	+4n(180°)
Coherent noise II		+(0°)		+(0°)	+2(0°)	+4n(0°)
	−(180°)		−(180°)		−2(180°)	−4n(180°)
Fourth sequence						
Coherent noise I	+(0°)	−(0°)	−(180°)	+(180°)	0	0
Coherent noise II		−(0°)	−(0°)		−2(0°)	−8n(0°)
	+(180°)			+(180°)	+2(180°)	+8n(180°)

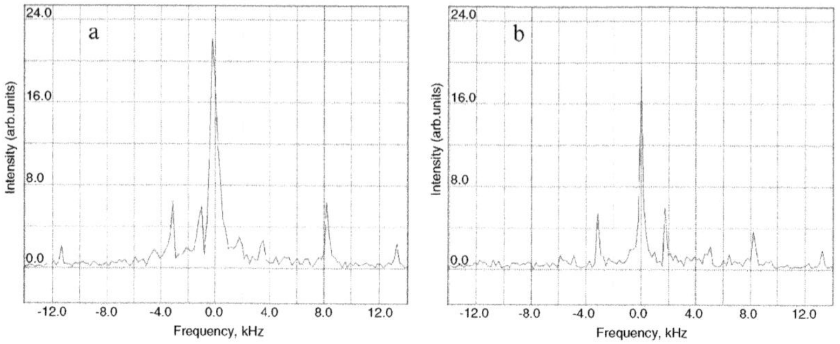

Figure 9. The NQR signal in a powdered sample of RDX ($C_3H_6N_6O_6$) and sand obtained with: (a) conventional SSFP pulse sequence consisting three blocks of sequences

$$(\theta_{0^\circ} - \tau - \theta_{0^\circ} - \tau -)_{2n} - \tau_1 - (\theta_{180^\circ} - \tau - \theta_{180^\circ} - \tau -)_{2n} - \tau_1 - (\theta_{0^\circ} - \tau - \theta_{180^\circ} - \tau -)_{4n}$$

(b) modified SSFP pulse technique consisting of four blocks of pulse sequences

$$(\theta_{0^\circ} - \tau - \theta_{0^\circ} - \tau -)_{2n} - \tau_1 - (\theta_{180^\circ} - \tau - \theta_{180^\circ} - \tau -)_{2n} - \tau_1 - (\theta_{0^\circ} - \tau - \theta_{180^\circ} - \tau -)_{4n}$$
$$- \tau_1 - (\theta_{0^\circ} - \tau - \theta_{0^\circ} - \tau - \theta_{180^\circ} - \tau - \theta_{180^\circ} - \tau -)_{4n}$$

The experimental conditions are: $\tau_1 = 40$ ms, $\tau = 0.7$ ms and n = 20.

4.3. NQR SIGNAL ENHANCEMENT

A long spin-lattice relaxation time (T_1) in some explosive types significantly reduces the efficiency of applied multi-pulse sequences. Examples of these explosives include PETN, AN and TNT. A quadrupolar spin-system once excited by an RF pulse or pulse sequence recovers to thermal equilibrium with the lattice in a time T_1. To take PETN as a specific example, a long T_1 time of about 32 s means that the ^{14}N NQR signal which decays throughout the application of a first pulse sequence over a few seconds cannot be efficiently re-excited by the application of a second pulse sequence until around a T_1 time has elapsed. In most applications this period is excessive and so detection results are generally based on the single application of a pulse sequence. Being able to effectively manipulate the recovery time allows a pulse sequence to be repeated on a much shorter timescale resulting in major improvements in the effectiveness of such techniques.

The cross polarisation methods[21,22] can be used to reduce the recovery time after excitation of a quadrupole spin-system (Q spins) in hydrogen (P spins)) containing solids. The recovery time reduction can be achieved by using contact between the proton and quadrupolar spin-systems, when the energy level separations in these systems are made the same and level crossings take place. According to the spin-temperature concept,[23,24] for the technique to be effective the proton system should be much "cooler" than the RF pulse sequence "heated" quadrupolar spin-system. It is important to point out that efficiency of the method is not critically dependent on the uniformity of the magnetic field. Removing uniformity requirements lowers the hardware cost of generating a static magnetic field considerably and makes the use of these techniques over large inspection volumes or stand-off feasible.

Cross-polarisation is used to enhance the signal-to-noise ratio in both nuclear magnetic resonance (NMR) and NQR. In NMR experiments the signal enhancement is normally achieved by transferring magnetisation from abundant spins like protons ^1H to rare spins.[21,22] In the case of NQR, magnetisation is transferred to quadrupolar nuclei from ^1H nuclei previously polarised in a reference magnetic field.[25–27] Two important benefits can be obtained (i) polarization transfer resulting in increased NQR single shot signal response and (ii) a dynamic reduction in recovery time of the NQR system allowing scan repetition on a much shorter timescale. Both enhancements offer a leap forward for QR explosive detection capability.

Let us consider the sample containing two spin-systems, namely magnetic nuclei, usually protons (P) with spin-lattice relaxation time T_{1P} and quadrupolar nuclei (Q) with spin-lattice relaxation time T_{1Q} (Figure 10). We assume that spin-systems P and Q are connected via dipole-dipole interactions and relaxation times T_{1P} and T_{1Q} are long enough for the experiment.

The typical cross-polarisation experiment is clearly illustrated in Figure 11. A basic approach of this sort of experiment is to initially polarize the P spins using a static magnetic field so that the P energy levels have much greater energy separation than the Q levels. Given time to equilibrate the P levels will have relative occupation numbers determined by the Boltzman distribution. Thus the relative population difference between the two P levels, hence polarisation, will correspondingly be much greater than the Q levels of interest. By reducing the external magnetic field adiabatically the P level splitting is reduced such that the proton and quadrupolar energy levels separations equalise allowing a transfer of polarisation. This results in a net polarisation flow from protons to the quadrupolar nuclei. This effect can also be explained through the concept of spin temperatures, where energy flows from the "hot" Q spin-system to the "cold" P spin-system "cooling" the Q spin-system. Applying conventional pulse detection techniques soon after removing the magnetic field, the NQR response can be improved according to the ratio of the proton NMR to NQR frequency.[25] Thus the Q signal enhancement becomes greater the higher the initial polarization for P system. It should be noted that the efficiency of cross-polarisation also depends on the strength of the dipole–dipole interactions between the Q and P spins and the relaxation times T_{1P} and T_{1Q}.

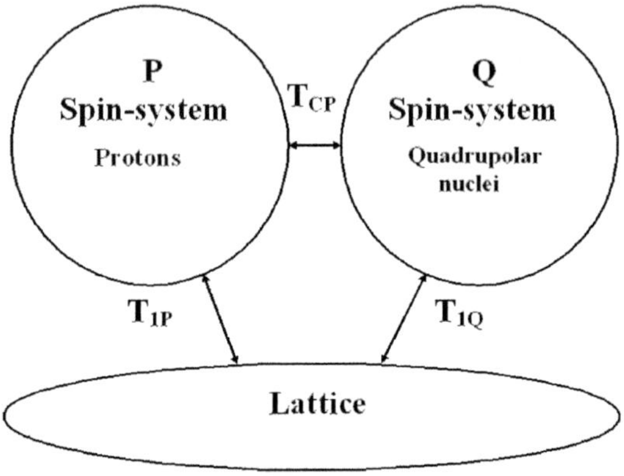

Figure 10. Proton (P) spin-system, quadrupolar (Q) spin-system and lattice. Spin systems P and Q are connected with a lattice and this connection may be by the spin-lattice relaxation time T_{1P} and T_{1Q} accordingly. Connection between P and Q spin-systems via dipole–dipole interactions may be characterised by the cross-correlation time T_{CP}.

The use of the cross-polarization effect between quadrupolar nuclei and protons has been shown to significantly enhance the NQR signal level using the same NQR pulse sequence for a long T_1 substance, TNT[26] and PETN.[27] The cross-polarization concept for NQR has been shown to effectively lower the waiting time required between repeated pulse sequences.[27]

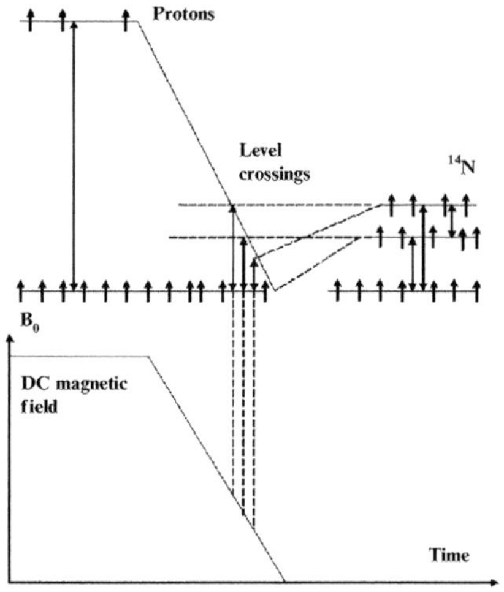

Figure 11. Cross-polarizations NQR experiment. The static magnetic field initially polarizes the P spins using a static magnetic field so that the P energy levels have much greater energy separation than the Q levels. By reducing the external magnetic field adiabatically the P level splitting is reduced such that the proton and quadrupolar energy levels separations equalise allowing a transfer of polarisation.

The cross polarisation method can be very efficient for reducing the relaxation time of Q spins previously excited (or "heated") by a sequence of RF pulses. In order to achieve the relaxation time reduction the contact between the "cold" P spin-system and the "hot" Q spin-system can also be used. During the contact the Q spin-system mainly relaxes via a cross-polarisation "mechanism" described by a cross-polarisation time (T_{CP}). During this time the P and Q spin-systems come to a common spin temperature. An important condition for the method to be effective is that T_{CP} must be much shorter than T_{1P} and T_{1Q} of the substance.

4.3.1. *Signal enhancement*

As was mentioned above, the cross-polarization method offers higher signal the greater the initial polarization of the P spins. The higher polarization is achieved by applying a strong magnetic field during the polarization time. The optimal time for this polarization depends on a number of conditions; in general it will be around 5 T_{1p}. A series of experiments were performed to determine the dependence of the ^{14}N NQR signal intensity for a sample of PETN with the ^{1}H polarization time. The results of these measurements are presented in Figure 12a and b. In these figures NQR signal intensities are plotted as a function of both the polarization time and applied DC magnetic field. Note that all these measurements commenced when Q spins were fully relaxed. Figure 12a shows the dependence of I_p/I on the polarization time for different values of B, where I_p and I is the total intensity of NQR signal obtained with and without the ^{1}H polarization respectively. This figure, in fact, demonstrates the enhancement of the ^{14}N NQR signal in PETN which can be achieved by using the cross-polarization method. Another plot presented in Figure 12b is the dependence of I_p/I on the same parameters as in Figure 12a.

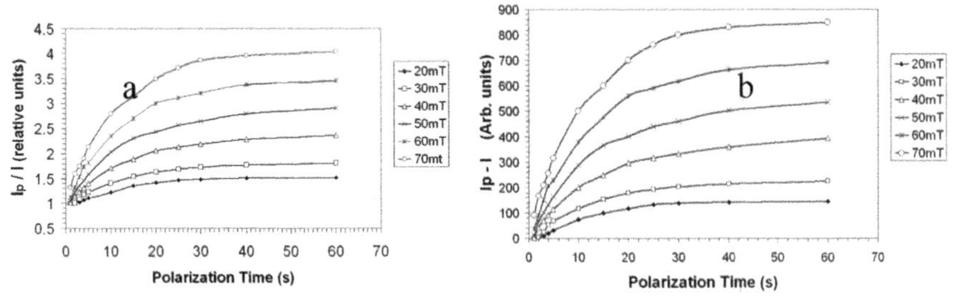

Figure 12. The dependence of (a) I_p/I and (b) $I_p - I$ on the polarization time for values of B.

Figure 13 shows the dependence of NQR signal intensity in PETN on the magnetic field. As would be expected the trend is linear with field magnitude given the initial P spin level separation is greater than the Q spin level separation.

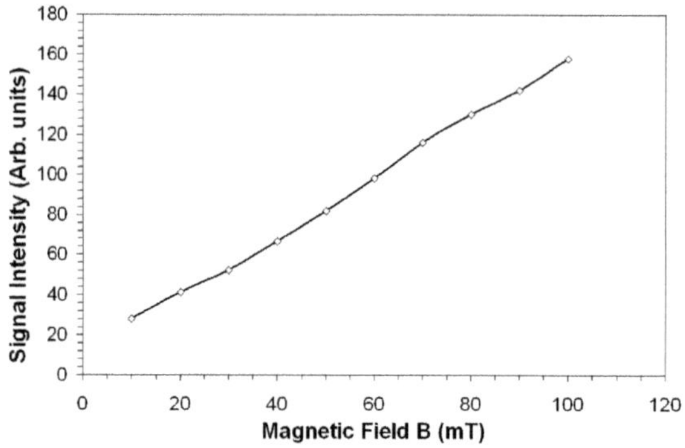

Figure 13. The dependence of NQR signals intensity in the PETN sample on magnetic field *B*. The proton polarization time is 60 s.

4.3.2. *Relaxation time reduction*

Several experiments were performed in order to investigate the recovering process of NQR signals in the Q spin-system.[27] The sample of PETN previously excited by the first pulse sequence was placed in magnetic field to polarize P spins. After polarisation the sample was moved back into the coil where NQR signal was detected using the second pulse sequence. Note that both pulse sequences used are the same. The results of these experiments are presented in Figure 14 and demonstrate the shortening of the signal recovery time. Figure 14a shows the dependence of the signal intensity on the waiting or polarization time for different values of *B*. The dependence of I_p/I_0 on the waiting time and *B*, where I_0 and I_p is the total intensity of NQR signal obtained with first and second pulse sequence accordingly is shown in Figure 14b.

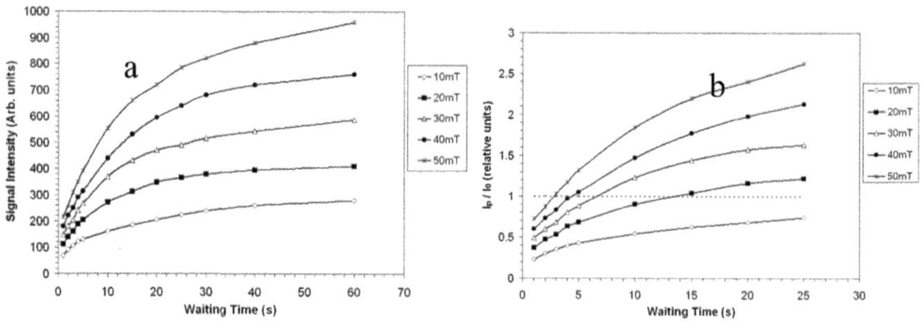

Figure 14. The dependence of (a) NQR signals intensity and (b) I_p / I_0 in the PETN sample on the waiting (or polarization) time for different values of B . I_0 and I_p is the total intensity of NQR signal obtained with first and second pulse sequence accordingly.

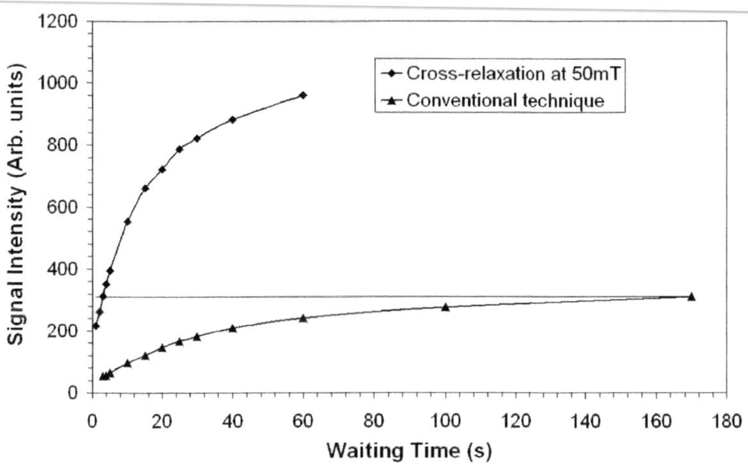

Figure 15. Comparison of the cross-polarization (at 50 mT magnetic field) and conventional detection technique. In each case the sequence is repeated twice with varying waiting times between each application. The intensity derived from second application sequence is shown.

One can see that using cross-polarization method the waiting time between two pulse sequences to obtain the same intensity can be reduced considerably. This fact is clearly demonstrated in Figure 15 where the cross-polarization detection technique is compared with the regular used technique. It can be seen that the cross-polarization method can effectively reduce the time between pulse sequences and in the case of PETN this reduction can be up to 60 times in this example using a small magnetic field.

4.4. EXTENSION OF EXPLOSIVE DETECTION CAPABILITIES

At the moment the NQR scanners for transport security are designed to be able to detect RDX, HMX and PETN based explosive materials including any plastic, powder, sheet and bulk explosives. Detection technique for other explosives such as ammonium nitrate (AN), black powder (KNO_3 component) and TNT has also been developed and will be used in subsequent NQR scanner systems [28].

As an example we present some of interesting NQR parameters for AN and AN based explosives obtained recently. Pure AN (NH4+NO3−) is chemically stable and has a relatively low sensitivity.[23] Confinement and/or the presence of contaminants may cause it to explode as has occurred in various accidental explosions. The AN has been used in conjunction with a fuel (ANFO) for terrorist purposes (ANFO used by IRA, ETA, FARC, by Palestinian extremists and in Oklahoma City, US, 1995) mainly in vehicle bomb attacks.

AN has a melting point of 169°C, below which it can exist in five different crystallographic forms (phases I–V) which are temperature dependent. The stable form at room temperature is AN-IV, but the phase transition to AN-III occurs at only 32°C.

TABLE 3. NQR parameters vs. temperature for AN (spectral line ν_+).

Temperature (°C)	NQR frequency (kHz)	Line width (Hz)	T_1 (s)
5	502.01	215	17
10	500.60	210	16
15	499.19	200	15
20	497.64	195	14.0
25	496.12	190	13
28	494.98	210	12.5
30	494.14	215	12.0

TABLE 4. NQR parameters vs. temperature for AN (spectral line ν_-).

Temperature (°C)	NQR frequency (kHz)	Line width (Hz)	T_1 (s)
5	421.70	215	25
7	421.90	215	24
9	422.10	215	23.5
10	422.20	215	23
12	422.38	215	22.5
15	422.62	215	21.5
17	422.80	210	21
20	423.05	200	20
23	423.30	195	19
25	423.47	190	18
28	423.68	210	17.0
30	423.80	215	16.5
32	423.90	215	16
33	424.06	215	16
34	424.19	215	15
35	424.30	215	15
36	424.40	215	14

The NQR parameters for AN sample are presented for spectral line ν_+ and ν_- in Tables 3 and 4 respectively. The relative intensities of NQR signal for PETN, AN and AN based explosives obtained by using the two-pulse Hahn sequence are presented in Table 5.

It should be noted that the detection of NQR signals in TNT sample is quite complicated by the existence of at least two polymorphic forms, monoclinic and orthorhombic, which have different NQR properties. Further complications are caused by the multi-line spectrum and a very small NQR signal-to-noise ration. Nevertheless this substance can be detected with the NQR technique.

5. Current and potential applications

Figure 16 demonstrates the current and potential NQR applications. At the moment the most important application of NQR method is transport security and aviation security in particular. In this area the NQR machines are used for screening baggage and shoes.

TABLE 5. Relative intensities of NQR signal for PETN and AN compounds.

Substance	Line and frequency (kHz)	Intensity (arb. units)
PETN	ν_+ 890	1.0
PETN	N_- 495	0.6
AN	ν_+ 497	1.1
AN	ν_- 424	0.8
ANFO	ν_+ 497	1.0
ANFO	ν_- 424	0.7
RIOGEL+	ν_+ 497	0.4
RIOGEL+	ν_- 424	0.2

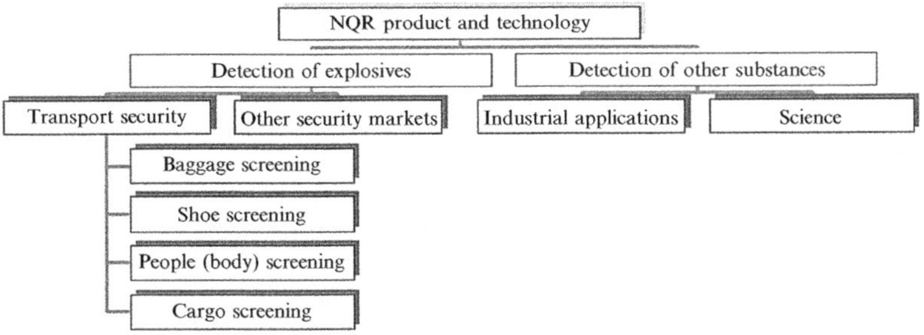

Figure 16. Current and potential NQR applications.

Figure 17 is a general block diagram illustrating the Baggage Scanner which is an NQR system for detecting a resonance signal in the specimen.[29] As shown in this figure, the system comprises Coil and Shield Assembly, System PC 1, Receiver Subsystem 2, RF Pulse Generator and Event Control Unit 3 and Power Amplifier (PA) 4. The Coil and Shield Assembly consists of the Coil System, Conveyor Unit and Position Sensor, Q-switch, Preamplifiers 6, RF Monitor Probe, Tuning and Matching Circuits. The system contains two different matching circuits and preamplifier channels designed to operate in the high frequency band and the low frequency band, normally below 1 MHz. Separate amplifiers and isolation networks are used for high and low frequency. Care has been taken to optimise the receive sensitivity by noise matching the front end of the receive amplifiers.

The Coil System includes a single low loss resonant 160-l coil, fixed capacitors and tune circuit. The resonant coil is tuned to the appropriate frequency for irradiation and detection of a particular explosive. This coil geometry was chosen because it has been shown to be a highly efficient means of detecting an NQR response from a target material.

The RF Pulse Generator and Event Control Unit generate RF pulses which are amplified by the Power Amplifier (PA) and transfer the pulses to the Coil System. These RF pulses can excite NQR signals in the specimen under investigation which is located in the coil. The transmit signal is

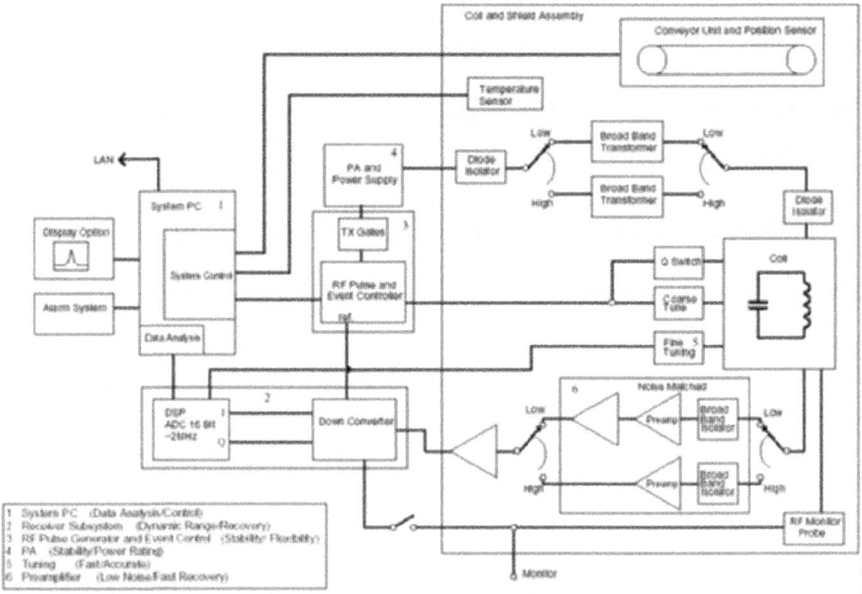

Figure 17. General block-diagram of the NQR Baggage Scanner.

efficiently matched and coupled into the coil by broadband networks. The transmit signal is generated digitally by a programmable pulse generator and event controller. This controller also sends signals to the demodulator in order to provide synchronous detection capabilities. The isolation diodes reduce the coupling from the noisy transmitter during the receive period.

The tuning of the coil is accomplished via electromechanical means. The tuning is broken up into two functional groups, firstly, course tuning to select the type of explosive and secondly, fine tuning is used to compensate for changes in the tuning frequency caused by luggage items. Extremely efficient isolation of the channels used to transmit and receive signals is required because a single coil is used for both transmitting the energizing pulse and receiving the NQR response. These responses are typically many orders of magnitude different in intensity. In this design this isolation is accomplished by appropriate placement of isolation diodes and broadband switching networks.

It is highly desirable to remove the energy from the coil as soon as possible so that the small NQR response can be observed. Coil damping is particularly important in the lower frequency ranges where "ringing" persists for a longer period of time. At low frequencies the coil damping is accomplished by two separate means, firstly, the Q-switch shown and secondly by circuits within the amplifier block.

The signal from the front end amplifiers is sent to an analog quadrature demodulator before being digitised by an ADC/DSP data acquisition card. The ADCs are 16 bit sigma delta converters. The ADCs typically run at a conversion frequency of 1–2 MHz and with over sampling; this arrangement offers a high dynamic range. The importance of high stability and dynamic range comes from issues associated with the removal of spurious responses from piezo-electric and magneto-acoustic sources within the luggage under examination. Overall system control and network interfacing is accomplished by an on board System PC. The System PC generates control signals, which are transferred to RF Pulse Generator and Event Control Unit and prescribe parameters for RF pulses. The range of functions controlled by the System PC includes control of the conveyors, alarm options, optical sensors, thermal sensors, logging of system information and data, downloading of control scripts to the pulse generator and DSP and some signal processing and display functions.

To move the luggage under investigation, Conveyor Unit with Position Sensor is used. This Conveyor Unit comprises means for exact positioning of the sample in the coil. The System PC generates control signals, which are transferred to Conveyor Unit and Position Sensor.

Besides its specifically determined NQR frequency, each target substance also has characteristic relaxation parameters. This means that for the optimum detection of each target substance it is desirable to use an optimum pulse sequence. Such a pulse sequence is determined for each target substance beforehand and entered into the memory of the System PC and later used in the detection procedure. Similarly, parameters for the signal processing of each target substance are entered into the memory of the System PC and later used in the process of detection.

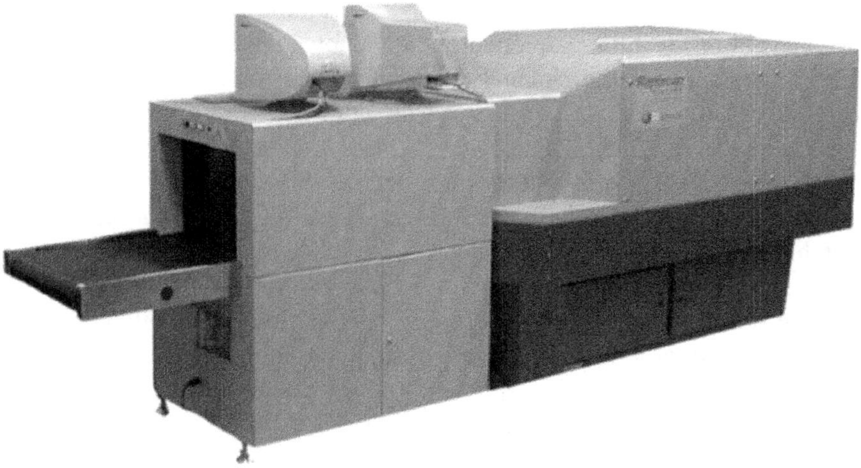

Figure 18. QXR 1000 Baggage Scanner.

One way to improve the performance of the Checkpoint systems is to combine different technologies that are based on different physical principles that provide different, but overlapping performance envelopes enable a combined system to "fill the gaps" left by single technology solutions. NQR technology has many benefits that are complimentary with other detection technologies such as single and multi-view, dual energy transmission x-ray, x-ray diffraction, CT and trace detection as it can increase the accuracy of automated detection.

Modern baggage scanners,[1] such as QXR 1000 baggage scanner, utilises the NQR technology combined with dual energy x-ray and countermeasures detection in a single, integrated platform (Figure 18). These machines provide improved checkpoint detection of a broader range of aviation treats. A greatly enhanced checkpoint screening solution is achieved by combining the automatic detection of explosives of NQR method with an existing x-ray system. These units require no additional screeners, or checkpoint lines.

The shoe scanners developed by GE Security contain NQR and metal detection means. These machines provide highly accurate automated explosives and weapon detection.

There are also follow-on opportunities in the explosive detection market to screen cargo, people and mail. Moving in these directions is the next phase of expansion of NQR technology:

Transportation – rail, subways, buses & cargo trains, etc.

Commerce – freight, parcels, mail, mailrooms and delivery services

Facility & Event Protection – corporate foyers, cultural & entertainment venues, government buildings etc.

Critical Infrastructure – utilities, power plants, water treatment facilities

6. Conclusion

NQR is a very promising explosive detection technique for the transport security. Recent advances in instrumentation and detection technique have shown that NQR has excellent potential in this field. NQR technology has many benefits that are complimentary with other detection technologies such as x-ray and trace detection as it can increase the accuracy of automated detection. We believe that the new generation of NQR based explosives detection machines will be designed to operate much more effectively in the real world conditions.

References

1. Rudakov T N, Rainer T J, Hayes P A et al., Detection of explosives by quadrupole resonance method: new aspects for security In: Schubert H, Kuznetsov A (eds), Detection and Disposal of Improvised Explosives, Springer, New York, pp. 191–204 (2006)
2. Hirshfeld T, Klainer S M, Short range remote NQR measurements, J. Molec. Struct. 58, 63–77 (1980)
3. Rudakov T N, Magnetic Resonance: Nuclear Quadrupole Resonance, Instrumentation. In: Lindon J C, Tranter G E, Holmes J L (eds), Encyclopedia of Spectroscopy and Spectrometry, Academic Press, London, vol. 2, pp. 1663–1671 (1999)
4. Gerothanassis I P, Methods of avoiding the effects of acoustic ringing in pulsed Fourier transform nuclear magnetic resonance spectroscopy, Prog. NMR Spectros. 19, 267–329 (1987)
5. Marino R A, Klainer S M, Multiple spin echoes in pure quadrupole resonance, J. Chem. Phys. 67, 3388–3389 (1977)
6. Ostroff E D, Waugh J S, Multiple spin echoes and spin locking in solids. Phys. Rev. Lett. 16, 1097 (1966)
7. Rudakov T N, Mikhaltsevitch V T, Flexman J H, Hayes P A, Chisholm W P, Modified multipulse technique for the effective detection of pure nuclear quadrupole resonance. Appl. Magn. Reson. 25, 467–474 (2004)
8. Ernst R R, Anderson W A, Application of Fourier transform spectroscopy to magnetic resonance, Rev. Sci. Instrum. 37, 93–102 (1966)
9. Freeman. R, Hill H D W., Phase and intensity anomalies in Fourier transform NMR, J. Magn. Reson. 4, 366–383 (1971)

10. Rudakov T N, Mikhaltsevich V T, Selchikhin O P, The use of multi-pulse nuclear quadrupole resonance techniques for the detection of explosives containing RDX, J. Phys. D: Appl. Phys. 30, 1377–1382 (1997)
11. Klainer S M, Hirschfeld T B, Marino R A, Fourier transform NQR spectroscopy. In: Marshal A G (ed) Fourier, Hadamard and Hilbert Transforms in Chemistry, Plenum, New York, pp. 147–181 (1982)
12. Buess M L, Garroway A N, Miller J B, NQR detection using a meanderline surface coil, J. Magn. Reson. 92, 348–362 (1991)
13. Rudakov T N, Belyakov A V, Modifications of the steady-state free-precession sequences for the detection of pure nuclear quadrupole resonance J. Phys. D: Appl. Phys. 31, 1251–1256 (1998)
14. Osokin D Ya, Pulsed spin-locking in nuclear quadrupole resonance of 14N, Zn. Eksp. Teor. Fiz. 84, 118–123 (1983)
15. Buess M L, Garroway A N, Yesinowsky J P (1994), US Patent No. 5365171
16. Smith J A S, Peirson N F (1996) Patent Int. Publication No. WO 96/26453
17. Rudakov T N, Hayes P A, Some aspects of quasi-steady state of nitrogen-14 quadrupolar spin-system, Chem. Phys. Lett. 405, 334–338 (2005)
18. Rudakov T N, Modification of SSFP technique for the effective detection of NQR signals. Phys. Lett. A 358, 322–326 (2006)
19. Ernst R R, Bodenhausen G, Wokaun A, Principles of Nuclear Magnetic Resonance in One and Two Dimensions. Clarendon press, Oxford (1987)
20. Rudakov T N, Belyakov A V, Modifications of the steady-state free-precession sequence for the detection of pure nuclear quadrupole resonance. J. Phys. D: Appl. Phys. 31, 1251–1256 (1998)
21. Mehring M, Principles of High Resolution NMR in Solids, 2nd edn, Springer, Berlin (1980)
22. Slichter C P, Principles of Magnetic Resonance, 3rd edn, Springer, Berlin (1990)
23. Abragam A, The Principles of Nuclear Magnetism, Clarendon Press, Oxford (1961)
24. Goldman M, Spin Temperature and Nuclear Magnetic Resonance in Solids, Clarendon Press, Oxford (1970)
25. Blinc R, Apih T, Seliger J, Nuclear quadrupole double resonance techniques for the detection of explosives and drugs, Appl. Magn. Reson. 25, 523–534 (2004)
26. Thurber K R, Sauer K L, Buess M L, Klug C A, Miller J B, Increasing ^{14}N NQR signal by H-N level crossing with small magnetic fields, J. Magn. Reson. 177, 118–128 (2005)
27. Rudakov T N, Hayes P A, Cross-polarisation method for improvement of ^{14}N NQR signal detectability. J. Magn. Reson, 183, 96–101 (2006)
28. Köhler J, Meyer R, "Explosives, Fourth revised and extended edition", Weinheim: VCH Verlagsgesellschaft mbH, (1993)
29. Flexman J H, Rudakov T N, Hayes P A, Shanks N, Mikhaltsevitch V T, Chisholm W P, The detection of explosives in airport baggage using the direct nuclear resonance method in NATO Science Series II Detection of Bulk Explosives, Schubert H, Kuznetsov A (eds.), Kluwer, Dordrecht, vol. 138, pp. 113–124 (2004)

DOUBLE RESONANCE DETECTION OF (MAINLY NITROGEN) NQR FREQUENCIES IN EXPLOSIVES AND DRUGS

JANEZ SELIGER*, VESELKO ŽAGAR
University of Ljubljana, Faculty of Mathematics and Physics Jadranska 19, 1000 Ljubljana, Slovenia and "Jozef Stefan" Institute, Jamova 39, 1000 Ljubljana, Slovenia

Abstract Various nuclear quadrupole double resonance techniques, useful in the detection of nuclear quadrupole resonance frequencies in explosives and drugs, are reviewed with a special emphasis on the detection of ^{14}N and ^{17}O. Various techniques, as for example,

- The Slusher and Hahn's technique
- The technique based on the proton T_1 dispersion
- The level crossing technique
- The solid effect technique
- The technique using multiple frequency sweeps
- The two-frequency irradiation technique

are discussed and a strategy of detecting the nuclear quadrupole resonance frequencies is presented together with some illustrative examples.

Keywords: nuclear quadrupole resonance, NQR; double resonance, magnetic field cycling, nitrogen, oxygen

1. Introduction

Nuclear quadrupole resonance (NQR) has become in recent years a recognized method for detection and identification of explosives and drugs. The method is based on the fact that the quadrupole atomic nuclei with the spin S equal one or larger (^{2}H, ^{14}N, ^{17}O, ^{35}Cl...) exhibit in zero external magnetic field molecular specific NQR frequencies. The degeneracy of the nuclear ground state is namely partially or completely removed by the presence of the

* E-mail: janez.seliger@fmf.uni-lj.si

strongly inhomogeneous local electric field caused by the electric charge distribution in the vicinity of the atomic nucleus. Even small changes in the electric charge distribution may produce a significant shift of the NQR frequencies.

Nitrogen and oxygen atoms are contained in a number of explosives and drugs. Their quadrupole nuclei, ^{14}N with the natural abundance 99.635% and ^{17}O with the natural abundance 0.037%, have the NQR frequencies in the MHz region. The ^{14}N and ^{17}O NQR frequencies can be used for the detection and identification of solid illicit substances.

A ^{14}N nucleus has in zero magnetic field three generally nondegenerated nuclear quadrupole energy levels (Figure 1). The NQR frequencies are labeled as $\nu_+ > \nu_- \geq \nu_0$:

$$\nu_+ = \frac{e^2qQ}{4h}(3+\eta)$$
$$\nu_- = \frac{e^2qQ}{4h}(3-\eta) \tag{1}$$
$$\nu_0 = \nu_+ - \nu_- = \frac{e^2qQ}{2h}\eta.$$

Here e^2qQ/h is the quadrupole coupling constant and η is the asymmetry parameter of the electric-field-gradient (EFG) tensor. The EFG tensor is a traceless second rank tensor composed of the second derivatives of the electrostatic potential with respect to the coordinates. It has three principal values labeled as V_{XX}, V_{YY} and V_{ZZ}, $|V_{XX}| \leq |V_{YY}| \leq |V_{ZZ}|$. The quadrupole coupling constant is in fact the largest principal value V_{ZZ} = eq of the EFG tensor at the position of the nitrogen nucleus multiplied by the electric quadrupole moment eQ of the nitrogen nucleus and divided by the Planck's constant h. The asymmetry parameter η is defined as $\eta = (V_{XX} - V_{YY})/V_{ZZ}$. It ranges between 0 and 1.

A ^{17}O nucleus has a spin I = 5/2 and in zero magnetic field three doubly-degenerated nuclear quadrupole energy levels (Figure 1). The energies E of the nuclear quadrupole energy levels are calculated from the secular equation:

$$x^3 - 7(3+\eta^2)x - 20(1-\eta^2) = 0 \tag{2}$$

Here an energy E is given as $E = (e^2qQ/20)x$, where x is a solution of the secular equation and e^2qQ is the quadrupole coupling constant multiplied by the Planck constant h. The three NQR frequencies are usually named as $\nu_{5/2-1/2} > \nu_{5/2-3/2} \geq \nu_{3/2-1/2}$.

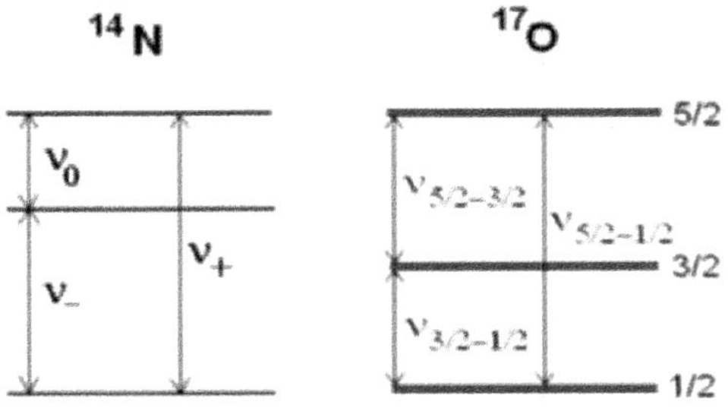

Figure 1. Nuclear quadrupole energy levels and resonance frequencies of ^{14}N and ^{17}O.

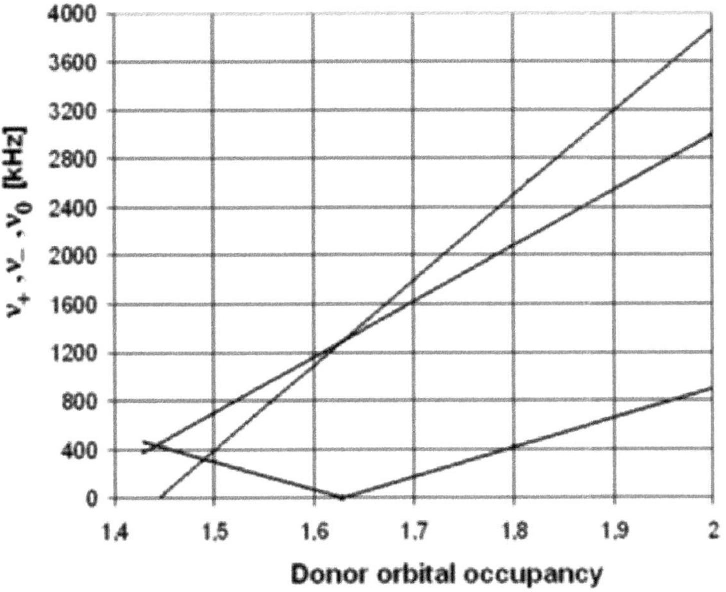

Figure 2. ^{14}N NQR frequencies in substituted pyridines.

They uniquely depend on the quadrupole coupling constant e^2qQ/h and the asymmetry parameter η, that are from the NQR frequencies determined in the following way. First the asymmetry parameter η is determined from the ratio $R = \nu_{3/2-1/2}/\nu_{5/2-3/2}$ which monotonously varies from $R = 0.5$ at $\eta = 0$ to $R = 1$ at $\eta = 1$. When η is known, the quadrupole coupling constant is calculated from any NQR frequency.

The sensitivity of the ^{14}N NQR frequencies to the changes in the electron charge distribution is illustrated in the case of substituted pyridines. Rubenacker and Brown[1] have shown that there exists a correlation between the donor orbital occupancy and the principal values of the EFG tensor. The ^{14}N NQR frequencies calculated on the basis of their analysis are illustrated in Figure 2.

The highest ^{14}N NQR frequency ν_+ ranges from about 3.9 MHz in case of a non substituted pyridine to about 0.5 MHz in case of a protonated nitrogen.

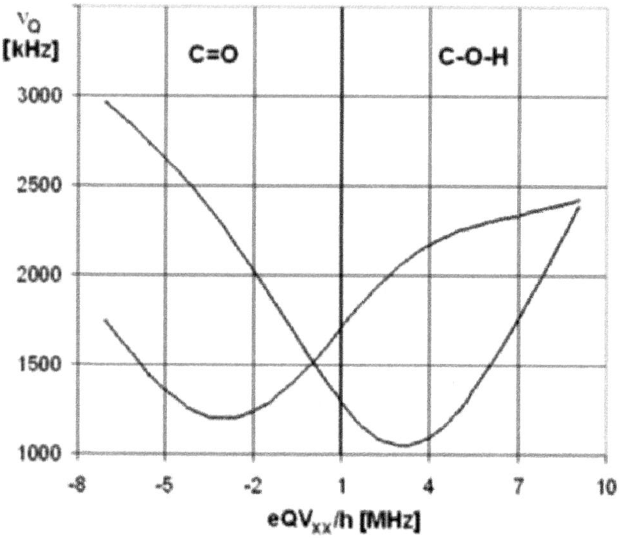

Figure 3. ^{17}O NQR frequencies $\nu_{3/2-1/2}$ and $\nu_{5/2-3/2}$ in C–O bonds as functions of the principal value V_{XX} of the EFG tensor. The principal axis X points perpendicular to the C–O bond.

It has been also shown that there exists a correlation between the principal values of the EFG tensor at the oxygen position in the C–O bonds.[2] The ^{17}O NQR frequencies as calculated on the basis of this correlation are presented in Figure 3.

A strong variation of the ^{17}O NQR frequencies depending on the local electric charge distribution is observed.

These examples illustrate the fact that the ^{14}N and ^{17}O NQR frequencies indeed strongly vary from substance to substance, and can be used to different-tiate between them.

The determination of either the NQR frequencies or the quadrupole coupling constant e^2qQ/h and the asymmetry parameter η in solids is possible by different means: solid state NMR, pulsed or cw NQR and nuclear quadru-pole double resonance (NQDR).

Solid state NMR measures the quadrupole shift of the resonance frequency. There is small number of NMR lines and the EFG tensor can be precisely determined in single crystals. On the other hand in polycrystalline samples we generally observe broad frequency distributions. Only the central line, corresponding to the $-1/2 - 1/2$ transition, exhibits a second order quadrupolar shift in the lowest order of the perturbation theory. This relatively narrow frequency distribution, which can be used for the determination of e^2qQ/h and η, occurs only in case of half-integer spin nuclei (^{17}O,...). In case of integer-spin nuclei (^{14}N,...) there is no central line and the sensitivity of detection is rather low.

Pulsed NQR measures the NQR frequencies the most precisely. In case of ^{14}N the search of the NQR frequencies represents generally a lengthy procedure, while in case of ^{17}O the sensitivity of pure NQR is often too low.

NQDR has in case of low NQR frequencies much higher sensitivity than pure NQR, while its resolution is somewhat lower. It can be used for the "rough" determination of the ^{14}N NQR frequencies in a short time, and for the determination of the ^{17}O NQR frequencies. In the continuation we present the NQDR techniques that are used for the measurement of the ^{14}N and ^{17}O NQR frequencies.

2. Nuclear quadrupole double resonance

NQDR is an indirect technique for the detection of the NQR frequencies. Two spin systems are needed to do a NQDR experiment: an abundant spin system with a strong NMR signal and the system of quadrupole nuclei. The abundant spin system is in most cases the proton spin system. We shall in the continuation assume that this is the case.

The NQDR experiment is based on magnetic field cycling. An experiment is schematically presented in Figure 4.

The experiment consists of three phases: polarization, mixing and detection. During the polarization phase the proton magnetization reaches its equilibrium value in a high magnetic field B_0. This phase lasts approximately $5T_1(B_0)$, where $T_1(B_0)$ is the proton spin-lattice relaxation time in the high magnetic field B_0. After the equilibrium proton magnetization is reached, the external magnetic field is adiabatically reduced to a low value B, when the mixing phase starts. The frequency separation Δν, $\Delta\nu = \nu_L$, of the proton Zeeman energy levels is during the adiabatic demagnetization reduced from $\Delta\nu = \gamma_H B_0/2\pi$ to $\Delta\nu = \gamma_H B/2\pi$. Here γ_H is the proton gyromagnetic ratio. During the adiabatic demagnetization the populations of the proton energy levels remain in most cases nearly untouched. The proton magnetization is at the beginning of the mixing phase nearly equal to the equilibrium proton magnetization M_0 in the high magnetic field B_0, $M_0 = CB_0/T_L$. Here C is the Curie constant and T_L is the temperature of the sample. This value of the proton magnetization is larger than the equilibrium proton magnetization M in the low magnetic field B, equal to $M = CB/T$. Due to the spin-lattice relaxation the proton magnetization relaxes during the mixing phase towards the equilibrium value M. The proton magnetization M(τ) is at the end of the mixing phase equal

$$M(\tau) = M + (M_0 - M)\exp(-\tau / T_1(B)) \quad (3)$$

Here $T_1(B)$ is the proton spin-lattice relaxation time in the low magnetic field B.

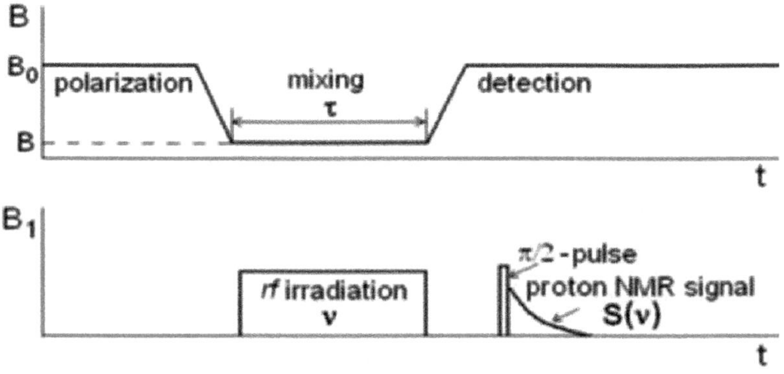

Figure 4. Schematic presentation of a NQDR experiment.

The process can also be described using the concept of spin temperature. During the adiabatic demagnetization the proton spin temperature reduces to the value $T = T_L(B/B_0)$. The "cold" proton spin system absorbs during the mixing phase energy from the crystal lattice, which is at a higher temperature T_L. The inverse proton spin temperature $\beta = 1/T$ exponentially approaches the inverse lattice temperature $\beta_L = 1/T_L$ with the time constant $T_1(B)$.

After the mixing phase is finished, the initial magnetic field B_0 is adiabatically restored and the proton NMR signal S is measured. It is proportional to M(τ). Measurement of S in dependence of τ at a fixed value of B is used in the field-cycling spectrometers to determine $T_1(B)$.

The application of a pulse of a radiofrequency (*rf*) magnetic field during the mixing phase may increase the proton relaxation rate and consequently decrease the proton NMR signal S in the detection phase. This happens when the frequency ν of the rf magnetic field matches a resonance (NQR) frequency of the quadrupole spin system. In such a way the NQR frequencies are indirectly measured by observing the proton NMR signal.

Various NQDR techniques used for the measurement of the ^{14}N and ^{17}O NQR frequencies under different experimental conditions are briefly described in the continuation.

2.1. THE SLUSHER AND HAHN'S TECHNIQUE

The general NQDR scheme (Figure 4) applies for the Slusher and Hahn' technique.[3] The final external magnetic field B in the demagnetization process is equal zero. The proton Zeeman order transforms into the proton dipolar order and the spin temperature of the proton dipolar system is at the beginning of the mixing phase equal $T = T_L(B_{loc}/B_0)$. The proton local magnetic field B_{loc} is the result of the dipole–dipole interaction and is typically equal to a few gauss.

The rf magnetic field applied during the mixing phase is strong and phase modulated. The time between two repetitive sudden 180° phase shifts of the rf magnetic field is equal τ^*. The rf magnetic field, when in resonance, induces in the interaction picture a frequency splitting $\Delta\nu$ of the quadrupole energy levels of the order of $\gamma_Q B_1/2\pi$. Here γ_Q is the gyromagnetic ratio of the quadrupole nuclei and B_1 is the amplitude of the rf magnetic field. This splitting is analogous to the frequency splitting of the energy levels in the rotating frame, as observed in conventional NMR. The splitting $\Delta\nu$ is proportional to the amplitude of the *rf* magnetic field and can be made comparable to the width $\delta\nu$, $\delta\nu \sim \nu_{loc} = \gamma_H B_{loc}/2\pi$, of the proton absorption line in zero external magnetic field (Figure 5). Simultaneous flip-flop transitions in both spin system, conserving the total energy, equalize the spin temperatures of the spin systems. The time needed for this process to be completed is called the cross relaxation time t_{CR}. Suppose that t_{CR} is shorter than the time t^* between two repetitive phase shifts of the rf magnetic field. Then both spin systems have a low spin temperature at the

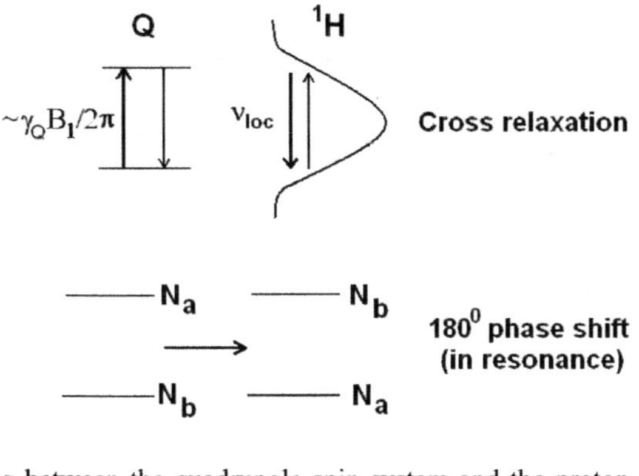

Figure 5. Coupling between the quadrupole spin system and the proton spin system in the Slusher and Hahn's technique and the effect of the sudden 180° phase shifts.

moment before the phase shift occurs. The phase shift of the rf magnetic field causes an inversion of the quadrupole energy levels in the interaction picture (Figure 5). The upper energy level is immediately after the phase shift more populated than the lower energy level. The spin temperature of the quadrupole spin system is negative. Downward transitions in the quadrupole spin system and upward transitions in the proton spin system equalize the spin temperatures. Then the next phase shift occurs.

This technique works well with the half-integer spin nuclei (^{17}O), but fails in case of integer-spin nuclei (^{14}N).[4] The reason is that the "quadrupolar precession" averages out the proton–nitrogen dipolar interaction to zero.

In the continuation we briefly discuss the application of the Slusher and Hahn's technique to ^{17}O. A more detailed discussion is given in reference[5].

Suppose we are dealing with a polycrystalline sample. Then the proton NMR signal S at the beginning of the detection period approximately equals

$$S(\nu) \cong S_0 \exp(-\tau / T_{1H}(0)) \exp(-W(\nu)\tau) \tag{4}$$

Here $T_{1H}(0)$ is the proton spin-lattice relaxation time in zero magnetic field and $W(\nu)$ is the double resonance contribution to the proton relaxation rate. The double resonance signal ΔS is the difference between the proton NMR signal S, $S=S_0\exp(-\tau/T_{1H}(0))$ measured in case of no *rf* irradiation, and the proton NMR signal $S(\nu)$. It depends on the relaxation rate $W(\nu)$. In resonance, when ν equals to a NQR frequency, say $\nu_{m,n}$, we have

$$W(\nu_{m,n}) = \frac{2}{9\tau^*} \frac{N(^{17}O)}{N(^{1}H)} \left(\frac{\nu_1}{\nu_{loc}} \right)^2 R_{m,n}(\eta) \tag{5}$$

Here $N(^{17}O)$ and $N(^{1}H)$ are the numbers of ^{17}O and protons in the unit cell respectively, $\nu_1 = \gamma_O B_1/2\pi$ and $R_{m,n}(\eta)$ is a factor depending on the transition (m,n) and asymmetry parameter η. For the 3/2–1/2 transition $R_{3/2,1/2}(\eta)$ monotonously decreases from 8 at $\eta = 0$ to 5.3 at $\eta = 1$. For the 5/2–3/2 transition $R_{5/2,3/2}(\eta)$ slightly increases from 5 at $\eta = 0$ to 5.3 at $\eta = 1$. For the 5/2–1/2 transition $R_{5/2,1/2}(\eta)$ is much smaller. It is equal zero at $\eta = 0$ and monotonously increases to 0.27 at $\eta = 1$.

The double resonance signal ΔS is in the limit of low $W(\nu_{m,n})$, i.e. when $W(\nu_{m,n})T_{1H}(0) << 1$, maximum at $\tau = T_{1H}(0)$, when it is equal

$$\Delta S = S_0 W(\nu_{m,n}) T_{1H}(0) / e \tag{6}$$

The strongest double resonance signal is thus observed when $\nu = \nu_{3/2,1/2}$, while the double resonance signal at $\nu = \nu_{5/2,1/2}$ is often not observed. The double resonance signal increases quadratically on increasing the amplitude B_1 of the rf magnetic field until the splitting $\Delta\nu$ of the quadrupole energy levels in the interaction picture does not become large as compared to ν_{loc}, when the cross relaxation time strongly increases. But even when $\Delta\nu$ is comparable to ν_{loc}, cross relaxation limits the sensitivity of double resonance.

As shown in reference[5], the cross relaxation reduces the double resonance contribution to the proton relaxation rate $W(\nu)$ to the value $W^*(\nu)$,

$$W^*(\nu) = \frac{1 - \exp(-\tau^*/t_{CR})}{1 + \exp(-\tau^*/t_{CR})} W(\nu) \tag{7}$$

Measurement of the double resonance signal DS at different values of τ^* can be used to determine the cross relaxation time t_{CR}.

As an experimental example we present the measurement of the ^{17}O NQR frequencies in TNT. The results are partially published in reference[6].

The double resonance signals as measured at T = −100°C are presented in Figure 6.

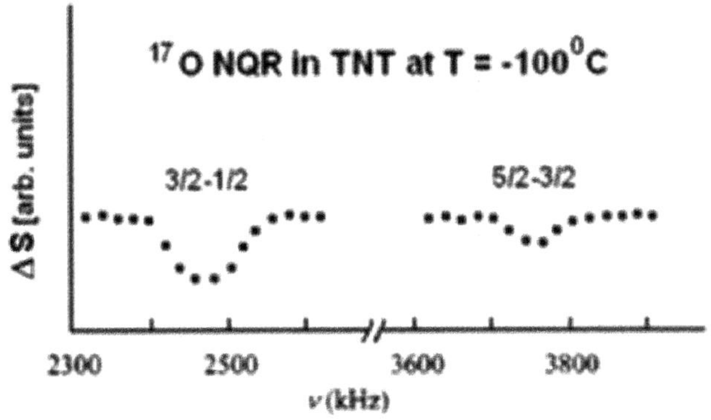

Figure 6. ^{1}H-^{17}O NQDR signals in TNT at T = −100°C.

The experiment was performed under the following experimental conditions. The duration of the polarization period in B_0 = 0.75 T was 45 s. The sample was within 0.1 s pneumatically transferred into a second magnet, where the external magnetic field was compensated to less than 50 μT. The mixing period lasted for 0.7 s. After the mixing period the sample was within 0.1 s transferred back into the first magnet, where the proton NMR signal was measured. During the mixing period we applied a phase modulated *rf* magnetic field with the amplitude B_1 = 4 mT. The time τ^* between two repetitive 180° phase shifts of the rf magnetic field was equal τ^* = 0.3 ms.

In order to determine the cross relaxation time between the ^{17}O and ^{1}H spin systems we measured the double resonance signal at two different values of τ^*: 0.3 and 3 ms. The results are presented in Figure 7.

Nearly no influence of the modulation time τ^* on the NQDR signal is observed. From the scattering of the experimental points we may estimate the difference $(S_0 - S)/S_0$ for the two measurements to differ less than a few percent. According to equation 7 this happens when 3 ms/t_{CR} < 0.5. The cross relaxation time t_{CR} is thus longer than 6 ms. This value is larger than t_{CR} = 0.8 ms as obtained for the nitro group in 2-nitrobenzoic acid.[5] There are two possible reasons for such a large difference: (a) the difference in the shortest O–H distance and (b) multiple slightly different ^{17}O NQR frequencies in TNT that are not resolved in the NQDR spectrum. In the later case to the NQDR line consists of resonance and off-resonance contributions and in case of the off-resonance contributions the cross relaxation time may be longer.

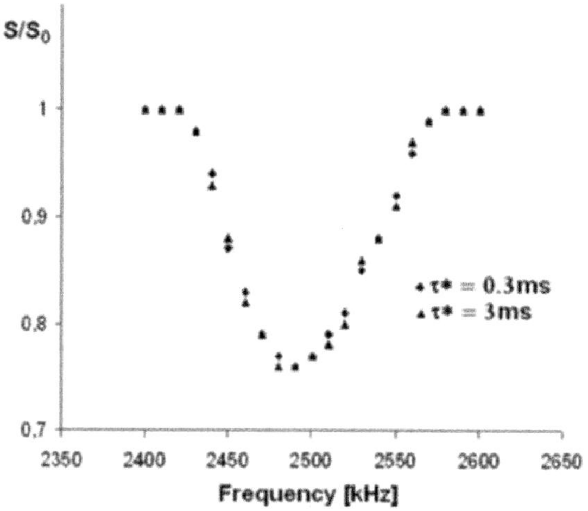

Figure 7. The normalized proton NMR signal S in TNT as a function of the frequency ν of the rf magnetic field measured at two modulation times τ^*.

2.2. LARMOR FREQUENCY DEPENDENCE OF PROTON T_1

This technique is used for fast measurement of low NQR frequencies. When the proton Larmor frequency matches a NQR frequency, the proton spin-lattice relaxation time may change, especially if the quadrupole spin system relaxes faster than the proton spin system.

In case of half-integer spin quadrupole nuclei this technique is used when the proton spin-lattice relaxation time in zero magnetic field is too short for the Slusher and Hahn's technique to be used. The resolution of this technique is low. The doubly degenerated quadrupole energy levels split in nonzero magnetic field B linearly with respect to B. The splitting depends on the orientation of the external magnetic field B with respect to the principal directions of the EFG tensor. In a polycrystalline sample we therefore observe broad structured dips in the Larmor-frequency dependence of the proton spin-lattice relaxation time T_1.

In case of ^{14}N (Figure 8) we generally observe three narrower dips in the Larmor-frequency dependence of proton T_1 at $\nu_L = \nu_0$, ν_- and ν_+.

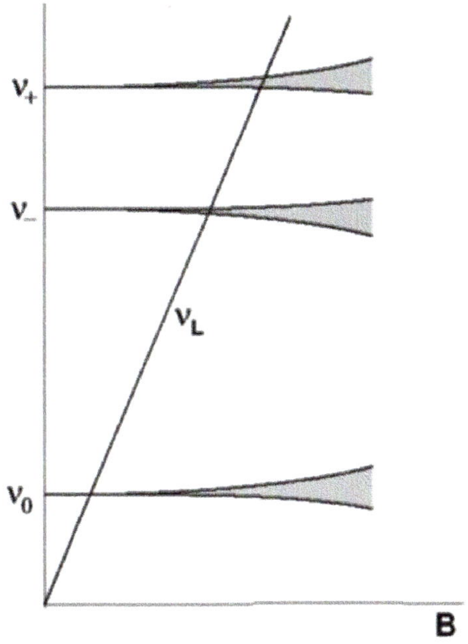

Figure 8. Schematic picture of proton (ν_L) and ^{14}N resonance frequencies in a low magnetic field B.

In case of a low ν_0 we often observe in addition a narrower dip at $\nu_L = \nu_0/2$.[7] This dip is associated with the fact that in solids the proton spin system – due to the dipole–dipole interaction absorbs energy also at multiples of the Larmor frequency. The quadrupolar dips are in case of ^{14}N relatively narrow because the energies of the nondegenerate nitrogen quadrupole energy levels in a nonzero magnetic field B shift quadratically with respect to B and in addition the ^{14}N magnetic moment is low as compared to the proton magnetic moment. The width $\delta\nu$ of the quadrupolar dip at $\nu_L = \gamma_H B/2\pi = \nu_0$ is of the order of $\delta\nu \sim (\gamma_N/\gamma_H)^2\nu_0 \approx \nu_0/200$. This width is in most cases smaller than the proton linewidth in magnetic field B. The widths of the dips around $\nu_L = \nu_-$ and $\nu_L = \nu_+$ are larger, especially when the asymmetry parameter η is low. The proton spin-lattice relaxation rate $W = 1/T_1$ may be for example in case when $\nu_L = \nu_0$ expressed as[8]

$$W = (1-\varepsilon)W_H + \varepsilon\left[W_0 + W_+W_- / (W_+ + W_-)\right] \tag{8}$$

Here $W_H = 1/T_{1H}(B)$ is the "pure" proton spin-lattice relaxation rate in the low magnetic field B, W_0, W_-, and W_+ are the nitrogen spin-lattice relaxation rates associated with the NQR transitions at the frequencies ν_0, ν_- and ν_+, and $\varepsilon = 2N(^{14}N)/3N(^1H)$.

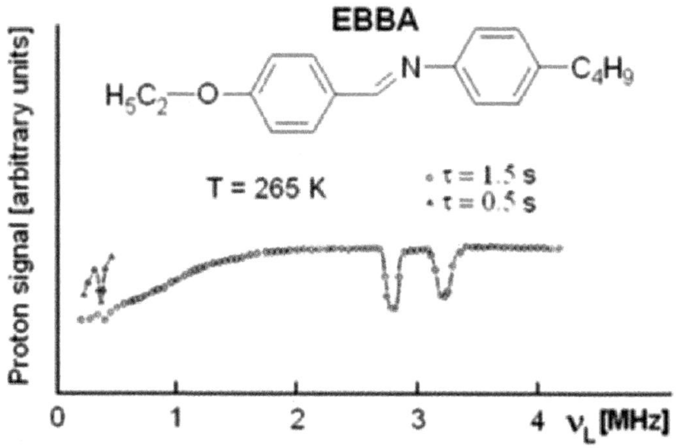

Figure 9. Determination of the ^{14}N NQR frequencies from the Larmor-frequency dependence of proton T_1.

In practice there is often no need to perform the real T_1 versus B measurement. In many cases it is enough to measure the proton NMR signal S as a function of the low magnetic field B at a fixed duration τ of the mixing period with no rf irradiation (Figure 4). As an experimental example we present in Figure 9 the measurement of the ^{14}N NQR frequencies in (p-ethoxybenzylidene)-p-n butylaniline (EBBA).[9]

The initial measurement was performed at fixed time τ = 1.5 s. Two dips are observed at around 2.9 MHz (ν_-) and 3.2 MHz (ν_+). The low-frequency dip at around 300 kHz (ν_0) is barely seen because of the low proton NMR signal, caused by fast proton spin-lattice relaxation below 1 MHz. In order to obtain a better expressed low-frequency dip, we shortened the duration of the mixing period to τ = 0.5 s and repeated the measurements in the low frequency region.

As seen from Figure 9 this technique works well when the nitrogen spin system relaxes fast. The low-frequency dip is narrow, while the high-frequency dips are significantly broader due to the presence of a higher magnetic field

B. We usually use this technique as the starting technique to get approximate values of the ^{14}N NQR frequencies and later improve resolution by other NQDR techniques.

2.3. LEVEL CROSSING AND SOLID EFFECT

The level crossing technique[10,11] is often used for the measurement of ^{14}N NQR frequencies. The basic experimental scheme presented in Figure 4 applies also in this case. Let's assume that the low magnetic field B is low enough that $\gamma_H B < 2\pi\nu_0$. Then during the demagnetization process the proton Larmor frequency becomes first equal to ν_+, and later to ν_- and ν_0 (Figure 8). We thus have three level crossings ($\nu_L = \nu_Q$, $Q = +, -, 0$) when the external magnetic field is reduced from B_0 to B and three level crossings when the external magnetic field is increased from B to B_0. If the rate at which the external magnetic field changes is not too high, the two spin systems exchange energy during each level crossing. The nitrogen spin system is at the beginning of the mixing period polarized by the proton spin system. The first set of level crossings decreases the proton magnetization. An additional change of the proton magnetization is caused by the second set of level crossings at the end of the mixing period. If during the mixing period we apply a rf magnetic field with the frequency ν equal to $\nu=\nu_+$, ν_-, or ν_0, the population of the nitrogen energy levels changes, and consequently, also the proton magnetization after the second set of level crossing changes. In general we observe a smaller proton NMR signal S at the end of the magnetic field cycle when $\nu = \nu_+$, ν_-, or ν_0.

There are two limitations for the application of the level crossing technique. First there should be a strong dipolar interaction between ^{14}N and ^{1}H in order to get a significant energy transfer between the spin systems during the level crossings. Second the ^{14}N spin lattice relaxation rates must be lower than $1/\tau$. The population of the nitrogen energy levels before the second set of level crossings is otherwise independent of the population induced by the first set of level crossings and the *rf* irradiation has nearly no effect.

The solid-effect transitions (Figure 10) are the simultaneous transitions between the energy levels of two spin systems caused by a strong rf magnetic field with the frequency ν equal to the sum or difference of the resonance frequencies of the two nuclear species. These transitions are allowed in solids due to the presence of the dipole-dipole interaction. The NQDR technique based on the solid effect is analyzed in references[12,13].

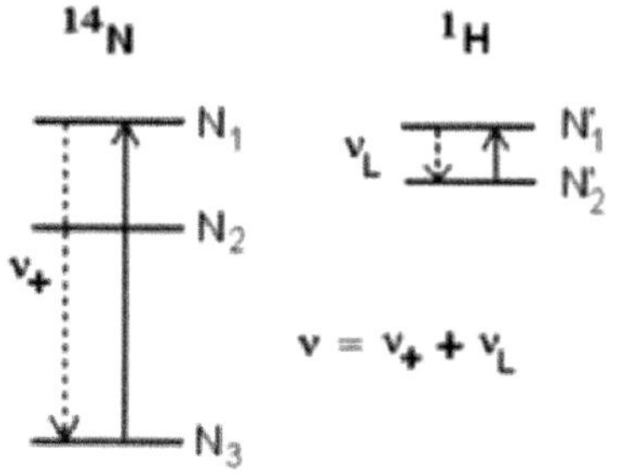

Figure 10. Solid-effect transitions in the system ^1H-^{14}N at $\nu = \nu_+ + \nu_L$.

The solid effect transitions tend to form an equilibrium in the two spin system. In the case shown in Figure 10 the equilibrium is reached when $N_1/N_3 = N_2'/N_1'$. If in the same system the rf magnetic field is applied at the frequency $\nu = \nu_+ - \nu_L$, the equilibrium is reached when $N_1/N_3 = N_1'/N_2'$. The simultaneous transitions in the two spin systems change the proton magnetization and consequently also the proton NMR signal S in the detection phase. Here we consider two limiting situations.

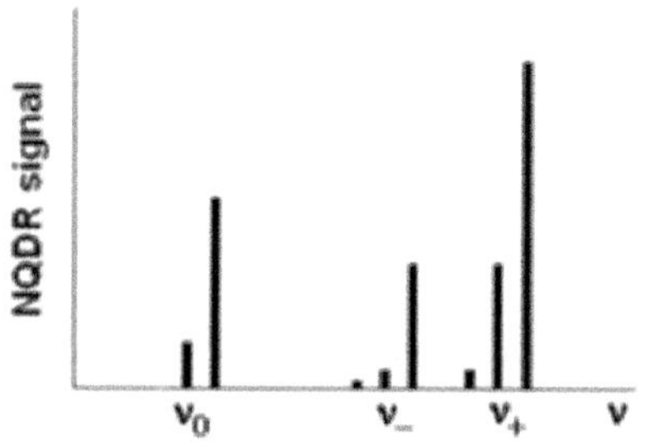

Figure 11. Relative intensities of ^1H-^{14}N NQDR signals in case of slow nitrogen spin-lattice relaxation.

When the ^{14}N spin-lattice relaxation rates are lower than the proton spin-lattice relaxation rate, the internal equilibrium of the two spin systems is indeed reached and both level crossing and solid effect contribute to the NQDR spectrum. The intensity of the NQDR lines in case of complete level crossing is shown in Figure 11.

As a realistic example we present the results of the measurement of the ^{14}N NQR frequencies in solid quinolinic acid using level crossing and solid effect.[13] The results are presented in Figure 12.

A nice triplet is observed at the highest ^{14}N NQR frequency ν_+. At lower frequencies overlaps of the lines may occur and in addition a direct absorption by the proton spin system is observed at multiples of the Larmor frequency.

Several spectra should be taken at different values of ν_L to clarify the NQDR spectra and to determine the NQR frequencies.

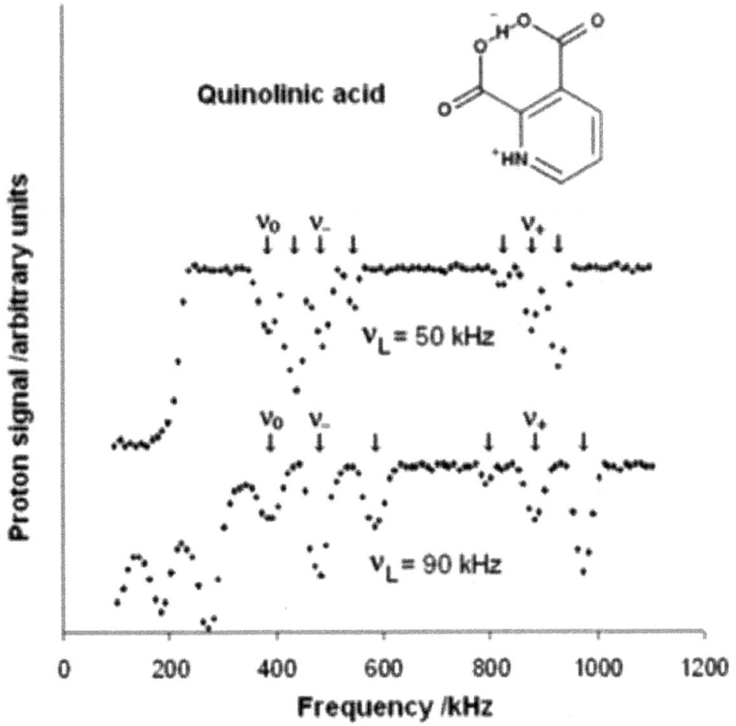

Figure 12. ^{1}H-^{14}N NQDR spectra in solid quinolinic acid using level crossing and solid effect.

When the spin-lattice relaxation rates of ^{14}N are high, the internal equilibrium in the proton and nitrogen spin system caused by the solid-effect transitions cannot be reached. An energy flow to the ^{14}N spin system occurs instead. In such a case a nearly symmetric doublet is observed around each NQR frequency ν_Q at the frequencies $\nu = \nu_Q \pm \nu_L$. The intensities of the two NQDR lines are higher than the ones observed in case of slow nitrogen spin-lattice relaxation, while the level crossing contribution at $\nu = \nu_Q$ is not observed.

The main limitation of the solid-effect technique is the proton–nitrogen dipolar interaction. If it is strong, as for example in case of a covalent N–H bond, or a strong N...H hydrogen bond, the solid effect lines are easily observed. If this is not the case, the solid effect lines are much weaker.

2.4. MULTIPLE FREQUENCY SWEEPS AND TWO-FREQUENCY IRRADIATION

When the low magnetic field B is properly chosen so that $\nu_L = \nu_0$, the proton and nitrogen spin systems exchange energy at this frequency. Expression 8 shows how nitrogen spin-lattice relaxation rates change the proton spin-lattice relaxation rate in this situation. Application of two rf magnetic fields at the frequencies $\nu_1 = \nu_-$ and $\nu_2 = \nu_+$ (Figure 13) acts similar as high spin-lattice relaxation rates W_+ and W_-.[8] Consequently the proton spin lattice relaxation rate (equation 8) strongly increases. As seen from equation 8 the application of a single frequency (ν_1 or ν_2) does not significantly influence the proton relaxation rate. The final situation is zero proton magnetization and not the equilibrium proton magnetization $M_0(B) = CB/T$ in the low magnetic field B as obtained in case of spin-lattice relaxation. In practice $B \ll B_0$ and the initial proton magnetization in the low magnetic field B is much larger than $M_0(B)$. Thus $M_0(B)$ can be approximated as zero.

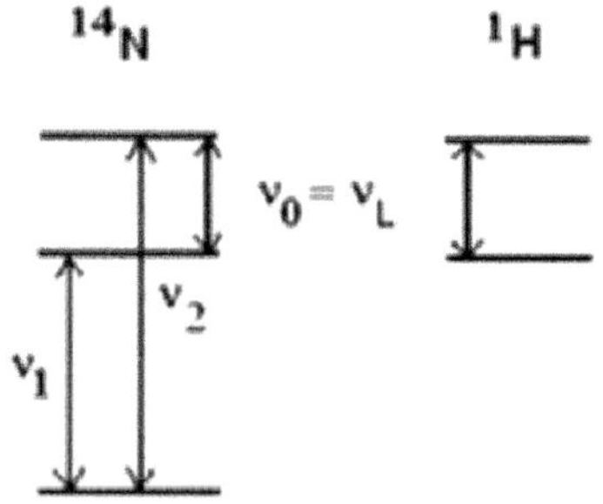

Figure 13. Two-frequency irradiation of ^{14}N coupled to ^1H.

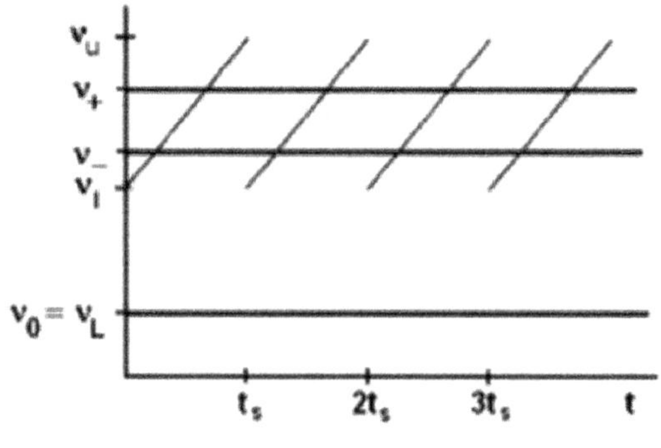

Figure 14. Application of multiple frequency sweeps of a rf magnetic field.

Another possibility of exciting the NQR transitions at $\nu = \nu_+$ and $\nu = \nu_-$ is the application of multiple frequency sweeps.[14] The situation is illustrated in Figure 14.

The low magnetic field B is chosen so that $\nu_L = \nu_0$. The sweep limits ν_l and ν_u are chosen so that each frequency sweep with the duration t_s covers both ν_+ and ν_-. When the frequency ν of the rf magnetic field passes a NQR frequency ν_Q, the rf magnetic field produces a change of the population of the two energy levels separated by ν_Q. In an ideal situation the population of the two energy levels inverts (adiabatic fast passage). One of the two energy levels is in both cases associated with the lowest frequency ν_0. Therefore after each sweep some flip-flop transitions in the coupled spin system occur to equalize the ratio of the populations of the proton energy levels and the ratio of the population of the nitrogen energy levels separated by ν_0. After each sweep the proton magnetization decreases. Multiple frequency sweeps are effective in the proton relaxation only when they cover both ν_+ and ν_-. The effect of multiple frequency sweeps is usually negligible if they cover only one NQR frequency.

The method of determination of ^{14}N NQR frequencies using multiple frequency sweeps and two-frequency irradiation is illustrated in Figure 15. As an example we choose diphenylamine.[15]

In the first part of the experiment (upper spectrum) we measured the Larmor frequency dependence of the proton NMR signal using multiple frequency sweeps with the frequency limits ν_l = 3 MHz, ν_u = 4 MHz and with the duration t_s = 10 ms. A dip is observed at ν_L = 460 kHz. So $\nu_0 \approx 460$ kHz. In the second part of the experiment we fixed B in the center of the dip ($\nu_L = \nu_0$) and varied the limits of the frequency sweeps. The proton NMR signal exhibits two steps when ν_l = 3 MHz and ν_u is varied. Additional two steps are observed when ν_u = 4 MHz and ν_l is varied. A step is observed when the sweep limit passes a NQR frequency. There are thus four NQR frequencies in the high frequency region and – within the experimental resolution – only one NQR frequency (unresolved doublet) in the low frequency region.

In the third part of the experiment we used the two-frequency irradiation. We fixed one frequency (ν_1) at ν_1 = 3,140 kHz and scanned the region between 3,550 and 3,750 kHz with the frequency ν_2. A dip was observed at approximately ν_2 = 3,600 kHz. Then we fixed ν_2 in the center of the dip and performed the scan with the frequency ν_1. The results are the lines labeled as A on the lower spectrum. In a similar way we obtained the lines labeled as B. In the diphenylamine crystal are thus two nonequivalent nitrogen positions: position A with the NQR frequencies ν_+(A) = 3,600 kHz, ν_- (A) = 3,140 kHz and ν_0(A) = 460 kHz and position B with the NQR frequencies ν_+ (B) = 3,700 kHz, ν_- (B) = 3,240 kHz and ν_0 (B) = 460 kHz.

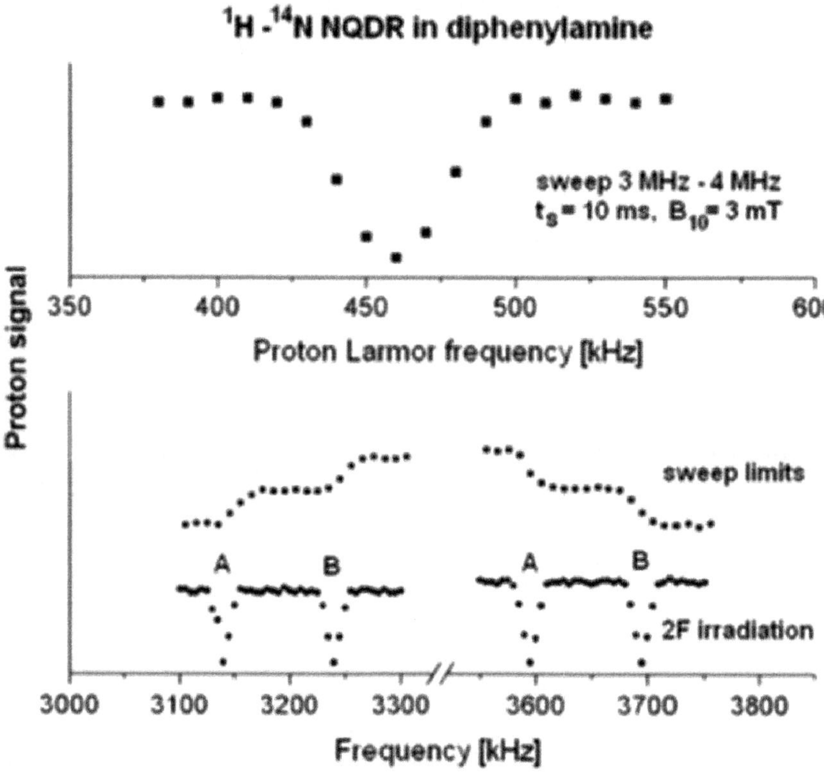

Figure 15. Determination of ^{14}N NQR frequencies in diphenylamine using multiple frequency sweeps and two-frequency irradiation.

3. Conclusions

The methods of determining low NQR frequencies by nuclear quadrupole double resonance are described and illustrated.

If a quadrupole nucleus has a half-integer spin (^{17}O, ^{39}K, …) it can usually be detected by the highly sensitive Slusher and Hahn's technique.[3] This technique works when the proton spin-lattice relaxation time in zero magnetic field is not too short. In case of ^{17}O the NQR frequencies $\nu_{3/2,\,1/2}$ and $\nu_{5/2,\,3/2}$ can usually be detected by this technique. The highest NQR frequency $\nu_{5/2,\,1/2}$ can often not be detected because of a rather low matrix element for this transition.

In case of a very short proton spin-lattice relaxation time in zero magnetic field the NQR frequencies can often roughly be determined from the Larmor frequency dependence of proton spin-lattice relaxation time. Sometimes also the solid effect technique can be used.

If a quadrupole nucleus has an integer spin (^{14}N, ^{6}Li, …), the sensitivity of the Slusher and Hahn's technique is strongly reduced due to spin quenching.[4] Several other techniques can de used to determine the NQR frequencies. The choice of the technique depends on the spin lattice relaxation of protons and the quadrupole nuclei.

The simplest technique for the determination of the ^{14}N NQR frequencies is the technique based on the Larmor frequency dependence of proton T_1.[7] This technique works well when the spin-lattice relaxation rates of the nitrogen nuclei exceed the spin-lattice relaxation rate of protons. In such a case a dip is observed in the Larmor frequency dependence of proton T_1 when the Larmor frequency matches a NQR frequency.

When the nitrogen spin system relaxes slow, the previous technique often gives no result. The level crossing technique[10,11] and the solid effect technique[13] can often be used in such a case. These two techniques require a strong proton-nitrogen dipolar interaction. The solid effect technique can also be used in a case when the nitrogen spin system relaxes fast. Its resolution is higher than the resolution of the technique using Larmor frequency dependence of proton T_1.

When the proton–nitrogen dipolar interaction is weak, but not only in this case, the technique using multiple frequency sweeps[14] and two-frequency irradiation[8] can be used. First the lowest ^{14}N NQR frequency ν_0 is determined by varying the low magnetic field while keeping the sweep limits constant. The upper NQR frequencies ν_+ and ν_- are then approximately determined by varying the frequency limits of the frequency sweeps. The two-frequency irradiation technique is finally used to improve the resolution. The resolution of this technique is typically a few kilohertz. This technique may be applied, when the nitrogen spin-lattice relaxation rates are moderate or low. When the nitrogen spin-lattice relaxation rates are high, the ^{14}N NQR frequencies can be determined either from the Larmor frequency of proton T_1 or using solid effect.

As already mentioned, the sensitivity of double resonance is high, while its resolution is moderate. If more precise data are needed, the NQR frequencies are first located by double resonance. Pulsed nuclear quadrupole resonance is later used to improve the resolution.

References

1. G. V. Rubenacker and T. L. Brown, Nitrogen-14 nuclear quadrupole resonance spectra of coordinated pyridine. An extended evaluation of the coordinated nitrogen model, Inorg. Chem. 19, 392–398 (1980).
2. J. Seliger, ^{17}O quadrupole coupling in C–O–H...O=C hydrogen bonds, Chem. Phys. 231, 81–86 (1998).

3. R. E. Slusher and E. L. Hahn, Sensitive detection of nuclear quadrupole interactions in solids, Phys. Rev. 166, 332–347 (1968).
4. G. W. Leppelmeier and E. L. Hahn, Nuclear dipole field quenching of integer spins, Phys. Rev. 141, 724–731 (1966).
5. J. Seliger and V. Žagar, Sensitivity of nuclear-quadrupole double-resonance detection of half-integer spin nuclei J. Magn. Reson. 194, 175–181 (2008).
6. J. Seliger, V. Žagar, R. Blinc, and F. Milia, Temperature dependence of ^{17}O and ^{14}N NQR frequencies in Commercial TNT, Appl. Magn. Reson. 29, 541–548 (2005).
7. J. Seliger, R. Blinc, H. Arend, and R. Kind, Proton-^{14}N double resonance study of the structural phase transitions in the perovskite type layer compound $(CH_3NH_3)_2CdCl_4$, Z. Phys. B 25, 189–195 (1976).
8. J. Seliger, V. Žagar, and R. Blinc, A new highly-sensitive ^{1}H-^{14}N nuclear-quadrupole double-resonance technique, J. Magn. Reson. A 106, 214–222 (1994).
9. J. Seliger, R. Osredkar, M. Mali, and R. Blinc, ^{14}N quadrupole resonance of some liquid crystalline compounds in the solid, J. Chem. Phys. 65, 2887–2891 (1976).
10. R. Blinc, M. Mali, R. Osredkar, A. Prelesnik, J. Seliger, I. Zupancic, and L. Ehrenberg, ^{14}N NQR spectroscopy of some amino acids and nucleic bases via double resonance in the laboratory frame, J. Chem. Phys. 57, 5087–5093 (1972).
11. D. T. Edmonds, Nuclear-quadrupole double-resonance, Phys. Rep. 29, 234–290 (1977).
12. J. Seliger, R. Blinc, M. Mali, R. Osredkar, and A. Prelesnik, Nuclear magnetic double resonance based on strong rf magnetic-field-induced coupling between spin systems, Phys. Rev. B 11, 27–36 (1975).
13. J. Seliger and V. Žagar, Measurement of the ^{14}N nuclear quadrupole resonance frequencies by the solid effect, J. Magn. Reson. 193, 54–62 (2008).
14. J. Seliger, V. Žagar, and R. Blinc, ^{1}H-^{14}N Nuclear quadrupole double resonance with multiple frequency sweeps, Z. Naturforsch. 49a, 31–34 (1994).
15. J. Seliger and V. Žagar, ^{14}N NQR Study of Diphenylamine, Z. Naturforsch. 63a, 88–92 (2008).

SIGNAL PROCESSING METHODS IN NQR

VADIM S. GRECHISHKIN[1*],
RUFINA V. GRECHISHKINA[2], HOON HEO[3]
[1]*Kaliningrad State University, St. A. Nevsky 14, Kaliningrad, 236041, Russia*
[2]*Baltic Military Naval Institute, 82 Sovetskiy av. Kaliningrad, 236036, Russia*
[3]*Korea University, Seoul, Korea*

Abstract To detect mines by the explosive itself, several new techniques have been proposed; one of the most promising is ^{14}N NQR. Design criteria allowing increase sensitivity of ^{14}N NQR are discussed. It is shown that combination of ^{1}H NMR with ^{14}N NQR leads to selective excitation or a better S/N; in this case, a strong magnetic field has to be switched on and off. Whereas to detect anti-tank mines strong magnetic field should be applied.

Keywords: NQR, explosive detection, spin electronics

1. Introduction

Nuclear Quadrupole Resonance (NQR) is very important for humanitarian demining, because this method has a very high success rate (97%) in field experiments in soil. Since the first experiments[1] of 16 years ago there has been a constant effort to develop improved systems using both results of experiments with mines and advanced techniques from digital electronics and signal processing.[2] On the whole, an NQR landmine detector is subject to three design criteria: firstly, that it should have high sensitivity and accuracy in signal detection with a low signal-to-noise ratio (SNR) for the separate NQR multiplets from the compounds to be detected in the remote NQR experiment; secondly, that it should be possible to detect simultaneously several compounds such as TNT, hexogen and tetryl; thirdly, that the phase stability of the system should be maintained when working with very broad NQR lines. The main difficulty in detecting TNT in mines, in contrast to other compounds of interest, is that in the manufacturing process, after

[*] E-mail: z3001@rambler.ru

J. Fraissard and O. Lapina (eds.), *Explosives Detection Using Magnetic and Nuclear Resonance Techniques*,

molten TNT has been poured into the plastic case of the mine and then cooled, the compound to be detected becomes a solid solution, with NQR spectra up to 100 kHz wide, which is much more difficult to detect than compounds giving narrow, well separated, lines.

2. NQR detector A

The system developed is an improvement of that in[3] and is specially designed for the simultaneous detection of two lines of TNT at 740 and 840 kHz, those of tetryl at 5,290 kHz and of hexogen at 5,192 kHz. Part of the system for TNT is shown in reference.[4] A quartz-crystal oscillator synchronizes a pulse generator that, in turn, handles gates 1 and 2, and oscillators 1 and 2 of the next stage, producing RF oscillations at 740 and 840 kHz (Figure 1).

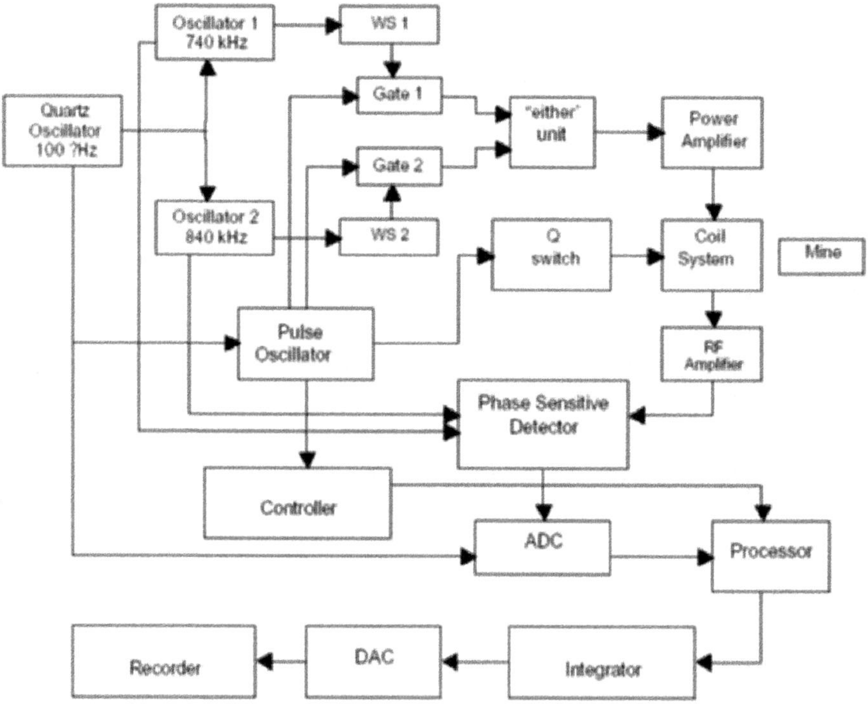

Figure 1. NQR detector A.

The system is an updated version of that in reference[5], and is distinguished from the latter by its higher phase stability and, in contrast to reference[6], provides accurate syncronization of oscillators 1 and 2 in the phase-sensitive detector.

3. Digital signal processing

As the NQR spectra of ^{14}N of TNT are up to 100 kHz wide and feature high multiplicity and low signal-to-noise ratio, the problem of detection is more complicated, which means that new approaches have to be applied. Often, the Fourier transformation usually used after preliminary filtering and time-averaging of the noisy NQR signals can be profitably complemented by novel techniques of spectral estimation.[7–9] According to reference[10], the Matrix Pencil Method can handle NQR signals with an SNR of 0.7 with close to 100% certainty of detection. To assist in the application of this method, the properly reworked version of the MATLAB program[11] was developed.

We solved the problem of TNT with the Matrix Pencil Method (ITMPM). For urotropin the success rate was 89% if SNR = 0.52; for TNT at 100 K the success rate was 100% for SNR = 0.8. It is possible to make a commercial NQR mine detector for all mines with TNT, RDX (Figure 2) and tetryl.[12,13] The equipment for baggage control is the same. If you have an NQR mine detector, you may use this device for baggage control in customs. We ran many tests in Kaliningrad, Novosibirsk, Cherepovez, Moscow, Perth (WA) and Poznan.

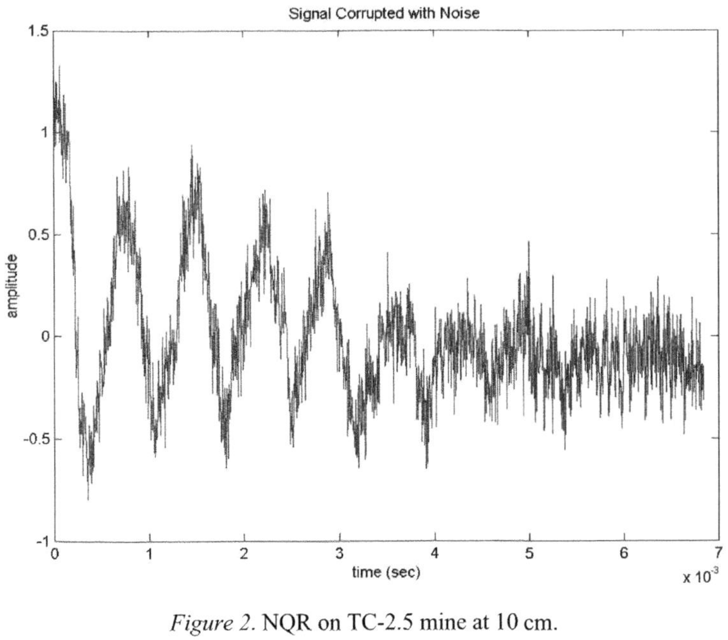

Figure 2. NQR on TC-2.5 mine at 10 cm.

During the crystallization of TNT the presence of pseudo-symmetry elements can provoke a deviation from the normal stacking sequence of the layers, leading to the formation of a twinned structure. Above and below the twin plane the molecular layers are arranged according to their normal monoclinic packing order, although the two halves of the crystal are now related by rotation of 180° about an axis normal to the (100) plane. Once twinning has occurred, additional molecular layers are added in the normal sequence until a second stacking defect returns the crystal to its original orientation. Random repetition of the stacking defect leads to the formation of a multiply twinned crystal. It is for this reason that in TNT we see broad NQR lines at room temperature. TNT is a problem for detection in an NQR mine detector, but stochastic synchronization and the MPM method can improve the sensitivity for its direct remote detection. This method is illustrated for RDX in mines.

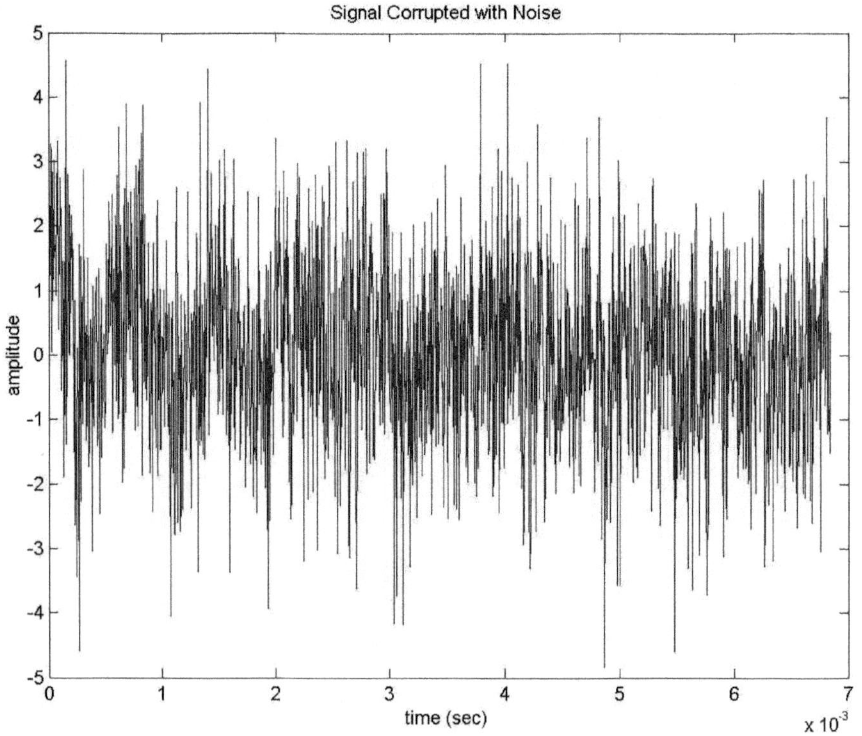

Figure 3. NQR of RDX with SNR = 0.05 for Marcov noise.

In a previous article[14] we described the application of our two-frequency method for the detection of explosives with short receiver dead-time, we constructed three perpendicular surface coils for TNT: two-crossed half-toroidal coils for vo and v- excitation and a ring-surface coil in the middle of the crossed coils for detection of the v+ NQR line. One may observe remote NQR with two-frequency excitation, using digital communications.[15]

If SNR < 1, then we use the ITMPM method. The excitation was made by PulseBlaster DDS ISA from SpinCore Technologies (Figures 4 and 5)

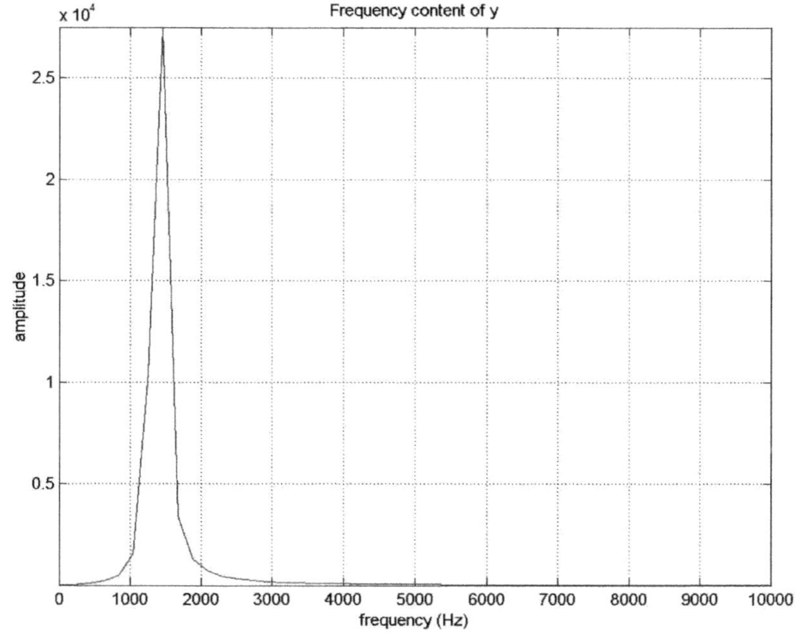

Figure 4. NQR after ITMPM transformation in real time with national instruments ADC 5112.

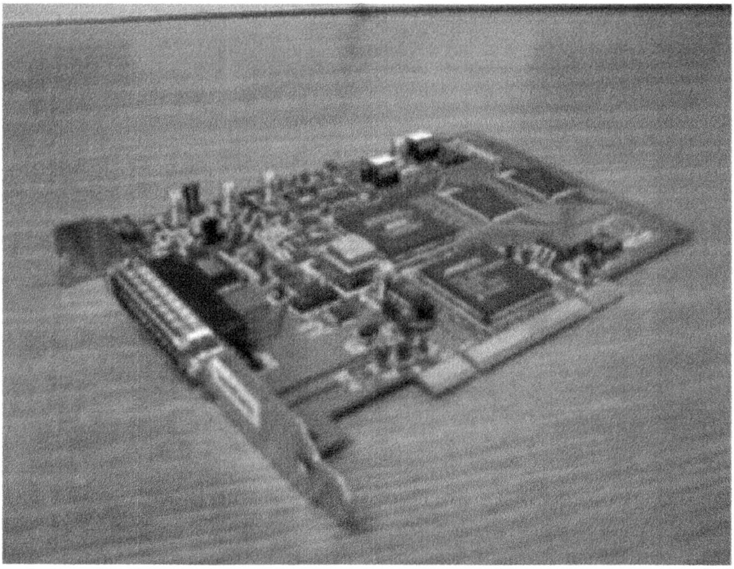

Figure 5. PulseBlaster DDS from SpinCore Technologies.

4. Dispersion of NQR signals

The NQR signal with noise is:

$$S(t) = A_c e^{-t/T_2} \cos[2\pi f_c t + \varphi(t)], \tag{1}$$

$$n(t) = x(t)\cos 2\pi f_c(t) - y(t)\sin 2\pi f_c(t) \tag{2}$$

where the T_2 relaxation time for RDX is 1.5 ms.

For noise we can write:

$$n(t) = n_c(t)\cos[2\pi f_c t + \varphi(t)] - n_s(t)\sin[2\pi f_c t + \varphi(t)] \tag{3}$$

where

$$\begin{aligned} n_c(t) &= x(t)\cos\varphi(t) + y(t)\sin\varphi(t) \\ n_s(t) &= -x(t)\sin\varphi(t) + y(t)\cos\varphi(t) \end{aligned} \tag{4}$$

After multiplication $s(t) + u(t)$ on reference signal in the synchron detector $c(t) = \cos\left[2\pi f_c t + \hat{\phi}(t)\right]$, we have:

$$e(t) = A_c e^{-t/T_2} \sin\Delta\varphi + n_c(t)\sin\Delta\varphi - n_c(t)\cos\Delta\varphi, \tag{5}$$

where $\Delta\varphi = \varphi - \hat{\varphi}$, the phase error.

We used for the phase dispersion of the NQR signal the formula in[15]

$$P(\Delta\varphi) = \frac{\exp(SNR \cdot \cos\Delta\varphi)}{2\pi I_0(SNR)}, \tag{6}$$

where I_0 is a Bessel function. With frequency offset we used the solution of the Fokker–Planck equation. For the compensation of phase errors we used the PAL-PAPS method. In Figure 1 we used a phase shift of 180° in channel v+, connected with the pulse generator. In the reference signal channel for v+ we used a phase shift of 270° connected with the pulse generator. Such a system gives the compensation of phase errors on 25°.

With such a program one may work with new parameters the noise, if SNR < 0.5. We are now testing such a possibility.

The colored noise can be seen in (Figure 3). This non-stationary noise is very bad for NQR mine detectors. We used the Kolmogorov–Fokker–Planck equation as a solution only for Marcov noise. We predicted the dispersion of the NQR line in the noise and displacement of the frequency offset to low frequencies. The signal noise file may be used for different types of noise (Figure 6).

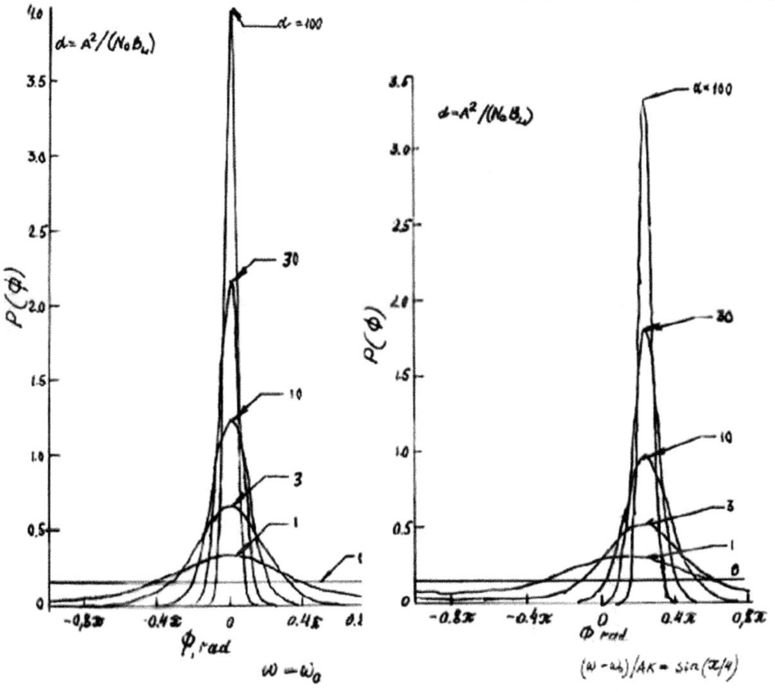

Figure 6. Dispersion of signal in the noise.

Figure 7. 1 kW amplifier from communication concepts.

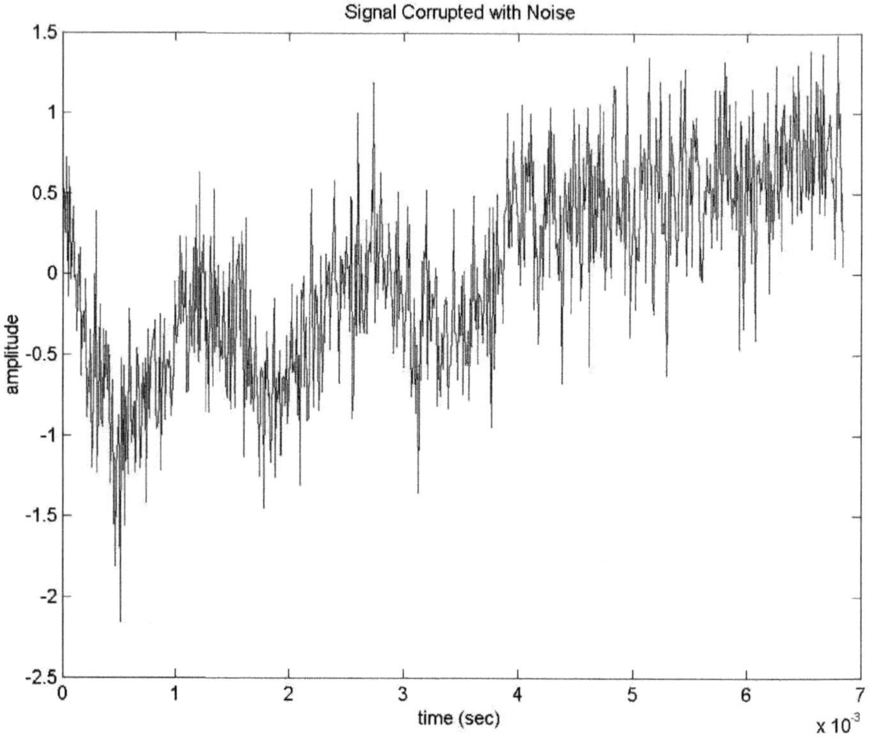

Figure 8. Remote NQR of TC-6 at 35 cm with a frequency of 5,192 kHz (10,000 accumulations).

We used amplifiers AN-779L and AR-347 from Communication Concepts (Figures 7 and 8).

5. NQR device with electronic field cycling for TNT mine detection (Detector B)

In Figure 9 is presented an NQR device with electronic field cycling for TNT mine detection, with: 1 – solenoid; 2 – saturating coil; 3 – receiving coil; 4 – mine; 5 – circuit for electronic cycling of magnetic field H_0; 6 – power pulse generator for electromagnetic field; 7 – pulse NMR spectrometer; 8 – programs; 9 – computer.

At room temperature NQR signals of TNT are usually very weak and are not detected by the direct NQR method. The suggested method of remote DNQR with electronic cycling of the magnetic field allows to increase the sensitivity by a factor of about 30.

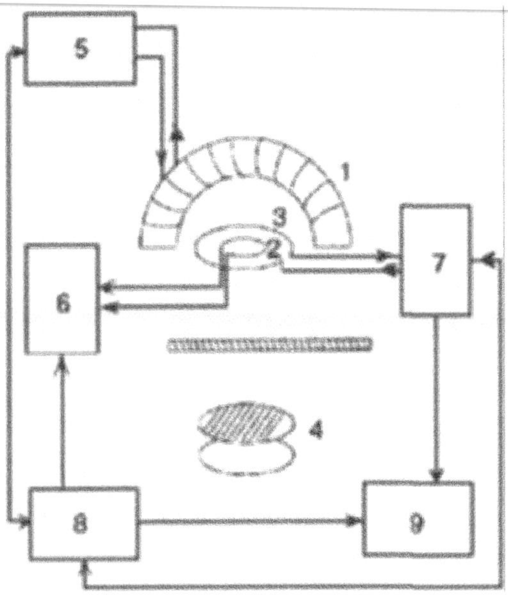

Figure 9. Remote DNQR.

The solenoid (1) works with the electronic magnetic field cycling circuit (5), which represents power pulse generator constant magnetic field, generating a current pulse I = 100 A with $H_0 = 0.25$ T.

The current pulse activates solenoid (1), producing a magnetic field. The ends of the solenoid are directed towards the ground (it stands upright). Trailing magnetic field between the ends of the solenoid is parallel to the Earth's surface. This magnetic field irradiates the mine and polarizes the magnetic moments of protons in the mine substance (4). The mine is irradiated at the NQR frequency of TNT, 841 kHz. The quadrupole system represents the totality of the nitrogen nuclei in the mine substance, and is saturated with the help of coil (2) from the power pulse generator (6) for the electromagnetic field (frequency 841 kHz). This results in leveling the population of the nitrogen nucleus and increasing the spin temperature of the quadrupole system. After turning on the magnetic field of solenoid (1) this high spin temperature passes to the proton system with energy level crossing. The receiving coil (3) induces a signal at the proton frequency. Receiving coil (3) works the pulse NMR spectrometer, and reducing the NMR signal conforms to appearance of NQR signal on frequency in zero magnetic field. The program block (8) produces current pulses to turn on the various elements of the device at the appropriate times. All elements of the scheme are conjugated through the program block (8), with each other and with the computer (9) (Figure 10).

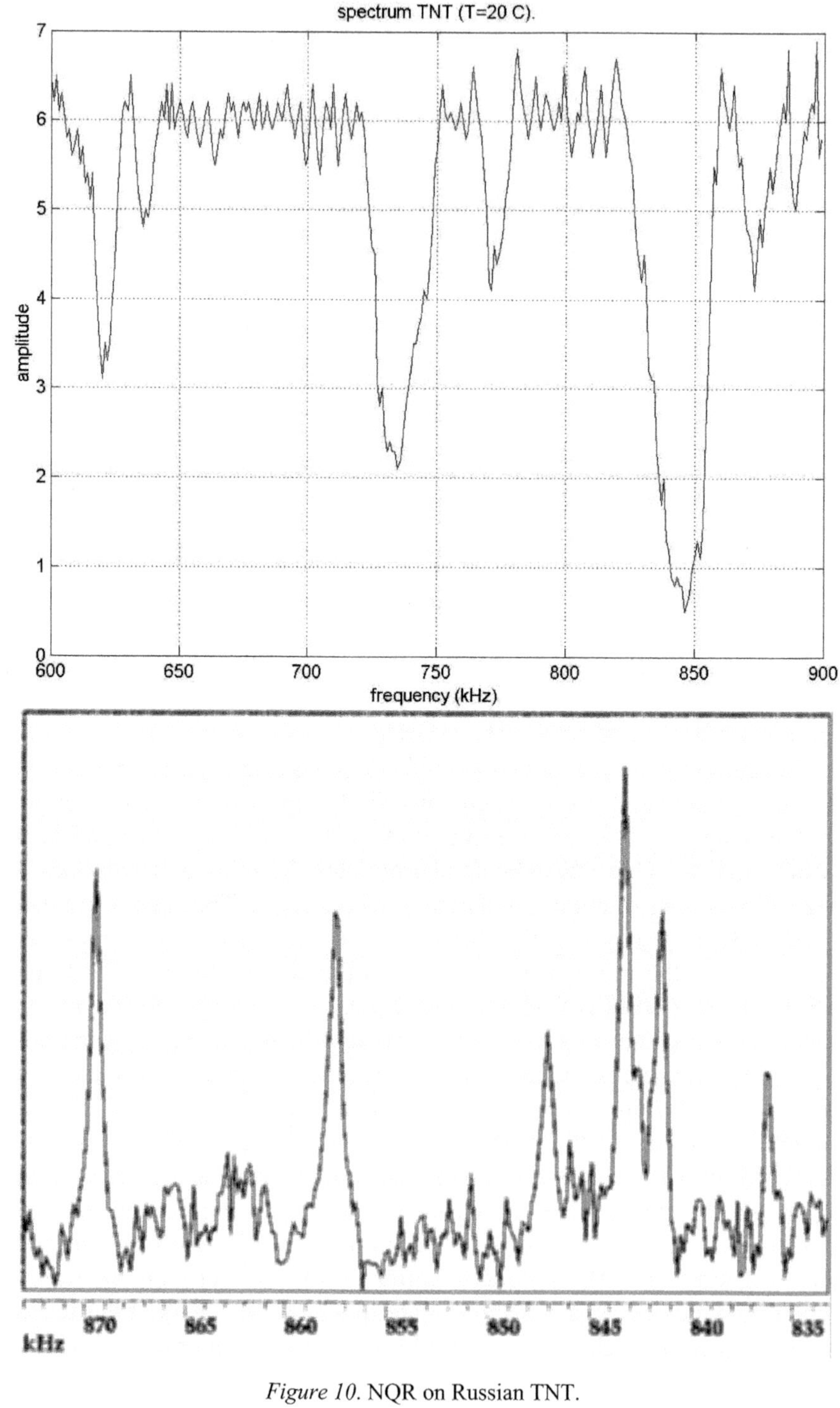

Figure 10. NQR on Russian TNT.

6. Conclusion

The traditional method of mine detection – metal detection by induction, and then manually probing and digging – does not work with mines with low or zero metal content, and forces deminers to treat every piece of metal in the soil as a potential mine. To detect mines by the explosive itself, several new technologies have been proposed; one of the most promising is nuclear quadrupole resonance with nitrogen-14 which is contained in practically all mine explosives. In NQR the quadrupole moment of a nucleus interacts with the electric field gradient of the surrounding molecule and crystal, and thus the energy level differences are molecule- and crystal-specific. In pure NQR, no external magnetic field is needed. One may, however, combine nuclear magnetic resonance (NMR), e.g. of protons, with NQR to achieve selective excitation or a better signal; in this case, a strong magnetic field has to be switched on and off (Figure 11).

With method 2, there is the problem that some anti-tank mines may be triggered by a strong magnetic field. One will have to analyse the scale of this problem (use of such mines in Cambodia) and propose solutions; these could involve, for example, first testing with a disposable permanent magnet on a moveable arm in front of a safe vehicle. Our NQR mine detector used in the 1984–1989 tests was homebuilt, but in the USA in 1999–2000 they used Tecmag equipment.

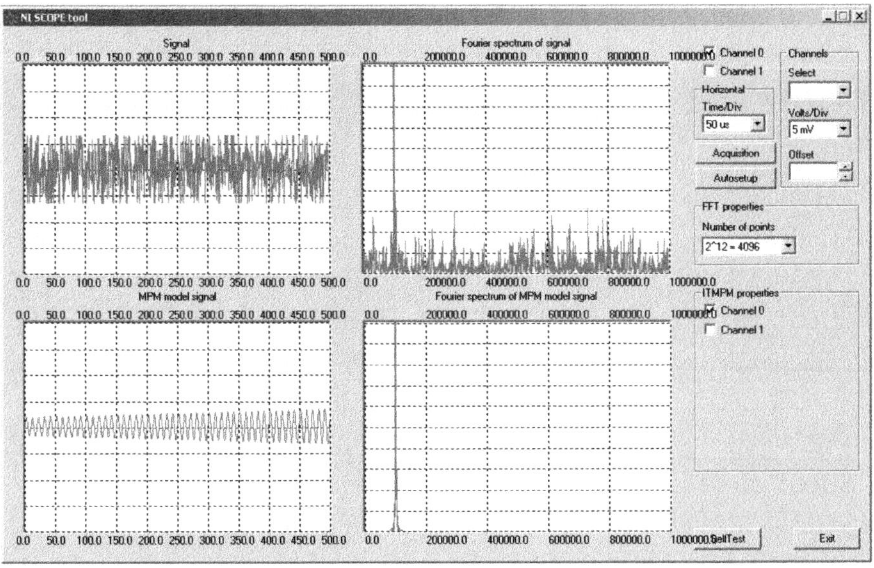

Figure 11. FFT and ITMPM with PulseBlaster in RDX.

References

1. V.S. Grechishkin, E.M. Shishkin, Org. Magn. Reson., 5, 1–4 (1973).
2. E.O. Azisov, V.S. Grechishkin et al., Izv. AN SU, Physics, 42, 2182–2185 (1978).
3. V.P. Anferov, V.S. Grechishkin, N.Y. Sinyavskii, Nuclear Spin Resonance (new methods), LGU press, Leningrad, 1990.
4. V.S. Grechishkin, N. Ja, Sinjavsky, Physics-Uspekhi 163(10), 95–120 (1993).
5. V.S. Grechishkin, N. Ja, Sinjavsky, UFN, 167, N4, 413–427 (1997).
6. V.S. Grechishkin, J. Molec. Struct. 83, 135–138 (1982).
7. V.S. Grechishkin, J. Molec. Struct. 83, 345–348 (1982).
8. V.S. Grechishkin, J. Molec. Struct. 83, 89–92 (1982).
9. V.S. Grechishkin, Solid State Phys. 33, 947–949 (1991).
10. V.S. Grechishkin, Appl. Phys. A58, 63–65 (1994).
11. V.S. Grechishkin, 16th NQR Congress, Leipzig, p. 16, 1999; NQR-DLME, Ljublyana, pp. 38–40 (2000).
12. V.S. Grechishkin, R.V. Grechishkina, NQR Mine-Detectors, Patent N2157002, 27.09.2000.
13. V.S. Grechishkin et al., Izv. Vuzov, Fizika, 6, 101–105 (2000).
14. V.S. Grechishkin, R.V. Grechishkina, Izv. Vuzov, Fizika, 8, 92–94 (2001).
15. V.S. Grechishkin, Probl. Math. Phys. Sci., 60–74 (2000); 72–81 (2001).

^{14}N NUCLEAR QUADRUPOLE RESONANCE SIGNALS IN PARANITROTOLUENE AND TRINITROTOLUENE. SPIN-LOCK SPIN-ECHO OFF-RESONANCE EFFECTS

ALAN GREGOROVIČ[1*], TOMAŽ APIH[1], JANKO LUŽNIK[1], JANEZ PIRNAT[2], ZVONE TRONTELJ[3]
T. Apih Institute Jožef Stefan, Jamova 39, 1000 Ljubljana, Slovenija
[2]Institute for mathematics, physics, and mechanics, Jadranska 19, 1000
[3]Faculty for mathematics and physics, Jadranska 19, 1000 Ljubljana, Slovenija

Abstract A simple, yet effective technique to enhance the ^{14}N NQR trinitrotoluene notoriously low sensitivity is the use of multipulse sequences. Here we investigate the off-resonance effects of the Spin-Lock Spin-Echo multipulse sequence, a predecessor of many advanced pulse sequences used for the same enhancement. Two samples have been used: paranitrotoluene, with a single ^{14}N site as a model compound for trinitrotoluene, and trinitrotoluene itself, with six ^{14}N sites. Our main focus has been the irradiation frequency dependence of the NQR signal, which is important when ^{14}N NQR is used for remote detection of explosives. The two related principal issues are: the target temperature uncertainty and the existence of multiplets with several closely spaced resonance frequencies. The first applies to any explosive, since in remote detection the temperature is only approximately known, whereas the second applies mainly to trinitrotoluene, with 12 resonance frequencies between 837 and 871 kHz. Our frequency dependent investigation shows that the signal intensity as well as the effective spin-spin relaxation time varies substantially with irradiation frequency in both samples. We provide a theoretical explanation of this variation which describes very well the observations and can be useful for increasing the reliability of remote detection signal processing.

[*] E-mail: alan.gregorovic@ijs.si

J. Fraissard and O. Lapina (eds.), *Explosives Detection Using Magnetic and Nuclear Resonance Techniques*,

Keywords: NQR, off-resonance SLSE pulse sequence, TNT, paranitrotoluene, PNT

1. Introduction

Multipulse sequences which create the spin-lock effect are often used in nuclear quadrupole resonance (NQR) experiments to increase sensitivity. By applying a train of radio-frequency (RF) pulses the magnetization M_{ss} is spin locked in the effective RF field and decays to equilibrium with a relaxation time T_{2eff}, which is much longer than a typical spin-spin relaxation time T_2. During one spin-lock decay one makes many acquisitions, $\propto T_{2eff}$, whereas without the spin-lock the number of acquisitions, $\propto T_2$, is much smaller. The first NQR sequence to create the spin-lock effect was spin-lock spin-echo (SLSE) described by Marino and Klainer[1] and shown in Figure 2. It consists of a preparatory pulse and a train of equally spaced RF pulses phase shifted by 90° relative to the preparatory pulse. Several versions of the original SLSE sequence were derived, commonly referred to as pulsed spin locking sequences (PSL), e.g. PAMS,[2] SORC,[3] PCPS,[4] to mention just a few. They all create the same spin-lock effect with some additional desirable properties but often at the expense of sensitivity.

The merit of any PSL sequence compared to the spin-echo (or single pulse) sequence is easily demonstrated by comparing the signal-to-noise (S/N) ratios of the two acquisition types. In a spin-echo pulse sequence a single signal with initial magnetization M_0 is acquired, whereas in a multi-pulse sequence one acquires and averages N_{acq} acquisitions every $t_{\mathrm{c}} << T_{2eff}$ of a quasi-steady state magnetization $M_{ss}(t) = M_{ss} \exp(-t/T_{2eff})$. The ratio of the respective S/N is

$$\frac{(S/N)_{PSL18mu}}{(S/N)_{ECHO}} = \frac{1}{\sqrt{N_{acq}}} \frac{M_{ss}}{M_0} \frac{1-e^{-N_{acq}t_c/T_{2eff}}}{1-e^{-t_c/T_{2eff}}} \tag{1a}$$

or if optimal number $N_{\mathrm{acq}} = 1.25\, T_{2eff}/t_c$ is chosen

$$\frac{(S/N)_{PSL18mu}}{(S/N)_{ECHO}} \sim 0.64 \frac{M_{ss}}{M_0} \sqrt{\frac{T_{2eff}}{t_c}} \tag{1b}$$

Although the value of $\sqrt{T_{2eff}/t_c}$ is usually much bigger than 1, the overall S/N gain might not be big if $M_{ss} << M_0$. The value of M_{ss} for SLSE on-resonance and optimal pulse lengths is M_0 and the gain in S/N (equation (1b)) is here indeed large, with typical values around 10.

The situation becomes much more complicated if the resonance ω_Q and irradiation frequency ω_{RF} don't coincide.[5] In a single echo experiment M_0

exhibits only a weak dependence on the frequency offset $\Delta\omega = \omega_Q - \omega_{RF}$ due to the finite pulse length. On the other hand, for PSL sequences the variation of M_{ss} is considerable even in the limit of infinite power and sharp pulses. In SLSE for example, the value of M_{ss} reaches M_0 at $\Delta\omega = 2\pi n/t_c$, where n is an integer. For other values of frequency offsets, its value is significantly smaller and exhibits a minimum every $\Delta\omega = 2\pi(n + ½)/t_c$. The M_{ss} offset dependence is well known,[3,6,7] although rarely taken into account. A far less known is the fact that T_{2eff} is offset dependent too and correspondingly influences the S/N as can be seen from equation (1b).

The above reasoning is very important for ^{14}N NQR detection of explosives. These have inherently a low S/N ratio and benefit significantly upon the use of PSL sequences. There are two problems associated with the off-resonance irradiation. The first problem is substantial temperature dependence of the NQR frequency and, to a lesser degree, other NQR parameters such as line widths and intrinsic relaxation times. Thus, whenever the sample temperature is only approximately known, as it is usually the case in remote detection, similar uncertainty exists for the resonance frequency. The second problem arises in laboratory as well as in remote detection because some explosives have multiple ^{14}N sites with closely spaced resonances. Here one can be on resonance only for a single ^{14}N site, whereas all the others are off-resonance. The existence of several closely spaced frequencies was recently used to increase the explosive detection probability[8] with promising results.

To elucidate these dependences we decided to investigate paranitrotoluene (PNT) and trinitrotoluene (TNT) with ^{14}N NQR SLSE multipulse sequence. TNT has been chosen because it is the most common explosive in anti-personnel landmines and also the most challenging for NQR detection.[9] On the other hand, the non-explosive PNT was used only as a TNT NQR-model compound. Namely, PNT and TNT share all major NQR parameters, apart the number of non-equivalent ^{14}N nuclei: PNT has only one and is as such suitable for analysis, whereas TNT has two crystallographic forms each with six ^{14}N non-equivalent sites.

In the following sections we first describe our experimental setup, then we derive a theoretical expression for the off-resonance steady state magnetization M_{ss} as in reference[10]. and compare it with our observations on PNT and TNT. We then outline the procedure for the derivation of the decay time T_{2eff} and present the results for two interactions: static homonuclear dipolar coupling and spin-lattice coupling. Finally we compare T_{2eff} theoretical predictions with our experimental PNT observations and comment on T_{2eff} in TNT.

2. Experiment

The measurements were carried out on our home built NQR spectrometer with the addition of a ^{1}H-^{14}N cross-polarization signal enhancement unit.

2.1. PARANITROTOLUENE

We have used ~70 g of laboratory grade PNT, which has a single ^{14}N site[11] with a quadrupole coupling constant e^2Qq/h = 1,340 kHz and asymmetry parameter $\eta = 0.408$.[12] There are three observable NQR transitions for $I = 1$ nuclei, with frequencies ν_+ = 1,141 kHz, ν_- = 868 kHz and ν_0 = 273 kHz. We have chosen to observe the ν_- transition, 887 kHz at room temperature, because it is in the range of ^{14}N TNT ν_+ frequencies[9] ~840–870 kHz. All measurements were done at room temperature.

2.2. TRINITROTOLUENE

The TNT sample was taken from an abandoned landmine and had a similar mass as PNT. TNT is stable in two crystallographic forms: monoclinic and orthorhombic, and each of these has six non-equivalent ^{14}N sites. Therefore we have up to 12 ν_+ frequencies in an uncharacterized TNT sample which for room temperature are listed in Table 1. Our sample of TNT turned out to be a composition of 30% monoclinic phase and 70% orthorhombic phase, whereas any arbitrary mix of the two phases can be found. Its spin-echo ^{14}N NQR spectrum at room temperature is shown in Figure 1.

TABLE 1. TNT ^{14}N ν_+ NQR frequencies at T = 300 K for the six nonequivalent sites of the two crystallographic forms.

Orthorhombic TNT ν_+ (kHz)	Monoclinic TNT ν_+ (kHz)
837.5	836.5
841.5	841.7
846.3	843.5
848.0	847.7
861.0	858.1
868.2	869.7

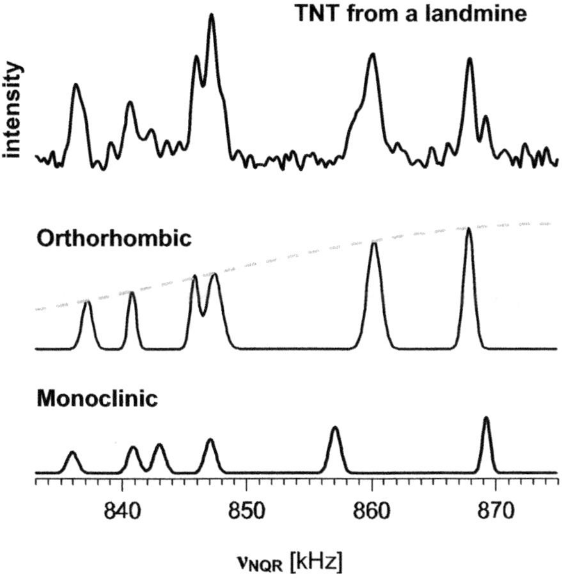

Figure 1. ^{14}N NQR spectrum of commercial TNT (blue line) obtained with a spin echo pulse sequence and its decomposition into orthorhombic (green line) and monoclinic (red line) component at T ≈ 304 K. Here the frequencies of the monoclinic and orthorhombic lines were taken from literature.[13,14] The effect of a filter centered at 875 kHz is shown by a dashed line.

2.3. ^{1}H–^{14}N CROSS-POLARIZATION

Because the ^{14}N NQR signal of a small (and also not so small) amount of TNT at room temperature is very hard to observe, we used signal enhancement through ^{1}H–^{14}N cross-polarization.[15] The cross-polarization unit consists of a transfer tube and a B_0 = 1.1 T permanent magnet. Every NQR acquisition began by placing the sample in B_0 and waiting for the establishment of a thermal equilibrium. For TNT this time is determined by ^{1}H $T_1(B_0)$ and was in our case ~30 s. After the equilibrium has been reached the sample was moved to the static-magnetic-filed-free coil for NQR inspection. At some point during this transfer the permanent magnet stray magnetic field reached a value where the ^{1}H and ^{14}N energy levels cross and the transfer of polarization from ^{1}H to ^{14}N took place.[16] The gain[15] of ^{14}N PNT magnetization in such a case is roughly $\gamma_{^1\mathrm{H}} B_0/\nu_Q \sim 50$ for $\nu_Q = 887$ kHz, where $\gamma_{^1\mathrm{H}}$ is the proton gyromagnetic ratio. Similar values are obtained also for TNT. Typically this cycle was repeated 1,024 times to further increase sensitivity and achieve an extremely good S/N.

2.4. PROBE

The coil was wound from 0.5 mm copper wire. Such coils have a high intrinsic Q-factor, which we intentionally lowered to ~36 with a resistor parallel to the coil. We have achieved two goals with such a scheme. First, the probe (resonant circuit) bandwidth was increased to 50 kHz, wide enough to detect all TNT frequencies without retuning. This way magnetization intensity variations could be attributed to probe unrelated sources. And second, a low-Q probe has also a short dead time, which in our case was less than 100 μs. This allowed us to use very small pulse spacing. On the other hand, we had to increase the number of signal averages to compensate for the smaller S/N of the low-Q probe.

2.5. SPIN-LOCK SPIN-ECHO MULTIPULSE SEQUENCE

The sample in the coil after cross-polarization is suitable for any conventional NQR experiment, but with an apparent spin T ~ 6 K. We have irradiated the sample with the SLSE sequence (shown in Figure 2): $90°_{\pm y}$–$(\tau - 90°_x - \tau)_N$, where $N = 4{,}096$, "90°" pulse lengths were $t_{90} = 33$ μs and $2\tau = 512$ μs, so that the repetition time was $t_c = 2\tau + t_{90} = 545$ μs. To remove the baseline offsets which arise from equipment imperfections and residual ringing we have cycled the preparatory pulse between $90°_{+y}$ and $90°_{-y}$ and subtracted the responses. The key component of a SLSE sequence is the establishment of a spin-lock state which happens in a time $<T_2$. On a short time-scale of about few t_c this state appears to be a steady state in the sense that after each $90°_x$ pulse, exactly the same magnetization is observed. However on a longer time-scale this magnetization decays with a decay time $T_{2eff} >> T_2$. The two processes together are usually referred as the establishment of a quasi-steady state.

The magnetization evolution between the pulses n and $n + 1$ has the form of an echo and is described by

$$s_n(t) = M_{SLSE} \exp\left(i\omega_Q t - \frac{nt_c}{T_{2eff}} \right) \tag{2a}$$

$$s_n(t) = \sum_{k=1}^{12} M_k \exp\left(i\omega_k t - \frac{nt_c}{T_{2eff,k}} \right) \tag{2b}$$

for PNT and TNT respectively. From each SLSE magnetization decay (a collection of all 4,096 $s_n(t)$) we extracted the two unknown parameters M_{SLSE} and T_{2eff} in the case of PNT and several M_k and $T_{2eff,k}$ for TNT. Here the resonance frequencies are exactly known; ω_k were taken from Table 1 and a predetermined frequency $\omega_Q/2\pi = 887$ kHz has been used for PNT. The experimental S/N and SLSE resolution $1/t_c$ did not allow for a reliable determination of individual linewidths which were therefore left out from equations (2a) and (2b). However, as we will see later, their effect becomes very noticeable in the case of TNT. In the following section we derive the expression for the single site intensity M_{SLSE} (a procedure described in reference[10]), which is later easily generalized to individual TNT intensities M_k.

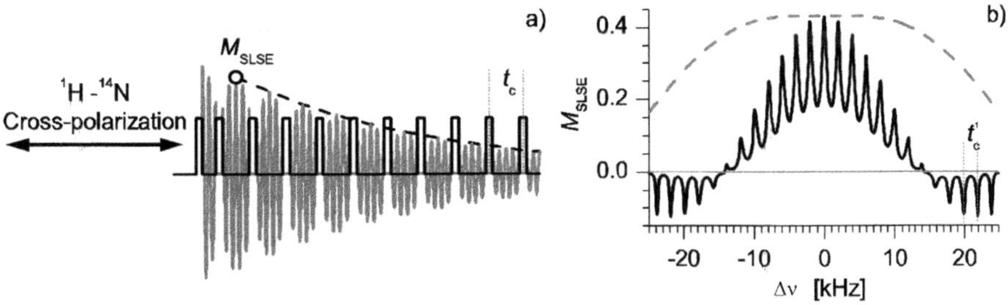

Figure 2. SLSE pulse sequence and the frequency offset dependence of the echo intensity: (a) the pulse sequence with RF pulses (black line), the spin echoes (gray line) and their decay to equilibrium (dashed line) and (b) the frequency offset $\Delta\nu = \nu_Q - \nu_{rf}$ dependence of the echo amplitude M_{SLSE} (black line), after the establishment of a common spin temperature for a single resonance line case. Here for comparison, the frequency offset dependence of the magnetization amplitude following a single pulse (dashed line) is also shown.

3. Steady state – M_{SLSE}

3.1. THEORY

The NQR spin-lock effect was first treated theoretically by Cantor and Waugh[17] for evolution times shorter than T_2 which confirmed the onset of a steady state. Theories for longer times which describe the quasi-steady state were developed by Provotorov and Khitrin[18] using the canonical transformation approach, and by Maricq[19] using the average Hamiltonian theory. The above approaches were all assuming an on-resonance case, which is unsuitable to describe TNT. Here we will re-derive the off-resonance SLSE steady state intensity as described in reference[10].

We consider a polycrystalline ensemble of equivalent $I = 1$ spins, which are evolving according to the spin Hamiltonian

$$\mathcal{H} = \begin{cases} \mathcal{H}_Q + \mathcal{H}_{RF} & ; \text{during RF pulses,} \\ \mathcal{H}_Q & ; \text{between RF pulses,} \end{cases} \tag{3}$$

and a pulse sequence $90°_{\pm y}-(\tau-90°_x-\tau)N$, where $\mathcal{H}_Q$ is the quadrupole Hamiltonian as the main interaction and $\mathcal{H}_{RF}$ is the interaction with the externally applied RF field. The spin operators will be written in terms of a fictitious spin[17] and are defined in Table 2.

TABLE 2. For each spin we define nine operators as in reference[17] and reliable them according to reference[20] to better reflect the fictitious spin behavior. The operators on the same line fulfill angular momentum commutation relations: $[S_{qx}, S_{qy}] = iS_{qz}$ and cyclic permutations of x, y, z (q can be replaced by p or r). For operations involving only operators from the same line the vector formalism can be used. To compact notation we thus define a vector $\vec{S}_q = (S_{qx}, S_{qy}, S_{qz})$ which satisfies $[\vec{S}_q, S_{rz} - S_{pz}] = 0$.

$S_{px} = \frac{1}{2} S_x$	$S_{py} = \frac{1}{2}(S_y S_z + S_z S_y)$	$S_{pz} = \frac{1}{2}(S_z^2 - S_y^2)$
$S_{qx} = \frac{1}{2} S_y$	$S_{qy} = \frac{1}{2}(S_z S_x + S_x S_z)$	$S_{qz} = \frac{1}{2}(S_x^2 - S_z^2)$
$S_{rx} = \frac{1}{2} S_z$	$S_{ry} = \frac{1}{2}(S_x S_y + S_y S_x)$	$S_{rz} = \frac{1}{2}(S_y^2 - S_x^2)$

The $\mathcal{H}_Q$ of $I = 1$ nucleus has three energy levels and all the transitions between them are allowed, so that there are three observable frequencies which we label as ω_p, ω_q and ω_r. In the principal axis system (PAS) the $\mathcal{H}_Q$ is

$$\mathcal{H}_Q = \omega_q S_{qz} + \frac{1}{3}\omega_{rp}(S_{qz} - S_{rz}) \tag{4}$$

with $\omega_{rp} = \omega_r - \omega_p$, while $\hbar$ is omitted throughout the paper. Similarly $\mathcal{H}_{RF}$ can be written as

$$\mathcal{H}_{RF} = 2\omega_1 \cos(\omega_{RF} t + \varphi)(\cos(\theta_p) S_{px} + \cos(\theta_q) S_{qx} + \cos(\theta_r) S_{rx}), \tag{5}$$

where $(\cos(\theta_p), \cos(\theta_q), \cos(\theta_r))$ defines the coil's orientation in PAS frame and ω_1 and φ are the external RF field strength and phase; $\varphi = \pm 90°$ for the first pulse and $\varphi = 0$ for the remaining pulses. We introduce the offset $\Delta\omega = \omega_q - \omega_{RF}$ as the difference between the resonant ω_q and the irradiation/observation frequency ω_{RF}. When ω_{RF} is close to other transition frequencies (ω_p and ω_r) the Hamiltonian in equation (4) can be written in another but equivalent form. The evolution of the density matrix ρ governed by the Liouville equation

$$\dot{\rho}(t) = i[\rho(t), \mathcal{H}], \tag{6}$$

will be monitored in the rotating frame defined by $R(t) = \exp(i(\omega_{RF} S_{qz} + ⅓\omega_{rp}(S_{rz} - S_{pz}))t)$ where the density matrix becomes $\rho^* = R(t)\rho R^{-1}(t)$. The rotating frame Hamiltonian $\mathcal{H}^*$ is piecewise time independent:

$$\mathcal{H}^* = \begin{cases} \Delta\omega S_{qz} \pm \omega_{1\theta} S_{qy} \equiv \mathcal{H}^*_{FIRST} & \text{; during the first pulse} \\ \Delta\omega S_{qz} + \omega_{1\theta} S_{qx} \equiv \mathcal{H}^*_{ON} & \text{; for remaining pulses} \\ \Delta\omega S_{qz} \equiv \mathcal{H}^*_{OFF} & \text{; for the period between pulses} \end{cases} \tag{7}$$

Here $\omega_{1\theta} = \omega_1 \cos(\theta_q)$ and we have omitted all rapidly oscillating terms.

The first pulse in the SLSE pulse sequence, the preparatory pulse $90°_{\pm y}$, is acting on a fully relaxed system described by $\rho_0^* = ⅓(1 - \mathcal{H}_Q/kT)$ in the high temperature limit (valid also for the cross-polarized system where spin $T \sim 6$ K). During the first pulse ρ_0^* evolves to

$$\rho^*(0) = \frac{1}{3}(1 - \frac{\omega_q}{kT}\vec{n}_0 \cdot \vec{S}_q - \frac{1}{3kT}\omega_{rp} S_\beta). \tag{8}$$

where $t = 0$ is defined at the end of the first pulse, $\vec{S}_q = (S_{qx}, S_{qy}, S_{qz})$, $S_\beta = S_{rz} - S_{pz}$. Here $\vec{n}_0 = (n_{0,x}, n_{0,y}, n_{0,z})$ is defined as

$$n_{0,x} = 2\frac{\omega_{1\theta}}{\tilde{\omega}_{1\theta}}\sin(\tilde{\omega}_{1\theta} t_{90°}), \tag{9a}$$

$$n_{0,y} = \frac{\Delta\omega\omega_1}{\tilde{\omega}_{1\theta}^2}(-1 + \cos(\tilde{\omega}_{1\theta} t_{90°})), \tag{9b}$$

$$n_{0,z} = 0, \tag{9c}$$

and $\tilde{\omega}_{1\theta} = \sqrt{\Delta\omega^2 + \omega_{1\theta}^2}$. We have already took into account the effect of the first pulse phase cycling by subtracting the evolution of ρ_0^* after a $90°_{+y}$ with the one after a $90°_{-y}$.

To find the intensity M_{SLSE} and also every M_k of TNT, we only need to know $\rho^*(t)$ at echo peaks, this is at discrete times 0, t_c, $2t_c$, $3t_c$, ... The evolution of the series $\rho^*(0)$ to $\rho^*(t_c)$, $\rho^*(2t_c)$, ... is in this case exactly described by the effective Hamiltonian

$$\mathcal{H}_{eff}^* = \Omega_{eff}\, \vec{n}_{eff} \cdot \vec{S}_q \tag{10}$$

defined by

$$\exp(i\mathcal{H}_{eff}^* t_c) = \exp(i\mathcal{H}_{OFF}^* \tau)\exp(i\mathcal{H}_{ON}^* t_{90°})\exp(i\mathcal{H}_{OFF}^* \tau) \tag{11}$$

and corresponds to the $\tau - 90°_x$-τ part of the pulse sequence. Here Ω_{eff} is found from

$$\cos(\frac{1}{2}\Omega_{eff} t_c) = \cos(\Delta\omega\tau)\cos(\frac{1}{2}\tilde{\omega}_{1\theta} t_{90°}) - \\ - \frac{\Delta\omega}{\tilde{\omega}_{1\theta}}\sin(\Delta\omega\tau)\sin(\frac{1}{2}\tilde{\omega}_{1\theta} t_{90°}) \tag{12}$$

while $\vec{n}_{eff} = (n_{eff,x}, n_{eff,y}, n_{eff,z})$ from

$$n_{eff,x} = \frac{\omega_{1\theta} \sin(\frac{1}{2}\tilde{\omega}_{1\theta} t_{90^\circ})}{\tilde{\omega}_{1\theta} \sin(\frac{1}{2}\Omega_{eff} t_c)} \tag{13a}$$

$$n_{eff,y} = 0 \tag{13b}$$

$$n_{eff,z} = \frac{\Delta\omega \cos(\Delta\omega\tau)\sin(\frac{1}{2}\tilde{\omega}_{1\theta} t_{90^\circ})}{\tilde{\omega}_{1\theta} \sin(\frac{1}{2}\Omega_{eff} t_c)} + \frac{\sin(\Delta\omega\tau)\cos(\frac{1}{2}\tilde{\omega}_{1\theta} t_{90^\circ})}{\sin(\frac{1}{2}\Omega_{eff} t_c)} \tag{13c}$$

One should be aware that equation (12) does not uniquely define Ω_{eff}, however all values satisfying equation 12 are equivalent because the spin evolution between nt_c is governed by $\exp(i\Omega_{eff}\, \vec{n}_{eff} \cdot \vec{S}_q^j t_c)$.

The SLSE steady state ρ_{ss}^* can now be defined as a state which is stationary under H_{eff}^*

$$[\rho_{ss}^*, H_{eff}^*] = 0. \tag{14}$$

An arbitrary initial state $\rho^*(0)$ is thus split in two parts: the steady state ρ_{ss}^* which commutes with H_{eff}^* and the remaining part $\rho_\perp^*$ which rotates around $\vec{n}_{eff}$. If the spins are uncoupled from each other then $\rho_\perp^*$ will rotate indefinitely, however when the spins become coupled then $\rho_\perp^* \to 0$. Although in the Hamiltonian (equation (3)) no such interaction is assumed, we shall include one in Section 4.2. We therefore assume here that $\rho_\perp^* = 0$. This mechanism has been treated rigorously in several articles.[17–19] In the present discussion, we just assume that such a state is formed soon after the preparatory pulse.

There are two commuting solutions of the equation (14)

$$\rho_\alpha^* \propto \vec{n}_{eff} \cdot \vec{S}_q \equiv S_\alpha \tag{15}$$

$$\rho_\beta^* \propto S_{rz} - S_{pz} \equiv S_\beta, \tag{16}$$

where $[\rho_\alpha^*, \rho_\beta^*] = 0$. Thus ρ_α^* and ρ_β^* correspond to two uncoupled heat baths with inverse spin temperatures α and β while the steady state ρ_{ss}^* is written as

$$\rho_{ss}^{*}=\frac{1}{3}(1-\alpha\Omega_{eff}S_{\alpha}-\frac{1}{3}\beta\omega_{rp}S_{\beta}). \tag{17}$$

We should emphasize here that ρ_{α}^{*} is not a function of α but rather a state associated with the temperature α and similarly for β.

The coefficients α and β are found by equalizing projections of $\rho^{*}(0)$ and ρ_{ss}^{*} onto states ρ_{α}^{*} and ρ_{β}^{*}:

$$\mathrm{tr}\{\rho(0)^{*}\rho_{\alpha}^{*}\}=\mathrm{tr}\{\rho_{ss}^{*}\rho_{\alpha}^{*}\}, \tag{18}$$

$$\mathrm{tr}\{\rho(0)^{*}\rho_{\beta}^{*}\}=\mathrm{tr}\{\rho_{ss}^{*}\rho_{\beta}^{*}\}. \tag{19}$$

and the two values are

$$\alpha=\frac{(\vec{n}_{0}\cdot\vec{n}_{eff})}{kT}\frac{\omega_{q}}{\Omega_{eff}}, \tag{20}$$

$$\beta=\frac{1}{kT}. \tag{21}$$

In our experiments the observed magnetization M is detected with the same coil used for RF irradiation so that in the laboratory system M is defined as

$$M=\cos(\theta_{q})S_{qx}, \tag{22}$$

whereas its steady state expectation value at echo peaks nt_c is

$$M_{ss}=\left\langle\cos(\theta_{q})\,\mathrm{tr}\{\rho_{ss}(S_{qx}\cos(\omega_{RF}nt_{c})-S_{qy}\sin(\omega_{RF}nt_{c}))\}\right\rangle_{\theta_{q}}$$

$$\equiv M_{SLSE}\cos(\omega_{RF}nt_{c}), \tag{23}$$

with only the in-phase component $\propto\cos(\omega_{RF}t)$

$$M_{SLSE}=-\frac{1}{6}\frac{\omega_{q}}{kT}\left\langle\cos(\theta_{q})n_{0,x}n_{eff,x}^{2}\right\rangle_{\theta_{q}} \tag{24}$$

In the above equations $\langle\cdots\rangle_{\theta_q}$ represents an average over all crystallite orientations θ_q.

The intensity M_k of the kth ^{14}N site in TNT is found by using the expression for M_{SLSE} with the appropriate frequency offset $\Delta\omega=\omega_k-\omega_{RF}$ for that site, whereas all the other parameters remain the same.

3.2. EXPERIMENTAL RESULTS FOR PNT

We have compared the prediction of the expression in equation (24) with ^{14}N experimental data on PNT as shown in Figure 3. The SLSE decay has been observed from $\Delta\nu = -15$ kHz to $\Delta\nu = 0$ kHz in steps of 470 Hz, where $t_c = 545$ μs. At each frequency the decay has been fitted with equation (2a) while assuming a very sharp resonance line. The thus extracted magnetization M_{SLSE} reaches its maximum on resonance $\Delta\omega = 0$ and exhibits an evident modulation with periodicity $1/t_c$. The other peaks correspond to $\Delta\omega = n2\pi/t_c$, where n is an integer. When moving away from resonance, the peaks become lower due to a finite 90° pulse length and obey the well known sinc dependence. The magnetization shows a minimum every $\Delta\omega = (n + ½)2\pi/t_c$, where n is an integer. Here, the magnetization is approximately 2/3 smaller than it is at the nearby maximum. It is immediately obvious from these results that the signal can be reduced by a factor of 3 just by missing the resonance for 1 kHz. In TNT for example, the temperature induced resonance variation of ν_+ lines[9] is 50-300 Hz/K and corresponds to a ΔT of 3-20 K for a 1 kHz offset. The target temperature in a remote detection can easily be miss-estimated by few degrees what will jeopardize the detection. In our measurement we have missed exact resonance by 190 Hz, which seemed negligibly small in the laboratory but had to be considered during the analysis. We should emphasize that although the match between the experimental points and the theoretical curve in Figure 3 is almost perfect, the curve is not a fit (neglecting minor refinements), but is rather calculated with equation (24) using the known NQR PNT frequency[12] and SLSE pulse sequence parameters.

For values of $\Delta\omega = n2\pi/t_c$, equation (23) slightly overestimates the observed values. We see two reasons for this. The first is our omission of the PNT linewidth in equation (2a), and the second is the omission of the homonuclear dipolar interaction in equation (3). Any additional small interaction $\mathcal{H}_\delta$ in Hamiltonian equation (3) will have appreciable impact on M_{ss} only when $|\mathcal{H}_{eff}| << |\mathcal{H}_\delta|$ which is the case at above conditions. When $\Delta\omega \rightarrow n2\pi/t_c$ then $\mathcal{H}_{eff} \sim 0$ and the interaction $\mathcal{H}_\delta$ will influence considerably the steady states ρ_α* and ρ_β* and eventually, when $\mathcal{H}_{eff} = 0$, produce a new steady state determined mainly by $\mathcal{H}_\delta$.

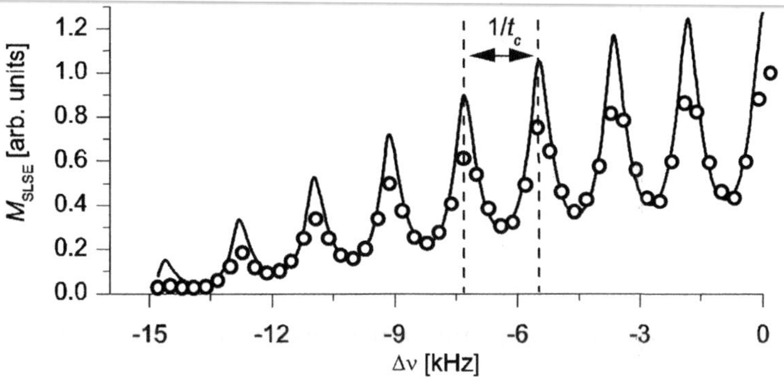

Figure 3. Comparison of PNT experimental (circles) and theoretical (solid line) in-phase steady state magnetization M_{SLSE} as a function of the frequency offset $\Delta\nu$. Experimental parameters are: ν_Q = 868 kHz, 90° pulse length 33 μs, τ = 256 μs, and $\omega_1 = 60 \cdot 10^3\ s^{-1}$. M_{SLSE} theoretical value is obtained from equation (24).

3.3. EXPERIMENTAL RESULTS FOR TNT

An analog experiment as for PNT has been performed for TNT with t_c = 545 μs. Here, ν_{RF} was varied from 837 to 868 kHz in steps of 250 Hz. At each step, the resonance line intensities M_k were determined by fitting the SLSE decay with equation (2b). The result of the analysis is shown in Figure 4b–f. Although all orthorhombic intensities have been fitted independently, the frequencies corresponding to M_3 and M_4 lie very close together and are showed as a sum. The 1/t_c fast modulation magnitude is here small compared with the one for PNT in Figure 3.

A comparison between the prediction of equation (24) and experiment is shown for M_5 in Figure 4e. There is considerable discrepancy where the reason is again the omission of the NQR linewidths from equation (2b). In our TNT sample the linewidths are of the order of 1 kHz and thus an appreciable fraction of the fast modulation frequency t_c^{-1} = 2 kHz. To correct for this, we have convolved the predicted M_k offset dependence (as in Figure 2) with Gaussian curve whose width is determined by the linewidths in spectrum Figure 1. Such a corrected simulation for each orthorhombic site is shown in Figure 4b–f. The agreement is now excellent, especially in view of the fact that there is practically only one adjustable parameter, a single vertical scaling of the six sinc functions. However, because of equipment imperfection, the scalings are allowed to differ, but by no more than 10%. All the other parameters are determined either by TNT or the applied pulse sequence.

4. Quasi-steady state decay – T_{2eff}

4.1. THEORY

As pointed out before, the magnetizations M_{SLSE} and M_k are steady if the spin Hamiltonian consists of $\mathcal{H}_Q$ and $\mathcal{H}_{\mathrm{RF}}$ only. On the contrary, M_{SLSE} and M_k decay to zero if an interaction between spins (usually dipolar) is included. We shall now include two interactions in our analysis which transform the steady state to a quasi-steady state: homonuclear dipolar interaction $\mathcal{H}_D$ and a single spin interaction with the lattice $\mathcal{H}_{SL}$. In our case they are small compared to $\mathcal{H}^*_{eff}$ (equation (10)) at practically all $\Delta\omega$, except for $\Delta\omega = n2\pi/t_c$ when $\mathcal{H}^*_{eff} = 0$, but we intentionally overlook such situations keeping in mind that at these points our result is only a crude approximation. We will first find some general equations about the system evolution under additional interaction and then list some properties of the individual interactions.

We start from the rotating frame with main interactions $\mathcal{H}^*_{ON}$ and $\mathcal{H}^*_{OFF}$, and a not yet specified perturbation $\mathcal{H}^*_{\delta}$. To remove the main rotating frame interactions, we make a transformation to the interaction representation $\rho^{**} = T(t)\rho^* T^{-1}(t)$ with a unitary operator $T(t)$ defined recursively as

$$T(t) = \exp(i\mathcal{H}^*_{OFF}t)\ ;\ 0 < t \le \tau$$

$$T(t) = \exp(i\mathcal{H}^*_{ON}(t-\tau))T(\tau)\ ;\ \tau < t \le \tau + t_{90^\circ} \tag{25}$$

...

The definition in equation (25) is necessary because in the rotating frame the RF interaction is piecewise time independent. Such transformation is also known as the toggling frame transformation.

Our aim is now to find the evolution of inverse temperatures α and β due to the additional interaction, while we neglect any perturbation of the steady states ρ_α^* and ρ_β^* due to $\mathcal{H}_\delta$. We expect the inverse temperatures α and β to change slowly with time, so that the evolution is sufficiently well determined with perturbation theory. Writing down the time dependent perturbation theory expression for the density matrix derivative[21] up to the second order we obtain

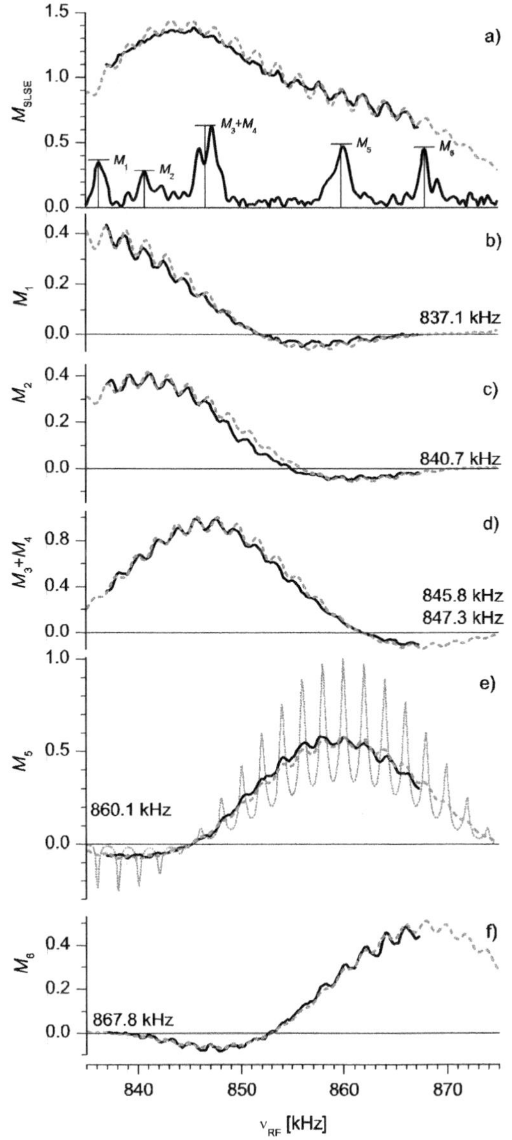

Figure 4. Comparison of a theoretical and experimental irradiation frequency dependence of the SLSE response in commercial TNT: (a) total echo amplitude (solid line) with a best fit (dashed line). For illustration a TNT spin-echo spectrum is also shown with labeled intensities M_k. In (b)–(f) contributions of various orthorhombic sites to the total echo amplitude (solid line) with the corresponding best fits (dashed lines) which have been corrected for the observed linewidths. In (e) also the prediction of the Equation (24) is shown, which does not include the observed linewidth (dotted line).

$$\dot{\rho}_{ss}^{**}(0) = \frac{\rho_{ss}^{**}(t_c) - \rho_{ss}^{**}(0)}{t_c} =$$

$$= \frac{i}{t_c} \int_0^{t_c} [\rho_{ss}^{**}(0), \mathcal{H}_\delta^{**}(t_1)] dt_1 - \frac{1}{t_c} \int_0^{t_c} \int_0^{t_1} [[\rho_{ss}^{**}(0), \mathcal{H}_\delta^{**}(t_2)], \mathcal{H}_\delta^{**}(t_1)] dt_2 dt_1 \quad (26)$$

where $\dot{\rho}_{ss}^{**}(0)$ stands for a cycle average derivative and $\mathcal{H}_\delta^{**} = T(t)\mathcal{H}_\delta^{*} T^{-1}(t)$.

Using $\rho_{ss}^{**}(nt_c) = \rho_{ss}^{*}(nt_c)$ which is valid for any integer n, we can insert equation (17) in equation (26) and select out only components ρ_α and ρ_β with traces in an analogous way as we have determined the initial spin temperatures in equations (18) and (19) to obtain

$$\Omega_{eff}\dot{\alpha} = -W_{\alpha\alpha}\Omega_{eff}\alpha - \frac{1}{3}W_{\alpha\beta}\omega_{rp}\beta \tag{27a}$$

$$\frac{1}{3}\omega_{rp}\dot{\beta} = -W_{\beta\alpha}\Omega_{eff}\alpha - \frac{1}{3}W_{\beta\beta}\omega_{rp}\beta. \tag{27b}$$

The decay rates are

$$W_{\mu\nu} = \frac{1}{t_c}\int_0^{t_c}\int_0^{t_1}\frac{\mathrm{tr}\{[[S_\nu, \mathcal{H}_\delta^{**}(t_2)], \mathcal{H}_\delta^{**}(t_1)]S_\mu\}}{\mathrm{tr}\{S_\mu^2\}}dt_2 dt_1 \tag{28}$$

where μ, ν stand for α or β. Due to the relationship $[\rho_\alpha, \rho_\beta] = 0$ only the second order term in equation (26) has a nonzero contribution to the transition rate $W_{\mu\nu}$.

The equations (27a) and (27b) are easily solved for the evolution of the inverse temperatures α and β which gives M_{SLSE} for a single crystallite orientation θ_q. In a powder however, many very similar magnetization decays have to be added together to form the bulk signal. The easiest way to characterize such a cumulative decay is to find its cumulative initial decay rate which is a powder average of individual decay rates weighted by their initial magnetization magnitude

$$\frac{1}{T_{2eff}} = \frac{\dot{M}(0)}{M(0)} = \frac{\left\langle \cos(\theta_q)\,\mathrm{tr}\{\dot{\rho}_{ss}^{*}(0)S_{qx}\}\right\rangle_{\theta_q}}{\left\langle \cos(\theta_q)\,\mathrm{tr}\{\rho_{ss}^{*}(0)S_{qx}\}\right\rangle_{\theta_q}}. \tag{29}$$

4.2. HOMONUCLEAR DIPOLAR AND SPIN-LATTICE INTERACTION

We need to explicitly specify the interactions $\mathcal{H}_\delta$ to proceed. The two main interactions are the ^{14}N–^{14}N homonuclear coupling, which in the laboratory frame is

$$\mathcal{H}_D = \sum_{\langle j,k\rangle}\mathcal{H}_D^{jk} = 4\sum_{\langle j,k\rangle}\sum_{\mu,\nu}\omega_{\mu\nu}^{jk}S_\mu^j S_\nu^k \tag{30}$$

and the spin-lattice coupling which includes the ^{1}H–^{14}N dipolar coupling and the dynamic part of quadrupole coupling $\mathcal{H}_Q$

$$\mathcal{H}_{SL} = 2\sum_{j,\mu}\omega_\mu(t)S_\mu^j. \tag{31}$$

Here the coupling constants $\omega_{\mu\nu}^{jk}$ are found from crystal structure

$$\omega_{\mu\nu}^{jk} = \frac{\mu_0 \gamma^2}{4\pi r_{jk}^5}(3(\vec{r}_{jk} \cdot \vec{e}_\mu)(\vec{r}_{jk} \cdot \vec{e}_\nu) - \delta_{\mu\nu} r_{jk}^2) \tag{32}$$

where the indexes μ and v are restricted to px, qx and rx only. The $\vec{e}_{px}$, $\vec{e}_{qx}$ and $\vec{e}_{rx}$ are the eigenvectors of the electric field gradient (EFG) tensor and $\vec{r}_{jk}$ is a vector connecting two ^{14}N nuclei. The spin-lattice couplings ω_μ are fluctuations associated with the fictitious single spin operator S_μ and are not known a priory. The index μ can be any of the px, qx, rx for heteronuclear dipolar interaction and py, qy, ry for quadrupole interaction. As usual, it is assumed that the interactions are time averaged to zero $\overline{\omega_\mu(t)} = 0$. It can be shown however, that for any inhomogeneous broadening $\overline{\omega_\mu(t)}$, $W_{\mu,\nu} = 0$ in equation (28). With reasonable assumptions equation (29) can be evaluated for both interactions,[10] however, the expressions contain huge amount of terms and offer no insight. In the following we summarize some T_{2eff} properties for each individual interaction.

For the homonuclear dipolar interaction all transition rates except $W_{\alpha\alpha}$ vanish. It thus follows from equation (27b) that the inverse temperature β is not changing with time, whereas the inverse temperature α mono-exponentially decays to zero. The two heat baths are effectively decoupled and T_{2eff} is thus a rather well defined quantity.

On the other hand, for the spin-lattice interaction all $W_{\mu\nu} \neq 0$, and there is a mix of the two heat baths corresponding to α and β:

$$W_{\alpha\alpha} = \frac{1}{16} c_0 + \frac{1}{16} c_\alpha w_{\alpha\alpha}(\omega_1, \Delta\omega, t_{90^\circ}, \tau) \tag{33a}$$

$$W_{\alpha\beta} = \frac{1}{16} c_\beta w_{\alpha\beta}(\omega_1, \Delta\omega, t_{90^\circ}, \tau). \tag{33b}$$

Here $c_0 = J_0 + 2J_q + J_p + J_r$, $c_\alpha = J_0 - J_q$, $c_\beta = J_p - J_r$ and depend only on sample characteristics (including T)

$$J_0 = \int_{-\infty}^{\infty} \omega_{qz}(0)\omega_{qz}(\tau)d\tau \tag{34a}$$

$$J_p = \int_{-\infty}^{\infty} (\omega_{px}(0)\omega_{px}(\tau) + \omega_{py}(0)\omega_{py}(\tau))\exp(i\omega_p \tau)d\tau \tag{34b}$$

$$J_q = \int_{-\infty}^{\infty} (\omega_{qx}(0)\omega_{qx}(\tau) + \omega_{qy}(0)\omega_{qy}(\tau))\exp(i\omega_q \tau)d\tau \tag{34c}$$

$$J_r = \int_{-\infty}^{\infty} (\omega_{rx}(0)\omega_{rx}(\tau) + \omega_{ry}(0)\omega_{ry}(\tau))\exp(i\omega_r\tau)d\tau, \tag{34d}$$

while the two functions $w_{\alpha\alpha}(\omega_1, \Delta\omega, t_{90}, \tau)$ and $w_{\alpha\beta}(\omega_1, \Delta\omega, t_{90}, \tau)$ contain only pulse sequence parameters. It is very likely that the two-exponential decay will be masked by the powder distribution. If this is the case, it is possible to evaluate T_{2eff} from equation (29):

$$\frac{1}{T_{2eff}} = \frac{1}{16}c_0 + \frac{1}{16}c_\alpha \frac{\left\langle \cos(\theta) n_{eff,x}^2 n_0 w_{\alpha\alpha} \right\rangle_\theta}{\left\langle \cos(\theta) n_{eff,x}^2 n_0 \right\rangle_\theta} + \frac{1}{16}c_\beta \frac{1+\eta}{3-\eta} \frac{\left\langle \cos(\theta) n_{eff,x} w_{\alpha\beta} \right\rangle_\theta}{\left\langle \cos(\theta) n_{eff,x}^2 n_0 \right\rangle_\theta} \tag{35}$$

4.3. PARANITROTOLUENE

The resonance frequency offset dependence of the relaxation time T_{2eff} for PNT is shown in Figure 5d for t_c = 545 μs. Similarly as the magnitude M_{SLSE}, also T_{2eff} shows the $1/t_c$ modulation, which however does not change appreciably with larger $\Delta\nu$ offsets. We have first tried to explain the relaxation only in terms of the homonuclear dipolar interaction, which is often held the T_{2eff} principal contribution. All parameters determining this relaxation are known for PNT (PNT's structure) so this is rather straightforward to evaluate. This contribution, for the used SLSE pulse sequence parameters, is shown in Figure 5a. As we see, the shape of this dependency is very similar to the observed one, but the homonuclear dipolar $1/T_{2eff}$ is roughly five times smaller than the experimental $1/T_{2eff}$. To better describe the experimental $1/T_{2eff}$, we have to include the spin-lattice interaction $\mathcal{H}_{SL}$ which we introduced in Section 4.2. In contrast with homonuclear dipolar interaction, there are no reports about the spectral densities J_0, J_p, J_q, and J_r for PNT, and we cannot evaluate the parameters c_0, c_α, and c_β in equation (4.2). We therefore use them as free parameters. The parameter c_α is associated with the function $w_{\alpha\alpha}(\omega_1, \Delta\omega, t_{90}, \tau)$ and for the used SLSE pulse sequence parameters is shown in Figure 5b. It has a similar $\Delta\nu$ dependence as the homonuclear dipolar interaction and the observed one. The parameter c_β is associated with the function $w_{\alpha\beta}(\omega_1, \Delta\omega, t_{90}, \tau)$ which is shown in Figure 5c. This $\Delta\nu$ dependence is now different from the observed one, and also from the one in Figure 5a and b. This term is responsible for the coupling of the two heat baths (see equations (27a) and (27b)) and also the origin of a spin-lock two-exponential decay nature. The third parameter c_0 is just the $\Delta\nu$ independent vertical offset of $1/T_{2eff}$. The best fit parameters of the equation (35) are $c_0 = 71\ s^{-1}$, $c_\alpha = 70\ s^{-1}$, and $c_\beta = -0.99\ s^{-1}$. Since $c_\beta << c_\alpha$ the decay should be pretty much mono-exponential, as indeed observed.

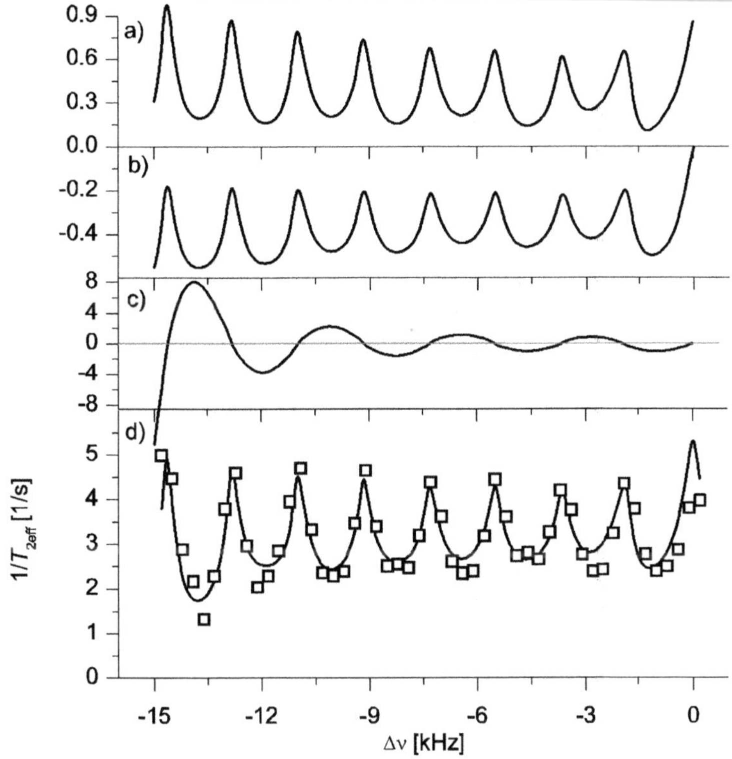

Figure 5. Theoretical and experimental relaxation rate T_{2eff} as a function of frequency offset $\Delta\nu$. In (a) theoretical homonuclear dipolar contribution with parameters $\omega_{qq} = 67.2\ \mathrm{s}^{-1}$ and $\omega_{rp} = 107.4\ \mathrm{s}^{-1}$, (b) spin lattice contribution with parameter $c_0 = c_\beta = 0$ and $c_\alpha = 1\ \mathrm{s}^{-1}$, in (c) spin lattice contribution with the parameter $c_0 = c_\alpha = 0$ and $c_\beta = 1\ \mathrm{s}^{-1}$ (d) (squares) the experimental value obtained at $\nu_q = 868$ kHz, $t_{90} = 33$ μs, $\tau = 256$ μs and $\omega_1 = 60 \cdot 10^3\ \mathrm{s}^{-1}$, (solid line) best fit with homonuclear dipolar and spin-lattice interaction with parameters $c_0 = 4.47\ \mathrm{s}^{-1}$, $c_\alpha = 4.39\ \mathrm{s}^{-1}$, and $c_\beta = -0.062\ \mathrm{s}^{-1}$.

4.4. TRINITROTOLUENE

In trinitrotoluene we observe similar behavior as in paranitrotoluene, but the extraction of 12 individual T_{2eff} is less reliable. It seems that the individual on-resonance- T_{2eff} vary in the range between 50 and 200 ms. We still observe the $1/t_c$ modulation which is smaller than in PNT, but not as much as we found for magnetization intensity. The relaxation behavior needs further studies and a better model equation (2b) which would include also individual linewidths.

5. Conclusions

We have shown that the off-resonance SLSE pulse sequence significantly influence the spin-lock state, in particular the magnitude of the spin-lock magnetization and the magnetization decay. We have experimentally observed and theoretically explained the $1/t_c$ modulation in $\Delta\nu$ dependence of M_{SLSE} and T_{2eff}. The modulation appears to be very strong in PNT as M_{SLSE} at $\Delta\nu = 0$ is almost three times bigger than at $\Delta\nu = 2/t_c$, whereas the decay time T_{2eff} diminishes only by a factor of ~2 so that the overall loss in S/N is not severe. We have also determined the three sample-specific parameters c_0, c_α, and c_β which enable us to find T_{2eff} under arbitrary pulse sequence parameters ω_1, τ, and t_{90}. In TNT the situation is somewhat different, and we don't observe a strong $1/t_c$ modulation. This seems to be due to larger linewidths of the ^{14}N NQR lines in TNT than in PNT. The detailed knowledge of frequency dependence of ^{14}N resonance amplitudes can be directly used for the increase of detection reliability in digital signal processing models, such as developed by Jakobsson et al.[8]

References

1. R.A. Marino, S.M. Klainer, J. Chem. Phys. 67(7), 3388 (1977)
2. D.Y. Osokin, J. Mol. Struct. 83(1–4), 243 (1982)
3. S.S. Kim, J.R.P. Jayakody, R.A. Marino, Z. Naturforsch. 47a, 415 (1992)
4. T.N. Rudakov, V.T. Mikhaltsevich, Phys. Lett. A 309, 465 (2003)
5. A.J. Blauch, J.L. Schiano, M.D. Ginsberg, J. Mag. Res. 139, 139 (1999)
6. V.T. Mikhaltsevich, T.N. Rudakov, Phys. Stat. Sol. (b) 241(2), 411 (2004)
7. K.L. Sauer, C.A. Klug, Phys. Rev. B 74(17), 174410 (2006)
8. A. Jakobsson, M. Mossberg, M.D. Rowe, J.A.S. Smith, IEEE Trans. Geosci. Rem. Sens. 43(11), 2659 (2005)
9. A.N. Garroway, M.L. Buess, J.B. Miller, J.B. Suits, A.D. Hibbs, G.A. Barrall, R. Matthews, L.J. Burnett, IEEE Trans. Geosci. Rem. Sens. 39(6), 1108 (2001)
10. A. Gregorovič, T. Apih, Relaxation during spin-lock spin-echo puise sequence in ^{14}N nuclear quadrupole resonance, J. Chem. Phys. 129, 214504 (2008)
11. J.V. Barve, L.M. Pant, Acta Crystallog. B 27, 1158 (1971)
12. Y. Hiyama, T.L. Brown, J. Chem. Phys. 75(1), 114 (1981)
13. R.A. Marino, R.F. Connors, J. Mol. Struc. 111, 323 (1983)
14. J.A.S. Smith, R.M. Deas, M.J. Gaskel, in International Conference on Requirements and Technologies for the Detection, Removal and Neutralization of Landmine and UXO, vol. 2, ed. by H. Sahli, A.M. Bottoms, J. Cornelis (Vrije Universiteit Brussels, Belgium, 2003), vol. 2, p. 715
15. J. Lužnik, J. Pirnat, V. Jazbinšek, T. Apih, A. Gregorovič, R. Blinc, J. Seliger, Z. Trontelj, Appl. Phys. Lett. 89, 123509 (2006)
16. M. Nolte, A. Privalov, J. Altmann, V. Anferov, F. Fujara, J. Phys. D: Appl. Phys. 35, 939 (2002)

17. R.S. Cantor, J.S. Waugh, J. Chem. Phys. 73(3), 1054 (1980)
18. B.N. Provotorov, A.K. Khitrin, JETP Lett. 34(4), 157 (1981)
19. M.M. Maricq, Phys. Rev. B 33(7), 4501 (1986)
20. V.L. Ermakov, D.Y. Osokin, Mol. Phys. 53(6), 1335 (1984)
21. C.P. Slichter, Principles of Magnetic Resonance, 3rd edn. (Springer, New York, 1996)

IDENTIFICATION OF LIQUIDS ENCOUNTERED IN CARRY-ON-LUGGAGE BY MOBILE NMR

JÖRG MAULER*, ERNESTO DANIELI, FEDERICO CASANOVA, BERNHARD BLÜMICH
ITMC, RWTH Aachen University, Templergraben 55, 52056 Aachen, Germany

Abstract Mobile NMR devices are suited for application at airports. They are low in maintenance, easy to use and offer safe operation.

Recently, great progress has been achieved in developing permanent magnets with high field homogeneity. Achieving high resolution proton spectroscopy would provide a fast and safe method to identify liquid explosives, without the need of cross checks with other techniques.

Keywords: mobile NMR, Halbach magnet, shim system, liquids, carry-on-luggage, bottled red wines

1. Objective

To date there is no procedure available to identify liquid and viscous explosives in carry-on-luggage at airports. NMR as a non-destructive and non-invasive method lends itself to be used for that. In this chapter suitable NMR-procedures are discussed.

A general requirement for such a procedure is fast detection performed with high accuracy. The procedure must end in a binary "Yes/No"-result identifying harmless from dangerous liquids without further need of interpretation by the security staff. The devices used for such measurements must offer ease of use, safe operation, and be low maintenance. This prohibits the use of liquid gases for cooling of super conducting magnets. Furthermore the magnet must provide a large enough bore for measuring liquids in bottle-size containers.

* E-mail: jmauler@mc.rwth-aachen.de

J. Fraissard and O. Lapina (eds.), *Explosives Detection Using Magnetic and Nuclear Resonance Techniques*,

All the criteria mentioned above point to mobile NMR technology as a possible alternative. During the last years considerable experience has been collected on how to perform relaxation measurements with mobile NMR equipment. The sensitivity of such a procedure is shown exemplarily by measurements of T_2-relaxation for differentiating different kinds of red wine (Figure 1).

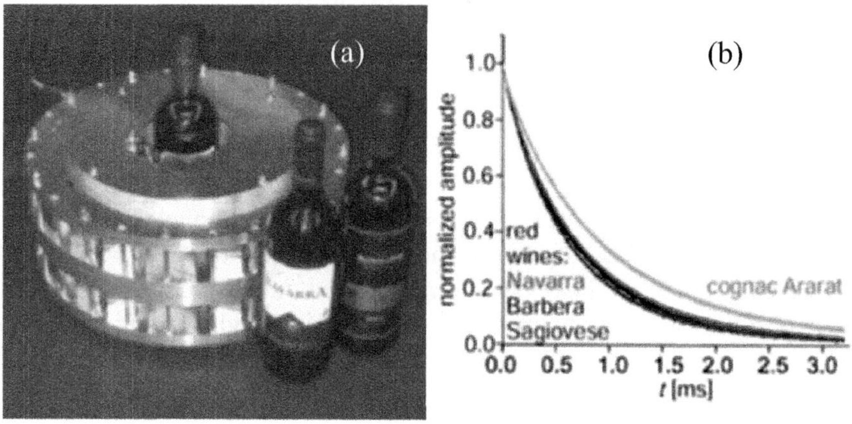

Figure 1. (a) Wine bottle in the Halbach magnet. (b) T_2-Relaxation-decays of bottled red wines.

The Halbach magnet used for the measurement of wine bottles is shown in Figure 1a and some representative data are shown in Figure 1b. If the spectrum of wines is limited, the samples can be differentiated based only on T_2 measurements. The design of a suitable magnet, a measurement and the data analysis protocol are discussed in the first part of this chapter.

Recently the proof of principle for spectroscopy in low fields generated by permanent magnets was shown,[1,2] but there are still some technical problems to be solved. Spectroscopy on liquids under stable conditions will be possible soon. The second part focuses on achieving sufficient field homogeneity with permanent magnets in order to run spectroscopy measurements. Based on their spectra, liquids will be recognizable with high specificity.

2. Constructing a Halbach magnet for relaxation measurements

Halbach cylinder magnets provide a strong and homogeneous B_0-field in the bore.[3] A cross section of an ideal Halbach magnet is shown in Figure 2a. The configuration can be expressed in terms of polar coordinates set in the center of the magnet. The magnetization vector continuously rotates, fulfilling that at a position with polar angle α, the magnetization vector is rotated by

2α. This leads to a superposition of the magnetic fields of the magnetization-vectors that produces a strong and homogeneous B_0-field in the central area of the magnet. In the case of an infinitely long cylinder, the field $B_0(\vec{r}) = c \ln(r_{out} / r_{in})$ inside the bore is perfectly homogeneous and the stray field outside is zero.

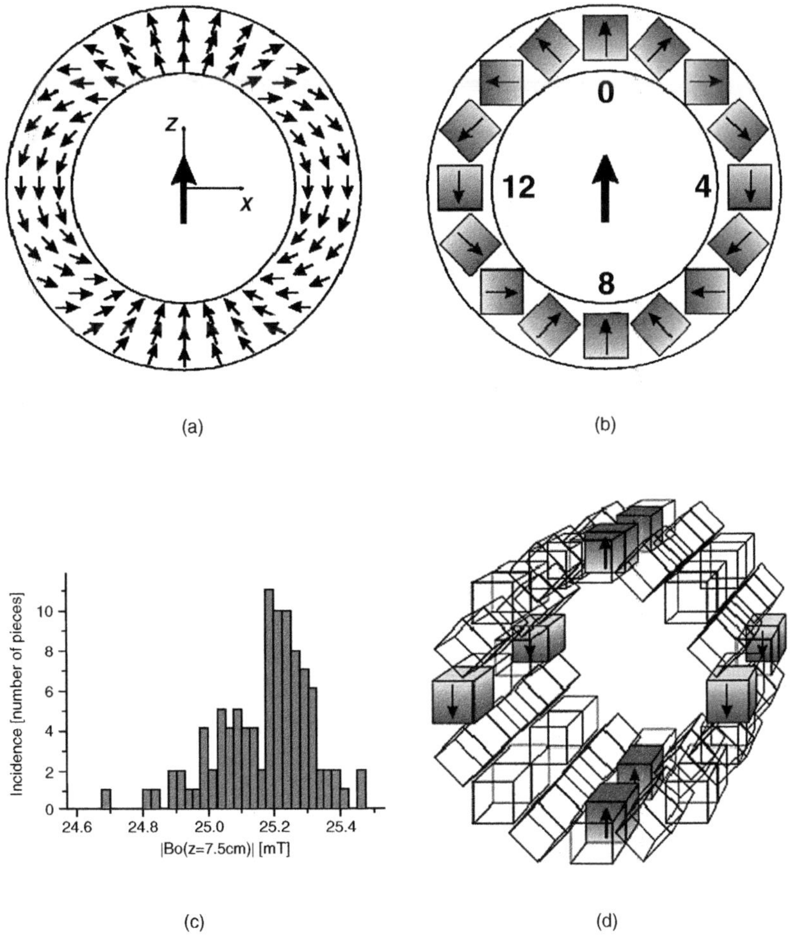

Figure 2. (a) Ideal Halbach magnet. The magnetization vector is rotated continuously. (b) Concept of a Halbach magnet built of cubes. (c) Distribution of magnetic field strengths of identical magnet blocks. (d) Arrangement of magnetic blocks that appropriates the ideal Halbach magnet.

For practical realization, the ideal Halbach magnet inevitably requires a simplification, the discretization of the ring. A discretization that uses identical blocks is discussed in reference[4]. As described there, a ring of the cylinder is substituted by a set of 16 magnetized cubes, sized 30 × 30 × 30 mm^3 (Figure 2b). Each magnet is rotated by an angle 2α to orient its magnetization along the direction required by the ideal Halbach magnet at the center position of the ring. The final alignment of the magnetization vector was equal to the direction one gets after integration over the cube volume in the ideal Halbach cylinder. Constructing the magnets from identical elements reduces the costs but must be paid in terms of magnetic field strength and homogeneity. Two sets of three rings were stacked to form a cylinder. The two sets were separated by a gap g = 93 mm in order to achieve a flat magnetic field profile along the long axis.

Typically, the remanent magnetization of a set of magnetic cubes varies within a certain distribution (Figure 2c). Mounting a Halbach magnet without any special strategy results in an additional loss in field homogeneity. Hence, before building the magnet, the magnetic field strength of all the cubes was measured and according to the position in that distribution the magnets were assigned to certain positions in the rings forming the Halbach cylinder. Generally the positions close to the center of the Halbach cylinder have a stronger impact on the field properties than the more peripheral positions. So the central positions are occupied by magnets taken from the center of the distribution. In order to avoid linear field variations, magnets at opposite positions with respect to the main axes of the magnet are filled with elements of preferably similar field strength. Positions with the antiparallel alignment of the magnetization vector (4 and 12 in Figure 2b) have a weaker impact on strength and homogeneity of the field than those aligned in parallel. So magnet cubes located further away from the center of the distribution are assigned to these positions. Exploiting this procedure helps us to reduce field inhomogeneities facilitating relaxometry experiments, but the achieved degree of homogeneity was not enough to resolve proton NMR spectra.

3. Relaxometry measurements

3.1. IDENTIFICATION EXCLUSIVELY BASED ON RELAXATION

The measurements were performed with a 2 cm diameter surface coil (Figure 3a). In order to fill the sensitive volume with liquid, the containers were tilted by 45°. The probe head was mounted in the Halbach magnet (Figure 3b), which was covered by a housing (Figure 3c).

The T_1 and T_2 relaxation values of 77 liquid and viscous samples from ten substance classes were measured and plotted in the T_1–T_2-space (Figure 4). T_2 could be determined with a four scan-experiment. For the T_1 measurements eight points of the saturation recovery curve were acquired, which led to a few seconds measurement time for each sample. This time might be decreased once the amount of points is reduced in a more optimized hardware setup.

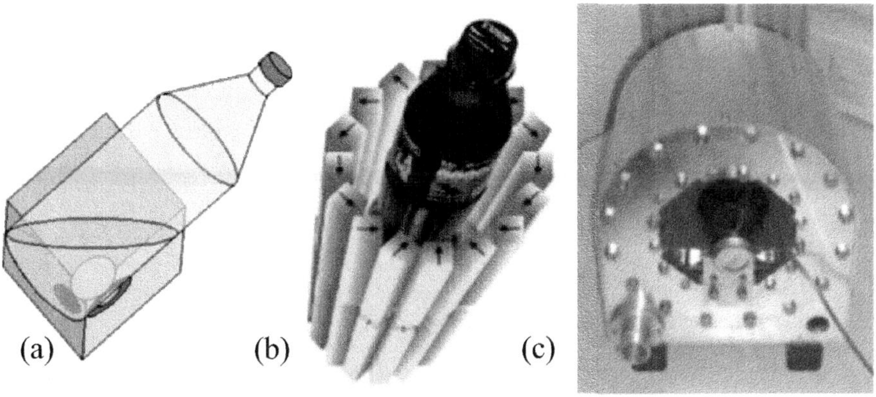

Figure 3. (a) Probe. (b) Magnet with bottle. (c) Real Halbach magnet.

The data in Figure 4a from the set of samples reveals two aspects: (1) the T_1–T_2-space will be filled more and more as the number of samples increases. (2) There are no disjoint, separable clusters of substance classes. The data points of the different sample classes are spread without any clear demarcation over the T_1–T_2-space.

The question is, to what extend a differentiation of liquid explosives and usual liquids can reliably be performed based on NMR relaxation times T_1 and T_2 with minimum time. To answer this question the error of the measurements must be evaluated and compared to the difference between the data points in the T_1–T_2 plot. Then, the measurement time can be decreased as long as the error does not make the identification impossible. Moreover, the option to measure relaxation times weighted by diffusion must be evaluated with respect to its tenacity to identify liquids. As the error of the measurement is too small to be visible on the scale of Figure 4a a zoom into this plot is displayed in Figure 4b. Obviously, the uncertainty is small enough to allow a differentiation of all the samples except the two alcoholic beverages on the left side. That implicates that for a limited set of samples, the identification can be performed based on relaxation data. This condition sounds rather restrictive, but compared to the current situation – liquids are rejected totally – the admission of a limited set of liquids is a considerable improvement. Particularly for real application the number of different liquids, passengers would like to take along, is limited.

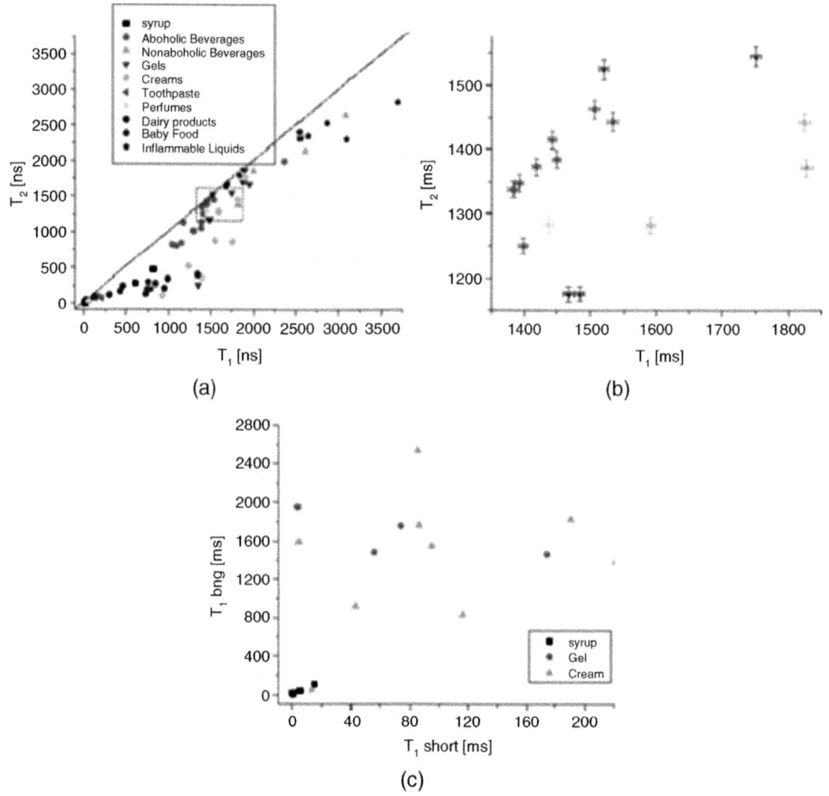

Figure 4. Relaxation measurements on 77 samples. (a) Total T_1–T_2-space. (b) Zoom into the T_1–T_2-space. (c) $T_{1,\text{short}}$ $T_{1,\text{long}}$ space.

In case of a 2- or n-component relaxation characteristic, the resultant parameter space can be extended to include further dimensions in order to differentiate liquids which are mapped onto the same points in the simple T_1–T_2-space. Syrup, gels, and creams revealed a 2-component T_1-relaxation (Figure 4c). So they can be identified in the three dimensional $T_{1,\text{short}}$–$T_{1,\text{long}}$–T_2-space. The equivalent can be performed with respect to the T_2-relaxation which leads to a 4-dimensional parameter space. But for some samples as in the case of the two alcoholic beverages in Figure 4b, this procedure does not help, because T_1 and T_2 are mono-exponential. Differentiation in this case would require the use of an additional method besides relaxometry.

3.2. IDENTIFICATION BASED ON THE COMBINATION WITH BAR CODE CROSS CHECK

Bottles expected to be transported through check points are expected to be commercial ones. An approach based on that and on ready-to-use technology is presented below. Its main idea is to check with NMR-measurements whether the content of a container has been altered. Therefore the bar-code available on every container can be read to identify the liquid. According to the bar-code the characteristic relaxation data can be retrieved from a database. The liquid is measured and the values are cross checked with the expected values taken from the database. In case they differ, the liquid was obviously exchanged or diluted and the container must be rejected.

The approach described above requires to define the products which are accepted to be taken into the database. Furthermore, the database must be updated regularly. A non-negligible number of liquids will change their composition from time to time. Wines even from the same vinery and other natural products will change their relaxation values every year. And there will be a significant amount of samples in which the recipe will be slightly modified within certain intervals of time. For instance it is imaginable that the food coloring is exchanged by the producer by a similar substance which shows the same color but a different relaxation value. Further on containers which are refilled by the passenger with a different liquid, will not pass the check. Finally one must cope with temperature shifts of the samples within a temperature range from refrigerator-level to summer-room-temperature in real application.

Thus relaxation measurements will not work in every case, but just for a limited set of liquids. NMR-Spectroscopy is a much more general approach. Possible solutions for low-field-NMR are discussed in the following section.

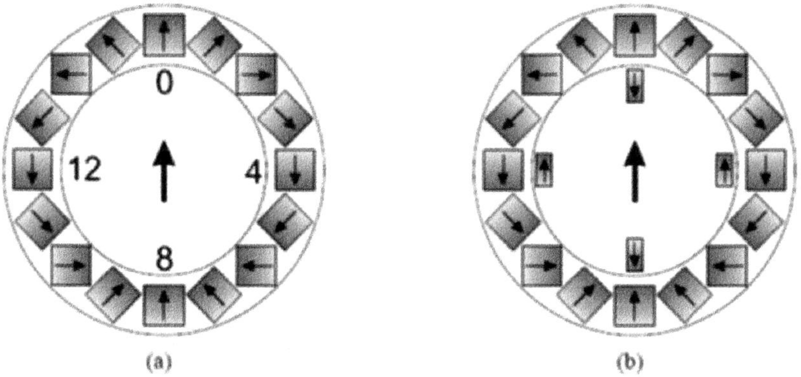

Figure 5. Implementing a shim system. (a) Halbach magnet. (b) Shim unit.

4. Shimming a Halbach array to achieve high-resolution spectroscopy

A Halbach magnet, built according to the procedure given in Section 2, does not provide the homogeneity needed to perform spectroscopy experiments. In order to enhance the homogeneity of that, a shimming approach is needed. Recently we implemented a shim system based on permanent magnets arranged in a Halbach configuration generating a weaker magnetic field antiparallel to the B_0-field.

As depicted in Figure 5a and b, small shim magnets were assembled at positions 0, 4, 8, and 12 with a magnetization vector antiparallel to the orientation of the magnet in the main cylinder. The shim unit comprises a total two four-element units, one for a set of three rings of the Halbach cylinder. These magnets can separately be shifted in three directions, x, y, and z. Decreasing the distance between two shim magnets leads to a higher quadratic component of the B_0-magnitude in the direction of the magnets. Shifting two opposite pairs of shim magnets, for instance $(0, \pm y_0, +z_0)$ and $(0, \pm y_0, -z_0)$ in the same direction generates corrective linear components, shifting them against each other leads to the generation of mixed terms.

B_0-Magnitude-field-maps measured for the unshimmed Halbach cylinder with a *RARE*-sequence are shown in the left column of Figure 6 and for the shimmed magnet in the right column. For all the three principal planes of the magnet (Figure 6 a and c: xz-plane, e and f: xy-plane, h and i: yz-plane), an improvement in field homogeneity facilitating spectroscopy in the center of the magnet could be achieved.

5. Spectroscopy measurements

Utilizing the shim unit described above, it was possible to enhance the spectral resolution up to two orders of magnitude (Figure 7). The improvement for a 6,000 mm^3 sample ($d = 20$ mm, $l = 20$ mm) is shown in the left column and for a 800 mm^3 volume sample ($d = 10$ mm, $l = 10$ mm) is displayed in the right column. The grey curves indicate the resolution for the Halbach cylinder without shim system, the black curves belong to the same sample volume and magnet but with the shim unit implemented. Zooming into Figure 7d reveals 0.85 ppm spectral resolution (Figure 7f). This resolution allowed us to resolve proton spectra of different solvents. Figure 7g shows the spectrum of acetic acid. Once the temperature shifts are manageable, spectroscopy performed in permanent magnets can be used to identify liquids in the carry-on-luggage at airports.

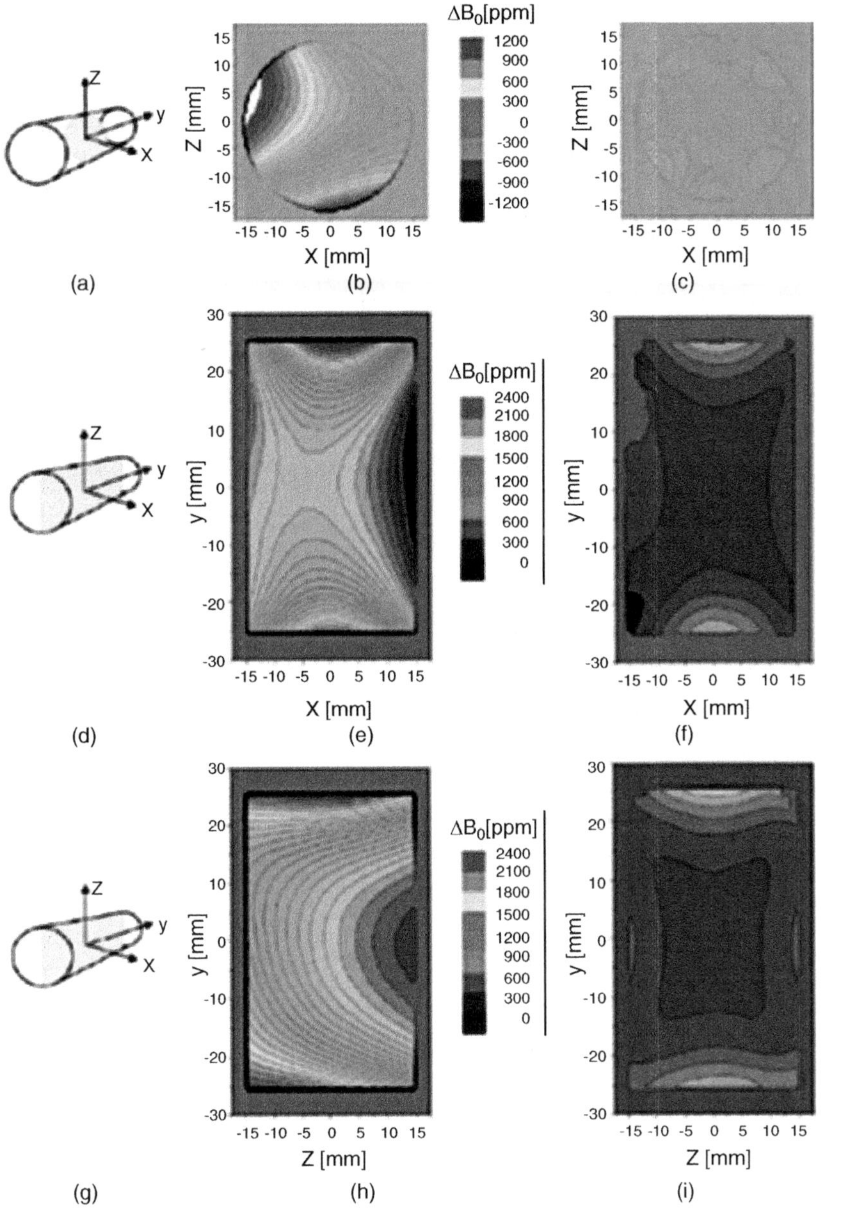

Figure 6. Improvements achieved by implementing a shim system. (a) *x–z*-plane. (b) *x–z*-field map without shim. (c) *x–z*-field map with shim. (d) *x–y*-plane. (e) *x–y*-field map without shim. (f) *x–y*-field map with shim. (g) *y–z*-plane. (h) *y–z*-field map without shim. (i) *y–z*-field map with shim.

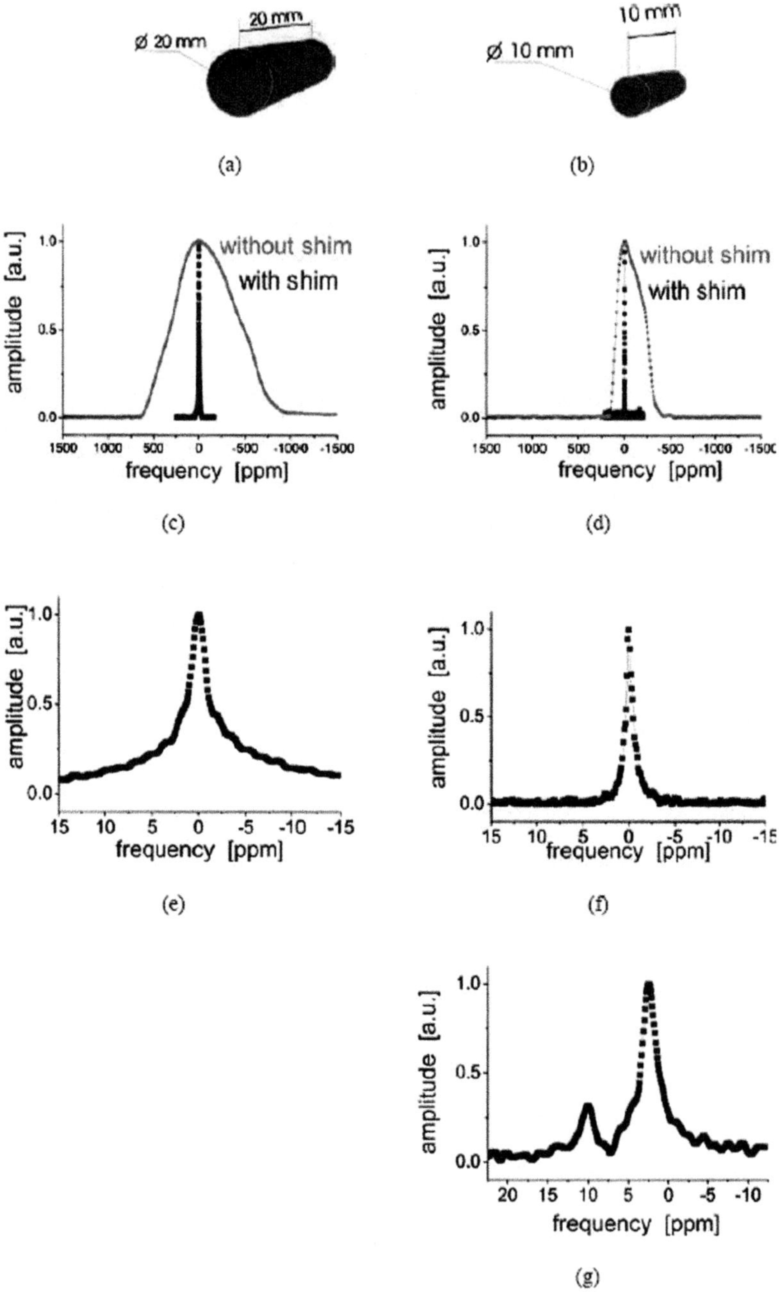

Figure 7. Spectroscopy performed in a Halbach magnet with a shim system. (a) V_b = 6,000 mm^3 (b) V_s = 790 mm^3. (c) Resolution in the large volume V_b. (d) Resolution in the small volume V_s. (e) Zoom into (c). (f) Zoom into (d). (g) Spectrum of acetic acid.

6. Summary

Mobile NMR devices are suited for application at airports. They are low in maintenance, easy to use and offer safe operation.

Identifying liquids based on relaxation data is possible only with certain restrictions, for instance the number of liquids must be limited. Therefore the help of an additional method is needed. One scenario could be the use of NMR together with bar-code scans to check whether the filling of a container has been altered. This can be achieved quickly and reliably with measurements by mobile NMR and a comparison with values from a database. That procedure would be independent of any further physical knowledge about the samples. But in addition to this, other measurements could be linked with NMR relaxation measurements, for example, microwave-spectroscopy.

Recently, great progress has been achieved in developing permanent magnets with high field homogeneity. Achieving high resolution proton spectroscopy would provide a fast and safe method to identify liquid explosives, without the need of cross checks with other techniques.

Acknowledgments This project has been supportet by the Ministerium für Innovation, Wissenschaft, Forschung und Technologie des Landes Nordrhein-Westfalen.

References

1. Perlo J., Demas V., Casanova F., Meriles C. A., Reimer J., Pines A., Blümich B.: High-Resolution NMR Spectroscopy with a Portable Single-Sided Sensor. Science 308, 1279 (2005).
2. Perlo J., Casanova F., Blümich B.: Ex Situ NMR in Highly Homogeneous Fields: ^{1}H Spectroscopy. Science 315, 1110–1112 (2007).
3. Halbach K.: Design of Permanent Multipole Magnets with Oriented Rare Earth Cobalt Material. Nucl. Instr. Meth. 169, 1–10 (1980).
4. Raich H., Blümler P.: Design and Construction of a Dipolar Halbach Array with a Homogeneous Field from Identical Bar-Magnets. Conc. Magn. Reson. 23B, 16–25 (2004).

THE TWO-FREQUENCY MULTIPULSE SEQUENCE IN NUCLEAR QUADRUPOLE RESONANCE OF N-14 NUCLEI

GEORGE V. MOZZHUKHIN[1,2*], BULAT Z. RAMEEV[1,3], NURCAN DOĞAN[1], BEKIR AKTAŞ[1]
[1]*Gebze Institute of Technology, 41400 Gebze-Kocaeli, Turkey*
[2]*Russian State University of Immanuel Kant, 236016 Kaliningrad, Russia*
[3]*Kazan Physical–Technical Institute, 420029 Kazan, Russia*

Abstract Combinations of two frequency pulses for NQR detection of explosives have been investigated. For the nuclear quadrupole system of ^{14}N nuclei with spin $I = 1$ we show that by this means it is possible to change the signal intensity of the transition detected. Theoretical and experimental studies of multiple-frequency sequences show that the proposed two-frequency sequences have considerable potential in the application of NQR for the detection of various materials.

Keywords: nuclear quadrupole resonance, NQR; two frequency pulses, two frequency pulses, nitrogen

1. Introduction

Multifrequency magnetic resonance techniques, a well known example of which is the Overhauser effect, have found wide application in research.[1] In nuclear magnetic resonance the related effect has also been experimentally demonstrated by Pound,[2] who observed a change in the signal intensity of one of the quadrupole transitions of ^{23}Na when RF power is applied to saturate another one. As regards pure nuclear quadrupolar resonance (NQR), two approaches to two-frequency methods have been developed. One of them is the double NQR–NQR method,[3] used for the detection of weak NQR signals of ^{23}Na, ^{39}K, ^{41}K, ^{131}Cs, ^{85}Rb and ^{87}Rb nuclei through the neighboring ^{35}Cl NQR signal. A development of the method was proposed

* E-mail: mgeorge@yandex.ru

by Weiss and Weiden.[4] The second approach proposed as the development of Pound's idea by Ainbinder and Grechishkin,[5] consists of the two-frequency excitation of the nuclear quadrupolar system with non-equidistant energy levels for nuclear spins I = 1, 2, 3, 5/2, 7/2, 9/2. In recent years, there has been a considerable renewal of interest in the methods of two- and three-frequency nuclear quadrupolar resonance;[6–10] this has been driven mainly by the possibility of developing further NQR detection methods for ^{14}N nitrogen nuclei with low resonance frequencies. The use of multifrequency NQR in quantum computations has also been proposed.[11,12]

Although nuclear two-frequency resonance was proposed by Ainbinder and Grechishkin even in 1966 (firstly for half-integer spins $I \geq 5/2$),[13] the application of this method was rather limited until the end of the twentieth century. This approach has been used mainly in two-frequency NQR for the identification of lines in complex NQR spectra. A first report of two-frequency methods for the NQR of nitrogen was devoted to the investigation of complex spectra of nitrogen-containing compounds with a small asymmetry parameter of the electric field gradient ($\eta \leq 0.2$).[14] Furthermore, using the density matrix method in the one-frequency approximation, different excitation sequences of two- and three-frequency NQR on ^{14}N nuclei have been also investigated.[15] It has been demonstrated that by excitation of the two-frequency echo the signals on the third non-irradiated transition can be observed. An experimental observation of two- and three-frequency NQR of nitrogen nuclei with a large electric field asymmetry parameter ($\eta \geq 0.2$) has been reported in reference[16].

It is well known that NQR can be effective for the detection of several types of explosives, such as 1,3,5-trinitroperhydro-1,3,5-triazine (cyclotrimethylene-trinitramine) $(CH_2–N–NO_2)_3$ (RDX), octahydro-1,3,5,7-tetranitro-1,3,5,7-tetrazocane $C_4H_8N_8O_8$ (HMX)[17] and trinitrotoluene $C_6H_2(NO_2)_3CH_3$ (TNT).[18] The presence of explosives is detected by the appearance of the NQR response of ^{14}N nitrogen nuclei. The NQR spectrum lines of RDX, HMX and TNT are between 0.6 and 5.3 MHz. However, the practical realization of NQR detection is a challenging task due to a very weak signal of the ^{14}N nuclei. Interference with external radiofrequency signals and internal noise of electronic equipment can mask the NQR signals or result in low detection rates.

The straightforward way to increase the NQR signal is to use multiple coherent averaging of NQR signals with multipulse sequences as well as digital methods of signal processing.[19] However, use of multiple averaging has limitations due to the need for a high rate of detection, while digital processing itself does not resolve the problem of the very low signal-to-noise ratio. Therefore, the application of more complicated techniques, including the double resonance approach, seems to be more promising.

As mentioned above, one of the possible techniques is double nuclear magnetic–nuclear quadrupolar resonance. Double NQR-NMR[20] provides a considerable gain in the detection sensitivity of weak NQR signals; however, the need for a strong and/or homogeneous magnetic field prevents the application of these methods outside laboratory conditions. Therefore the second approach, multifrequency NQR, turns out to be more practical although the gain in the signal-to-noise ratio (SNR) with this method is smaller than in double NMR-NQR.

As already mentioned, a renewal of interest in multifrequency NQR methods has been observed recently, in part related to the idea[21] of observing NQR signals on the third transition by applying two-frequency excitation, e.g. in the detection of explosives with resonance frequencies lower than 5 MHz. It has been suggested to use this scheme in low-frequency NQR as a possible way of overcoming the issues of probe ringing and spurious signals due to piezoelectric resonances. However, this NQR technique has some complications for practical realization due to the need for a complex system of three-frequency coils, special control system and complicated analysis of three-frequency signals. One has also to take into account the significant drift of NQR frequency with sample temperature. On the other hand, it is possible to realize the two-frequency NQR technique with recording of the secondary NQR signal *at one of the irradiated transitions*. Theoretically this technique is similar to the three-frequency NQR scheme mentioned above, while its practical realization is simpler.

It should be mentioned that the potential advantage of the application of two-frequency NQR is not limited to the increase in SNR but is also related to the possibility of improving the reliability of NQR detection of various explosives. In pulse NQR it is very important to recognize a resonance response of an explosive against the background of various RF noise and spurious signals, piezoelectric ringing from surrounding materials,[22] as well as the effects of interference of induction and echo signals.[23] The parallel recording of various NQR lines of ^{14}N increases the detection reliability by reducing false alarms. Moreover, two-frequency excitation with the use of multipulse sequences provides additional opportunities for signal recording.

In this work, we study various aspects of the two-frequency NQR technique for explosive detection. Our studies of the nuclear quadrupole system of ^{14}N nuclei with spin $I = 1$ show it is possible to enhance the signal intensity for the transition with the smaller frequency. We also demonstrate that the proposed two-frequency sequences show promise for use in the NQR detection of various compounds.

2. Theoretical model

The Hamiltonian for NQR energy levels of nuclear spin $I = 1$ is written as follows.[24]

$$\mathcal{H}_Q = \frac{e^2Qq}{4}[3I_z^2 - I^2 + \frac{\eta}{2}(I_+^2 + I_-^2)] \tag{1}$$

where $I_\pm = I_x \pm iI_y$, and I_x, I_y, I_z are the components of the nuclear spin operator in the principal axes system of the EFG tensor, η the asymmetry parameter of the EFG, q the EFG, and Q the quadrupole moment of the ^{14}N nucleus. Transition frequencies determined by the Hamiltonian are:

$$\omega_\pm = \frac{E_\pm - E_o}{\hbar} = \frac{eQq_{zz}}{4\hbar}(3 \pm \eta) \tag{2}$$

$$\omega_0 = \frac{E_+ - E_-}{\hbar} = \frac{eQq_{zz}}{2\hbar}\eta$$

where $\omega_+ = 2\pi\nu_+$ corresponds to the transition between the levels labelled as + and 0 (+↔0), $\omega_- = 2\pi\nu_-$ – to the transition (−↔0), $\omega_0 = 2\pi\nu_0$ – to the transition (+↔−), and $E_\pm = \frac{eQq_{zz}}{4}(1 \pm \eta)$, $E_0 = -\frac{eQq_{zz}}{2}$ are the eigenvalues of H_Q. Figure 1 shows the NQR energy level splitting.

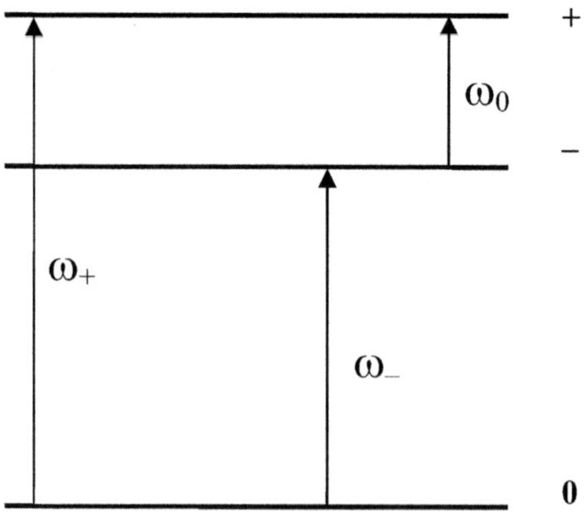

Figure 1. ^{14}N NQR energy levels.

Usually the calculations for multifrequency excitation in NQR are carried out for two cases:

The excitation length is close to the relaxation parameters of the NQR system. In this case we can use the rate equations for energy level populations,[24]

The duration of radiofrequency pulses and the period of the observation are smaller than the relaxation parameters of the NQR system. In this case it is possible use the solution for the density matrix without relaxation terms.[25]

2.1. THEORETICAL ANALYSIS WITH RATE EQUATIONS

The changes in intensity of NQR signals are conveniently analysed by means of population equations for energy levels N_m (where $m = 0, -1, +1$ are the values of the magnetic quantum numbers):

$$\frac{dN_m}{dt} = \sum_n (W_{mn} \cdot N_n - W_{nm} \cdot N_m) \tag{3}$$

where W_{mn} and W_{nm} are the probabilities of the direct and reverse transition.

The rate equations are applied to the case when a long saturating pulse with length τ ($\tau << \frac{1}{W_{nm}}, \frac{1}{W_{mn}}$) acts on one of the transitions, while the other transitions reveal a change in their signal intensities (in response to an exciting pulse with length $\Delta t << T_2$, where T_2 is the transverse spin relaxation time). The change in intensity may be measured by the change in the population differences as compared to the equilibrium populations:

$$\frac{A_i^s}{A_i^e} = \frac{n_i^s}{n_i^e} \tag{4}$$

where A_i^s, A_i^e are the amplitudes of the NQR signals on the i-transition ($i = 0, +, -$) with saturation on the adjacent transition and without it, respectively, n_i^s, n_i^e are the population differences with saturation on the adjacent transition and without it, respectively.

In RDX there is a transition with a frequency $\nu_- = 3{,}458$ kHz (the transition m = 0→m = −1). In the case of saturation at the frequency of $\nu_0 = 1{,}782$ kHz (the transition $m = -1 \rightarrow m = +1$) the signal intensity on transition ν_- increases by a factor of 1.2 in accordance with the expression given in reference[16]:

$$\frac{n_-^s}{n_-^e} = \frac{3}{3-\eta} \tag{5}$$

It is the ν_- transition that is usually used in RDX detection, since it reveals a small shift of the resonance frequency with temperature. The small temperature shift is an essential factor in searching for explosives.[22] A similar situation is realized in the NQR of HMX. Another possible advantage of two-frequency excitation is the use of the ν_0 transition for RDX and HMX detection. It is well known that the intensity of the NQR signal decreases with the transition frequency as a function of ν^2. However, in the case of mine detection, the influence of radiofrequency interference is largest just in the frequency range of ν_- and ν_+.[26] Therefore, observation in another frequency range, even at lower frequencies, provides an advantage in the signal/noise ratio. On the other hand, observation of the frequency ν_0 signals while saturating the frequency ν_- provides the following gain in the signal intensity[16]:

$$\frac{n_0^s}{n_0^e} = \frac{3}{4} \cdot \frac{(1+\eta)}{\eta} \tag{6}$$

Thus, the smaller the asymmetry parameter η, the higher the signal gain. However, the application of such a scheme requires that a number of conditions be satisfied. To solve equation (3) it has been assumed that the full saturation of a transition (resulting in the spin temperature $T_s = \infty$) is the result of the long radiofrequency pulse applied to the corresponding transition. The process of equalization of the populations depends on the relaxation constants and the strength of the radiofrequency magnetic field. Therefore, to estimate the applicability of equations (4) and (5) one has to calculate the spin temperature T_s of the saturated transition that can be reached under the real conditions of remote detection. The following expression may be used with a sufficient degree of accuracy[27]:

$$T_s = (1 + \gamma^2 \cdot B_1^2 \cdot T_1 \cdot T_2) \cdot T \tag{7}$$

where γ is the gyromagnetic ratio, B_1 the induction of the radiofrequency magnetic field, T_1 the longitudinal (spin-lattice) relaxation time, T_2 the transverse (spin-spin) relaxation time, and T the thermodynamic temperature of a sample.

Estimations of spin temperature show that it is harder to realize the two-frequency saturation experiments for RDX than for HMX and particularly TNT. The room-temperature relaxation parameters of RDX for the transition frequency ν_+ = 5,193 kHz are T_1 = 0.01 s, T_2 = 0.008 s,[28] while the relaxation parameters of TNT are $T_1 \sim$ 3–4 s, $T_2 \sim 10^{-3}$ s.[19] Therefore, for the same value of the radiofrequency field B_1 the spin temperature T_s of the saturated transition in TNT is two orders of magnitude higher than in RDX.

In remote experiments it is hardly possible to compensate this difference by increasing the amplitude of B_1.

However, two-frequency experiments on RDX allow to observe the so-called "capture effect", that is the effect of reducing the intensity of the spin-echo signal on frequency ν_+ under the influence of pulse excitation on a nearby transition with frequency ν_-.[29] This phenomenon is related to spin coherency break under the influence on adjacent transition.

TABLE 1. Dependence of transition signal on saturation frequency.[16]

Saturation frequency	$\frac{n_-^s}{n_-^e}$	$\frac{n_0^s}{n_0^e}$	$\frac{n_+^s}{n_+^e}$
ν_+	$\frac{2}{3}\cdot\frac{(1-\eta)}{(3-\eta)}$	$-\frac{3}{4}\cdot\frac{(1-\eta)}{\eta}$	
ν_-		$\frac{3}{4}\cdot\frac{(1+\eta)}{\eta}$	$\frac{2}{3}\cdot\frac{(1+\eta)}{(3+\eta)}$
ν_0	$\frac{3}{(3-\eta)}$		$\frac{3}{(3+\eta)}$

2.2. THEORETICAL ANALYSIS FOR MATRIX DENSITY

In this approach the pulse duration is considered to be much shorter than the relaxation time of the quadrupole system. Therefore, one can use the density matrix solutions, neglecting the relaxation terms. There are a number of theoretical investigations of the two-frequency signal on the base of the matrix density approach.[5,7,13–15,20,21] In the discussion below we provide a formulation that is adequate to analyse our experiments. We also assume the duration of the excitation pulses in the sequence to be negligibly small compared to the time intervals between the pulses. The time evolution of the signals is calculated using the density matrix approach in the resonance approximation.[1,27,28]

$$i\hbar\frac{\partial\rho}{\partial t}=[\mathcal{H},\rho]=\mathcal{H}\rho-\rho\mathcal{H} \tag{3}$$

where ρ is the density matrix, $i=\sqrt{-1}$, and $\mathcal{H}$ is the Hamiltonian of the NQR system:

$$\mathcal{H} = \mathcal{H}_{1x} + \mathcal{H}_{1y} + \mathcal{H}_Q \tag{4}$$

$$\mathcal{H}_{1x} = \gamma\hbar I_x H_x \sin\theta_x \cos\varphi_x \cos\omega_+ t$$

$$\mathcal{H}_{1y} = \gamma\hbar I_y H_y \sin\theta_y \sin\varphi_y \cos\omega_- t$$

$$\mathcal{H}_{1z} = \gamma\hbar I_z H_z \cos\theta_z \cos\omega_0 t$$

where γ is the gyromagnetic ratio; H_x, H_y and H_z are the components of the radiofrequency magnetic field; $\theta_x, \varphi_x, \theta_y, \varphi_y, \theta_z$ are the Euler angles to describe the axes of various coils of the multifrequency detector coils with respect to the principal axes of EFG. In the general case of the three-frequency NQR probe, there are three mutually orthogonal coils. Each of them is effective on its own frequency: the first one is on frequency ν_+ (I_x spin operator), the second operates on frequency ν_- (I_y) and the third on frequency ν_0 (I_z).

The solution of equation (3) is given as:

$$\rho(t) = \hat{U}(t)\rho(0)\hat{U}(t)^{-1} \tag{5}$$

where $\rho(0)$ is the initial (equilibrium) value of the density matrix operator, $\hat{U}(t)$ is the evolution operator.

In the equilibrium state the density operator is as follows:

$$\rho(0) = \frac{N}{2I+1}\exp\left(-\frac{\mathcal{H}_Q}{kT}\right) \approx \frac{N}{2I+1}\left(1 - \frac{\mathcal{H}_Q}{kT}\right) \approx \frac{N}{2I+1}\cdot\frac{\mathcal{H}_Q}{kT} \tag{6}$$

where N is the number of quadrupolar nuclei in the sample, T is the temperature of the sample.

The evolution operator $\hat{U}(t)$ has the following form:

$$\hat{U}(t) = D_n R_n \cdots D_2 R_2 D_1 R_1 \tag{7}$$

where R_n is the solution of the Shrödinger equation under the action of the nth excitation pulse, which means for the rotational operator of the nuclear magnetization during the RF pulse:

$$i\hbar\frac{dR_n}{dt} = \hat{\mathcal{H}}(t) R_n \tag{8}$$

while D_n is the solution of equation (8) for the time period between pulses n and (n + 1), i.e. under the effect of $\mathcal{H}_Q$ only (i.e. $\mathcal{H}_1 = 0$).

As an example of the rotational operators for different excitation pulses the following solutions are shown:

A. For the radiofrequency pulse on frequency ω_+:

$$R_x(t_{wx}) = \begin{pmatrix} \cos\alpha & 0 & i\sin\alpha \\ 0 & 1 & 0 \\ i\sin\alpha & 0 & \cos\alpha \end{pmatrix} \tag{8}$$

where $\alpha = \gamma H_x \sin\Theta \cos\varphi\, t_{wx}$

B. For the radiofrequency pulse on frequency ω_-:

$$R_y(t_{wy}) = \begin{pmatrix} 1 & 0 & 0 \\ 0 & \cos\beta & \sin\beta \\ 0 & -\sin\beta & \cos\beta \end{pmatrix} \tag{9}$$

where $\beta = \gamma H_y \sin\Theta \sin\varphi\, t_{wy}$

C. For the radiofrequency pulse on frequency ω_0:

$$R_z(t_{wz}) = \begin{pmatrix} \cos\gamma & -\sin\gamma & 0 \\ \sin\gamma & \cos\gamma & 0 \\ 0 & 0 & 1 \end{pmatrix} \tag{10}$$

where $\gamma = \gamma H_z \cos\Theta\, t_{wz}$

D. For the double-frequency radiofrequency pulse on frequencies ω_+, ω_-:

$$R_{xy}(t_{wxy}) = \begin{pmatrix} \frac{(\alpha^2\cos\sqrt{\alpha^2+\beta^2}+\beta^2)}{\alpha^2+\beta^2} & \frac{\alpha\beta(\cos\sqrt{\alpha^2+\beta^2}-1)}{\alpha^2+\beta^2} & \frac{i\alpha\sin\sqrt{\alpha^2+\beta^2}}{\sqrt{\alpha^2+\beta^2}} \\ -\frac{\alpha\beta(\cos\sqrt{\alpha^2+\beta^2}-1)}{\alpha^2+\beta^2} & \frac{(\beta^2\cos\sqrt{\alpha^2+\beta^2}+\alpha^2)}{\alpha^2+\beta^2} & -\frac{\beta\sin\sqrt{\alpha^2+\beta^2}}{\sqrt{\alpha^2+\beta^2}} \\ -\frac{i\alpha\sin\sqrt{\alpha^2+\beta^2}}{\sqrt{\alpha^2+\beta^2}} & \frac{\beta\sin\sqrt{\alpha^2+\beta^2}}{\sqrt{\alpha^2+\beta^2}} & \cos\sqrt{\alpha^2+\beta^2} \end{pmatrix} \tag{11}$$

E. For the double-frequency radiofrequency pulse on frequencies ω_0, ω_-:

$$R_{zy}(t_{wzy}) = \begin{pmatrix} \frac{\beta^2\cos\sqrt{\gamma^2+\beta^2}+\gamma^2}{\beta^2+\gamma^2} & \frac{-i\beta\sin\sqrt{\gamma^2+\beta^2}}{\sqrt{\gamma^2+\beta^2}} & \frac{-i\beta\gamma\cos\sqrt{\gamma^2+\beta^2}+i\beta\gamma}{\gamma^2+\beta^2} \\ \frac{-i\beta\sin\sqrt{\gamma^2+\beta^2}}{\sqrt{\gamma^2+\beta^2}} & \cos\sqrt{\gamma^2+\beta^2} & -\frac{\gamma\sin\sqrt{\gamma^2+\beta^2}}{\sqrt{\gamma^2+\beta^2}} \\ \frac{i\beta\gamma\cos\sqrt{\gamma^2+\beta^2}-i\beta\gamma}{\gamma^2+\beta^2} & \frac{\gamma\sin\sqrt{\gamma^2+\beta^2}}{\sqrt{\gamma^2+\beta^2}} & \frac{\gamma^2\cos\sqrt{\gamma^2+\beta^2}+\beta^2}{\beta^2+\gamma^2} \end{pmatrix} \quad (12)$$

2.3. TWO-FREQUENCY AND THREE-FREQUENCY SEQUENCES

In fact various multifrequency sequences can be applied. We show below only some of the possible sequences. For instance, in the sequence shown in Figure 2 the NQR signals on three frequencies are excited by two frequency pulses.

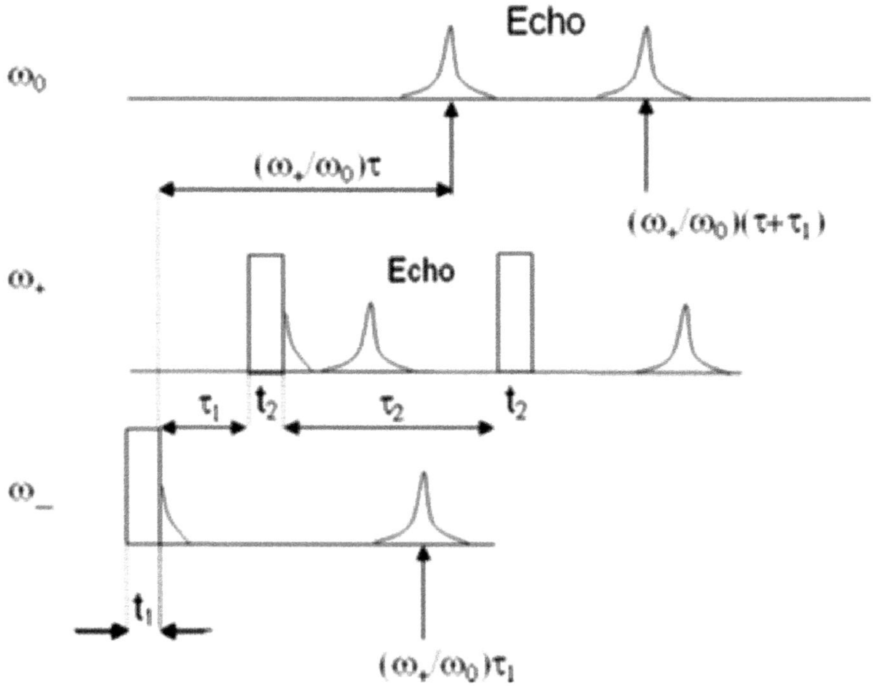

Figure 2. Three frequency signals excited by two frequency pulses.

In Figure 2 the signals after the third pulse are:

$$\langle I_x \rangle \propto \sin^2 \alpha (E_+ - D) \sin\alpha \cos\alpha \sin\omega_+ (t - 2\tau_2 - \tau_1) \tag{13}$$

where $D = E_o \cos^2 \beta + E_- \sin^2 \beta$, and $E_\pm$ have already been determined above (see below equation (2)).

$$\langle I_y \rangle \propto \omega_- \sin^2 \alpha \cos^2 \beta \sin\omega_- (t - \frac{\omega_+}{\omega_-}\tau_1) \tag{14}$$

$$\begin{aligned}\langle I_z \rangle &\propto \omega_- \sin\alpha \cos\alpha \sin\beta \cos\beta \sin\omega_0 (t - \frac{\omega_+}{\omega_0}\tau) + \\ &+ \omega_- \sin\alpha \cos\alpha \sin\beta \cos\beta \sin\omega_0 (t - \frac{\omega_+}{\omega_0}\tau - \frac{\omega_+}{\omega_0}\tau_1)\end{aligned} \tag{15}$$

An example of another sequence is shown in Figure 3:

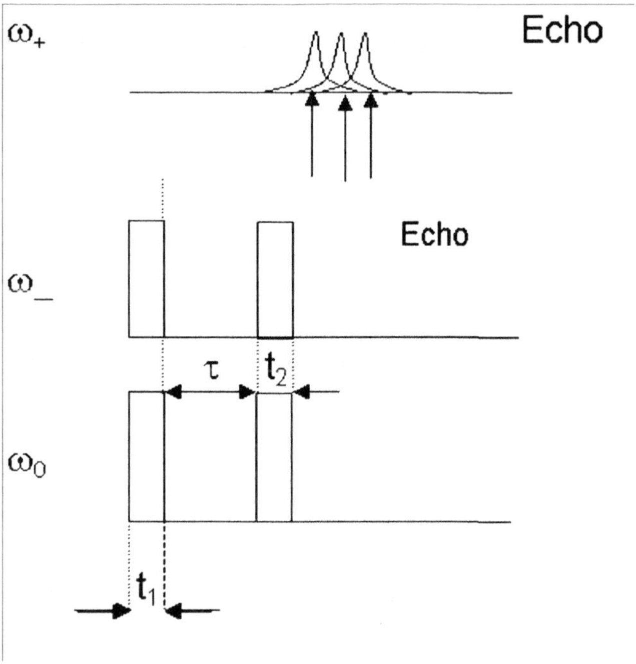

Figure 3. Sequence with two frequency pulses applied simultaneously. Only the signal on frequency ω_+ is presented.

The signal observed on frequency ω_+, taking into account also the relaxation parameters, is as follows:

$$\langle I_{x1}\rangle = \exp(-\frac{t-\tau}{T_2^+} - \frac{\tau}{T_2^0}) \cdot \sin\omega_+(t - \tau - \frac{\omega_0}{\omega_+}\tau); \quad (16)$$

$$\langle I_{x2}\rangle = \exp(-\frac{t-\tau}{T_2^+} - \frac{\tau}{T_2^-}) \cdot \sin\omega_+(t - \tau - \frac{\omega_-}{\omega_+}\tau); \quad (17)$$

$$\langle I_{x3}\rangle = \exp(-\frac{t}{T_2^+}) \cdot \sin\omega_+(t - 2\tau); \quad (18)$$

where T_2^+ is the transverse relaxation time on frequency ω_+, T_2^- is the transverse relaxation time on frequency ω_-, T_2^0 is the transverse relaxation time on frequency ω_0. The sequence shown in Figure 3 has been studied in reference[8]. It should be noted that in this work the authors observed one signals on frequency ω_+ instead the three neighbouring signals (see Figure 2).

Another example is the two-frequency sequence in Figure 4 where the signal on the transition with frequency ν_- is observed as the result of the excitation pulses on frequency ν_+.

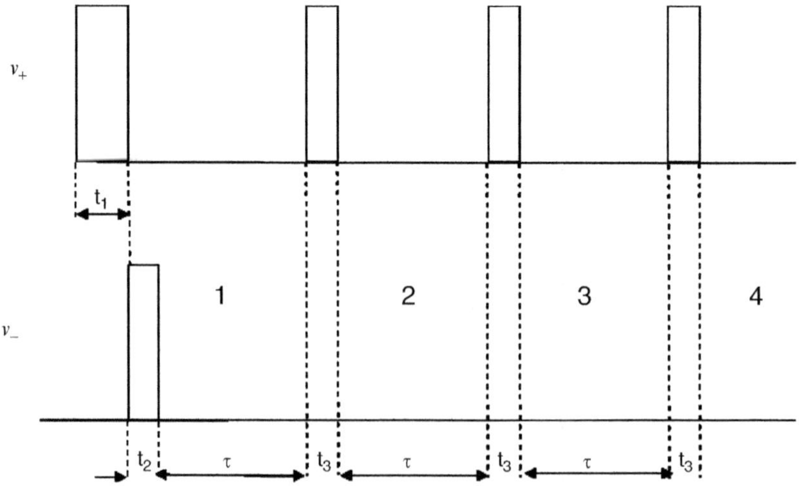

Figure 4. Two-frequency sequence with secondary NQR signals: sequence consists of the preparatory pulses on frequencies ν_+ and ν_-, followed by pulses on frequency ν_+. Secondary NQR signals are detected on ν_- in observation windows between the pulses on ν_+.

The solutions for the signals on ν_- for the consecutive observation windows (Figure 4) are the following.[30]

1. In the first observation window after irradiation by a pulse with frequency ν_+ (pulse duration is t_1) and a pulse with frequency ν_- (pulse duration is t_2), the signal detected on frequency ν_- is:

$$\langle I_y \rangle_1 \approx \frac{\hbar\omega_-}{kT} \sin(2\gamma H_y t_2 \sin\Theta_y \sin\varphi_y) \times \\ \times \left[1 - \frac{\omega_+}{\omega_-} \sin^2(\gamma H_x t_1 \sin\Theta_x \cos\varphi_x)\right] \sin\omega_-(t - t_1 - t_2), \tag{19}$$

where k is the Boltzmann constant, T is temperature. One could compare equation (5) with the solution for the free induction decay signal (i.e. after a single pulse on frequency ν_-) to see the appearance of an additional factor that decreases the signal magnitude in the first window:

$$\langle I_y \rangle \approx [1 - m \cdot \sin^2(\gamma H_x t_1 \sin\Theta_x \cos\varphi_x)] \langle I_{y0} \rangle, \tag{20}$$

where $m = \frac{\omega_+}{\omega_-}$.

2. In the second observation window after irradiation by a pulse with frequency ν_+ and duration t_3, the signal detected on frequency ν_- is:

$$\langle I_y \rangle_2 \approx -\frac{\hbar\omega_+}{kT} \sin(2\gamma H_x t_1 \sin\Theta_x \cos\varphi_x) \sin(\gamma H_y t_2 \sin\Theta_y \sin\varphi_y) \cdot \\ \sin(\gamma H_x t_3 \sin\Theta_x \cos\varphi_x) \sin\omega_-(t - m\tau - t_1 - t_2 - t_3) \tag{21}$$

3. In the third observation window, after the pulse repetition with the same parameters (ν_+ and t_3) the signal detected on frequency ν_- is:

$$\langle I_y \rangle_3 \approx -\frac{1}{2} \cdot \frac{\hbar\omega_+}{kT} \sin(2\gamma H_x t_1 \sin\Theta_x \cos\varphi_x) \sin(\gamma H_y t_2 \sin\Theta_y \sin\varphi_y) \cdot \\ \sin(2\gamma H_x t_3 \sin\Theta_x \cos\varphi_x) \sin\omega_-(t - 2m\tau - t_1 - t_2 - 2t_3) \tag{22}$$

4. In the fourth observation window, the signal is given as follows:

$$\langle I_y \rangle_4 \approx -\frac{1}{2} \cdot \frac{\hbar\omega_+}{kT} \sin(2\gamma H_x t_1 \sin\Theta_x \cos\varphi_x) \sin(\gamma H_y t_2 \sin\Theta_y \sin\varphi_y) \cdot \\ \sin(2\gamma H_x t_3 \sin\Theta_x \cos\varphi_x) \cos(\gamma H_x t_3 \sin\Theta_x \cos\varphi_x) \sin\omega_-(t - 3m\tau - t_1 - t_2 - 3t_3) \tag{23}$$

5. In the fifth observation window, after pulse repetition with the same parameters (ν_+ and t_3) the signal detected on frequency ν_- is:

$$\left\langle I_y \right\rangle_5 \approx -\frac{1}{2}\cdot\frac{\hbar\omega_+}{kT}\sin(2\gamma H_x t_1 \sin\Theta_x \cos\varphi_x)\sin(\gamma H_y t_2 \sin\Theta_y \sin\varphi_y)\times$$
$$\times\sin\left(2\gamma H_x t_3 \sin\Theta_x \cos\varphi_{xR}\right) \tag{24}$$
$$\times\cos^2\left(\gamma H_x t_3 \sin\Theta_x \cos\varphi_x\right)$$
$$\times\sin\omega_-\left(t-4m\tau-t_1-t_2-4t_3\right),$$

Thus the pulse sequence applied on frequency ν_+ with the repeat time τ results in the appearance of secondary echo signals after each excitation pulse in the time instants $n\cdot\frac{\omega_+}{\omega_-}\tau$, where the pulse number count n starts only after the action of the first two-frequency pair of pulses. We have to note that the effect of transverse relaxation will decrease the amplitude of signals as a function of n. This effect starts to be important only for higher n values and it has not been taken into account in our calculations.

2.4. TWO-FREQUENCY COMPOSITE PULSES

The action of two frequency pulses produces a signal on the third transition and changes the signal intensity of the other transitions. As one of the possible approaches in the multifrequency NQR experiments, we can consider the creation of such a pulse sequence that gives the maximum signal on the third non-irradiated transition. Interest in two/three-frequency NQR was initially motivated by the aim of excluding the influence of transient processes in the detection devices by recording the signal on the third non-irradiated transition, while the other two transitions are under the radiofrequency excitation.[7,8] However, for practical applications one has also to take into account the significant drift of the NQR frequency with sample temperature. It is obvious that it is very difficult to control all three NQR frequencies. The mutual influences of three resonance circuits create an additional problem in the experiment. Therefore, it is highly desirable to simplify such a system as much as possible while preserving its potential advantages. We analysed the action of different types of composite pulse R_{ij} (Figure 5) on the frequencies ν_i and ν_j, where i and j stands for x, y, z. The signs x, y, z are another way of noting +, –, 0, i.e. ω_x, ω_y and ω_z correspond to the frequencies ω_+, ω_- and ω_0.

There are six possible types of composite pulse R_{xy}, R_{yx}, R_{xz}, R_{zx}, R_{yz}, R_{zy}. The action of the composite pulse R_{yz} results in the signal of the free induction decay on frequency ω_- with coherence transfer from the transition with frequency ω_+:

$$R_{yz} = R_y R_z$$

$$\langle I_y \rangle \propto \sin 2\beta \cdot (\omega_+ \sin^2 \gamma + \omega_- \cos^2 \gamma) \sin \omega_- t \quad (25)$$

If one applies the radiofrequency pulse on frequency ω_0 with angle $\gamma = \pi/2$, it is possible to get the signal intensity in the receiver coil, which is proportional to $\omega_+\omega_-$ instead of ω_-^2 for the single ω_- frequency case.

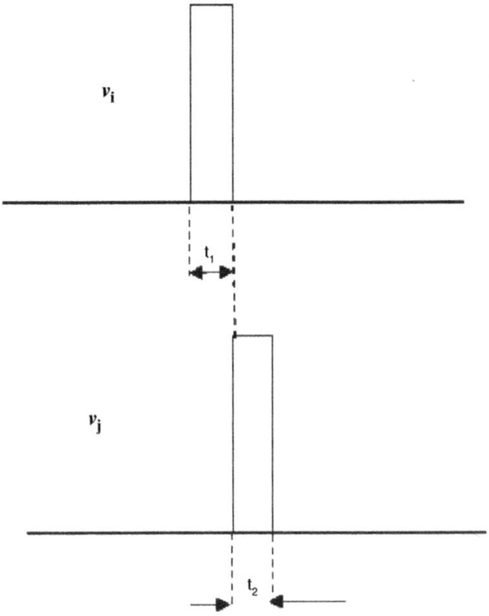

Figure 5. Two-frequency composite pulse R_{ij}.

3. Experimental procedures and samples

NQR experiments have been performed on a two-frequency pulse mode spectrometer with an operating frequency range of 0.5–18 MHz. The spectrometer consists of: (1) *Apollo Tecmag* NQR/NMR console (0.1–100 MHz) with two-channel transmitter and one-channel receiver modules; (2) two *Tomco BT-00500-Beta* power amplifiers with output power up to 500 W; (3) two-frequency detector unit. The detector unit includes a home-made low-noise single-channel preamplifier and a system of two orthogonal saddle coils, each of which is connected to a serial resonance circuit tuned to its own resonance frequency. The quality factors Q of the resonance circuits of the detector unit used in the NQR measurements of $NaNO_2$ are

74 for the frequency ν_- and 60 for ν_+, while the Q-factor of the detector used for RDX is about 100.

The RDX sample was in the form of about 400 g of a powder placed inside a cubic container with 100 mm edges. The room temperature NQR parameters of RDX are as follows. The asymmetry parameter $\eta = 60.24\%$, and the resonance frequencies are $\nu_+ = 5{,}048$ kHz (transition between the NQR energy states $0\rightarrow+1$) and $\nu_- = 3{,}359$ kHz ($0\rightarrow-1$). The relaxation parameters on frequency ν_+: longitudinal time $T_1 = 10$ ms, transverse time $T_2 = 8$ ms, and $T_2^* = 0.5$ ms (T_2^* is the relaxation time related to the linewidth;[1] $T_2^* = 1/\pi\Delta\nu_{1/2}$; $\Delta\nu_{1/2}$ is the full width at half height of the resonance line). The relaxation parameters on frequency ν_-: longitudinal time $T_1 = 10$ ms, transverse time $T_2 = 4$ ms, and $T_2^* = 0.4$ ms.

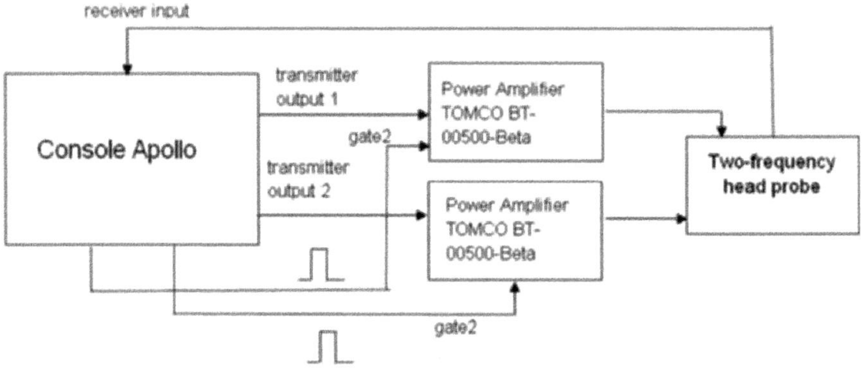

Figure 6. Block diagram of two-frequency NQR spectrometer.

The sodium nitrite sample consisted of 20 g of powder placed inside a Teflon ampule. The room temperature NQR parameters of this substance (see, e.g.[31]) are as follows. For the resonance frequency $\nu_+ = 4{,}640$ kHz the longitudinal relaxation time is $T_1 = 90$ ms, the transverse time $T_2 = 5.3$ ms, and $T_2^* = 2.0$ ms. For the resonance frequency $\nu_- = 3{,}601$ kHz the longitudinal relaxation time is $T_1 = 280$ ms, the transverse time $T_2 = 3.3$ ms, and $T_2^* = 2.6$ ms.

4. Two-frequency NQR experiments on ^{14}N nuclei

4.1. APPLICATION OF TWO-FREQUENCY NQR METHOD FOR SIGNAL CAPTURE

Firstly the two-frequency method was applied to assign definitely the adjacent NQR transitions. Application of multifrequency NQR to attribute the spectral lines is based on a change of the signal intensity for saturated adjacent transitions. The last phenomenon is sometimes called the "capture" effect in NQR.[29] There are two methods of realizing this effect. The first version described below consists in applying the saturating pulse to change the initial energy level populations of the adjacent transitions.[16] The results of these experiments on RDX at room temperature have been considered in reference[29]. The first method can be used to increase the detection reliability of RDX, which is a common explosive substance. For example, at room temperature one of the resonance frequencies of RDX (3.359 MHz) is very close to that of methylurea $C_3H_8N_2O$ (3.3503 MHz), which may cause mistakes in the detection of RDX. Moreover, it is known[25] that for some complexes the formation of crystal hydrates significantly decreases the resonance frequency (e.g. for the crystal hydrates of hexamethylenetetramine, the decrease is 3–8.5%).[32] This shows that it is possible to find compounds with resonance frequencies that mask those of explosives. Therefore, multifrequency NQR may be usefully applied to increase the reliability of identification of a given compound.

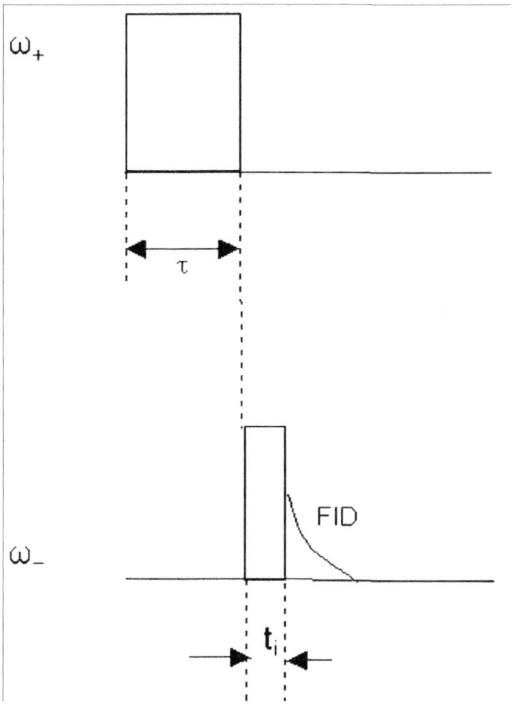

Figure 7. Sequence used to study the "capture" effect.

Figure 7 shows the sequence applied, while Figure 8 shows the amplitude variations of the NQR signal under two-frequency excitation. We observed the attenuation of the signal by nearly 40% of the single-frequency signal by increasing the transmitted pulse power. With further increase in the transmitter power, the signal remained unchanged. (It should be noted that, depending on the transition saturated, the signal intensity on the detected frequency could be increased as well.) However, according to the results of theoretical calculations (see Table 1),[16] the signal decrease is expected to be as high as 89%. The difference may be attributed to the incomplete saturation of the irradiated transition ω_+ due to the very small relaxation parameters of RDX (see equation (7)). Figure 9 shows the dependence of the signal magnitude at ω_- on the saturated frequency ω_+. Figure 10 shows the dependence of the relative signal amplitude on the saturation pulse duration.

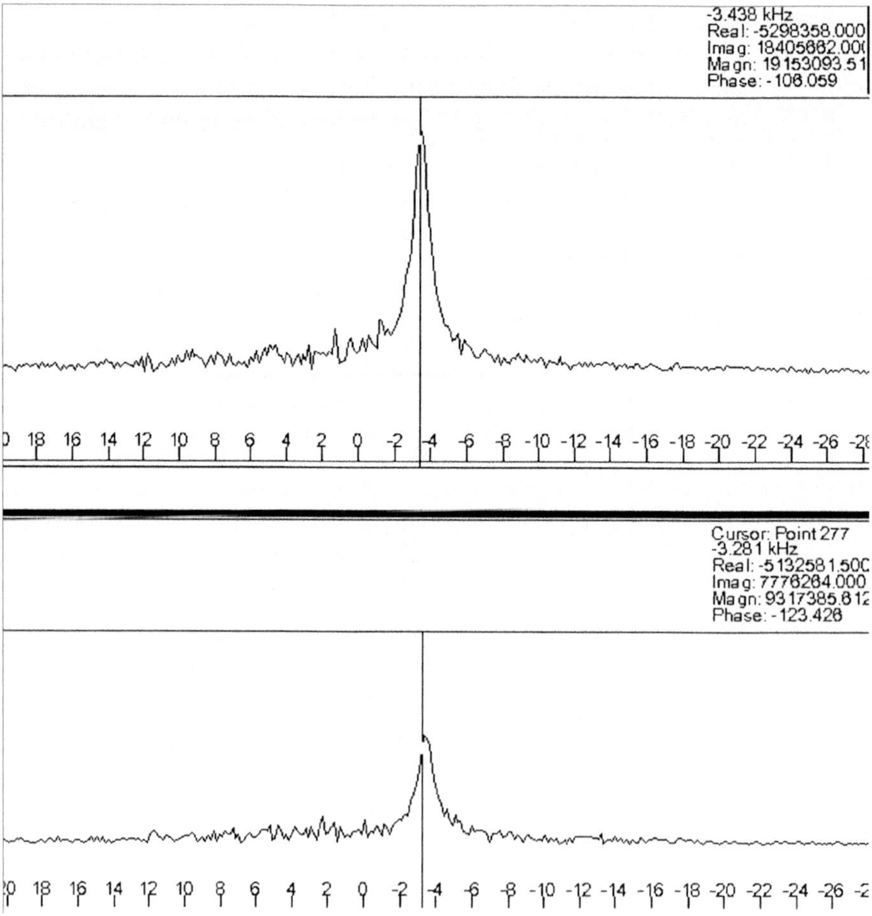

Figure 8. Top – one-frequency signal on $\omega_- = 3{,}362$ kHz; bottom – two-frequency signal on $\omega_- = 3{,}362$ kHz, while the 400 μs saturation pulse is applied at $\omega_+ = 5{,}049$ kHz.

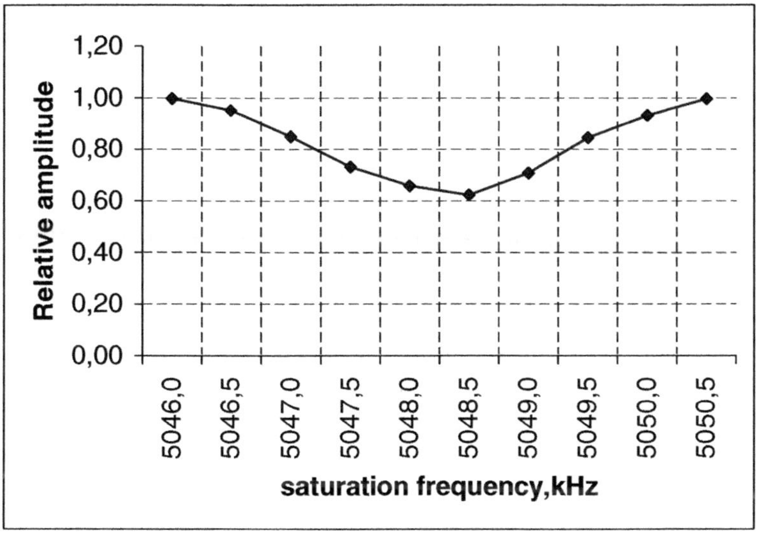

Figure 9. Signal magnitude on ω_{-} = 3,358 kHz as a function of ω_{+} frequency of saturation pulse.

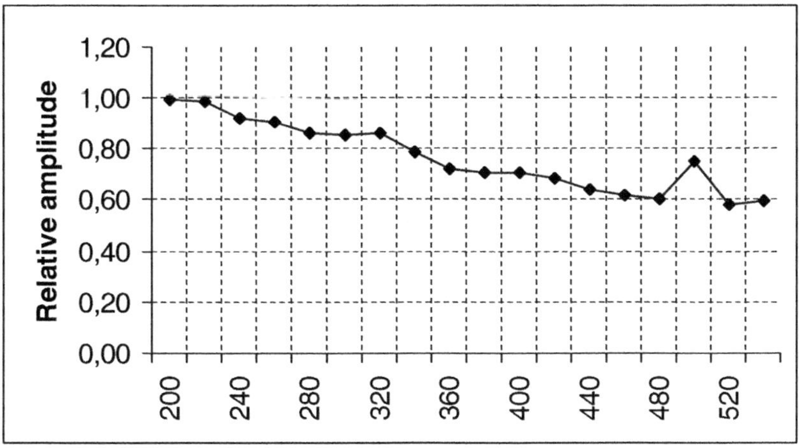

Figure 10. Dependence of NQR signal on frequency ω_{-} = 3,359 kHz on saturation pulse duration (in microseconds) on ω_{+} = 5,048.5 kHz.

Thus, the effect demonstrated above allows one to identify unambiguously the NQR lines in multi-component mixtures. This approach may be also used for the signal amplification on the detected transition.

4.2. TWO-FREQUENCY EXPERIMENTS WITH MULTIPULSE SEQUENCE

An assignment of the various lines in the NQR spectra of mixtures of different chemical compounds is one of the possible applications of the "capture" method. On the other hand, the NQR detection of chemical compounds is mainly based on the use of multipulse sequences. Assignments can also be made by means of two-frequency sequences. Figure 11 shows one of such two-frequency sequences that can be used for data verification. This sequence consists of two stages.[7] In the first stage one records the one-frequency signal using the spin locking spin echo sequence (SLSE) with averaging in n-windows between pulses (in Figure 11 it is the sequence on frequency ω_+). In the second stage one acquires the ω_+ signal applying an additional pulse on frequency ω_-. As a result, an important decrease in the signal echo intensity is observed (Figure 12).

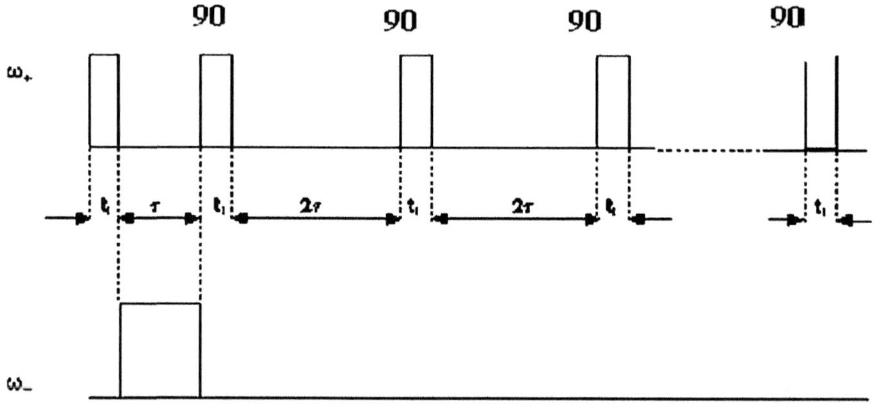

Figure 11. Two-frequency multipulse sequence $t_i^{\pm 90^\circ} - (\tau - t_i - \tau)^n$ that can be used to verify NQR lines.

The analysis of the above results on the two-frequency multipulse experiment is as follows: (i) the echo observed in the SLSE sequence is the result of action of the preparatory pulses; (ii) the pulses $(\tau - t_i - \tau)^n$ maintain the echo response in the rotation frame; (iii) an additional pulse on the second transition frequency destroys the echo magnitude. Application of the two-frequency preparatory pulses produces new coherences, which differ from the one-frequency coherence. In the above experiment the application of the second frequency pulse destroys the detected echo signal. However, there are schemes where the two-frequency multipulse sequences allow also to "lock" the signal on the observed transition. In Figure 4 is an example of such a two-frequency sequence that is discussed below.

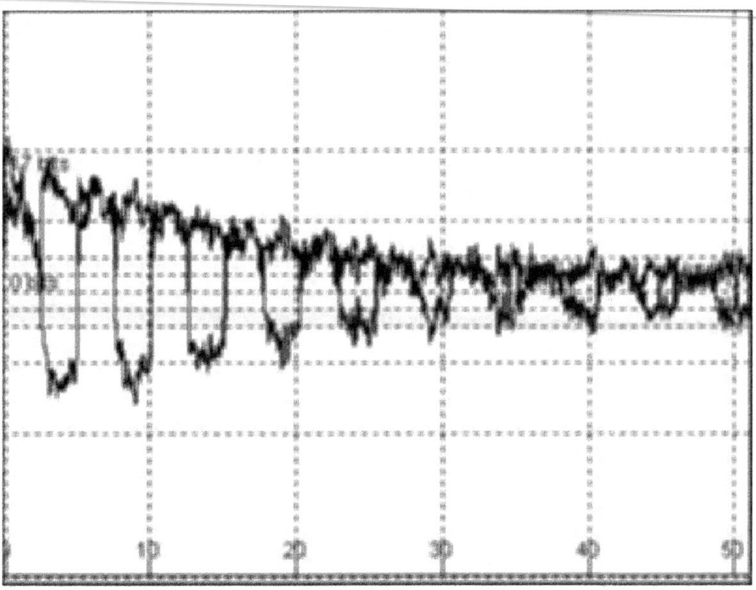

(a)

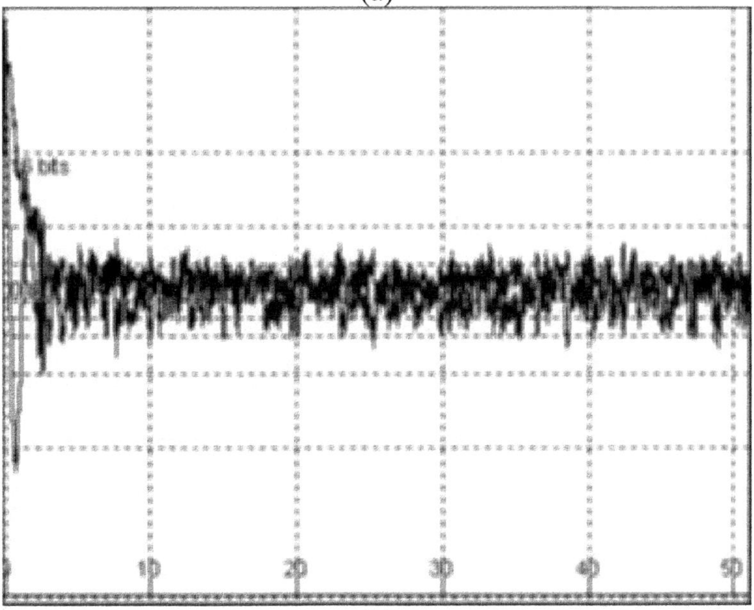

(b)

Figure 12. Experimental NQR signals recorded with SLSE sequence $t_i^{\pm 90^\circ} - (\tau - t_i - \tau)^n$ on frequency $\nu_+ = 4{,}633$ kHz for $NaNO_2$ at room temperature. Duration of observation windows is 2.56 ms, pulse length $t_i = 175$ μs, time between pulses $2\tau = 2.85$ ms. (a) One-frequency NQR signal; (b) same sequence with second-frequency pulse on $\nu_- = 3{,}590$ kHz (pulse length 300 μs). Tick labels on timescale are in milliseconds.

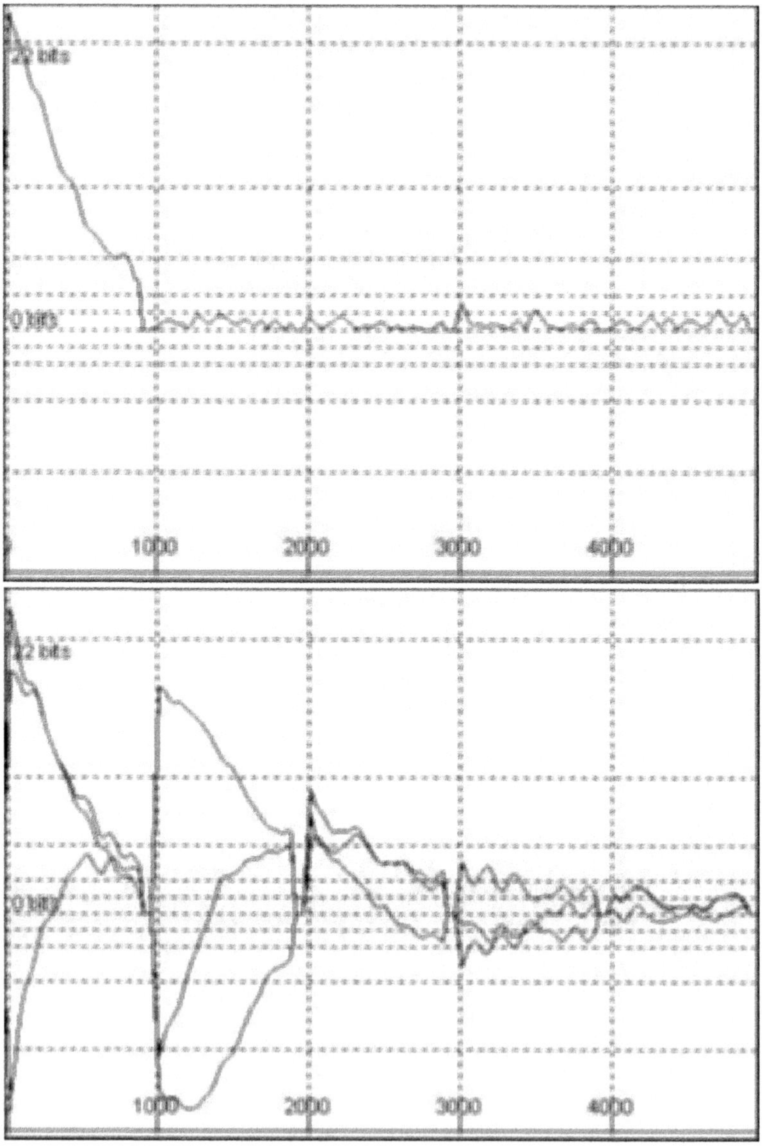

(a)

(b)

Figure 13. NQR signals in two-frequency multipulse sequence, recorded for RDX on transition ν_{-} = 3,359 kHz. First pulse duration is t_1 = 300 μs, length of all other pulses $t_2 = t_3$ = 250 μs, time interval between the pulses 1,100 μs, observation window time 1,000 μs. There are five observation windows; number of scans is 1,000. (a) Single-frequency FID signal on ν_{-} = 3,359 kHz (magnitude of the signal); (b) secondary echo signals (magnitudes, real and imaginary parts, of the signal are shown). Tick labels on the timescales are in microseconds.

It is remarkable that after the second exciting pulse the signal magnitudes (equations (19)–(24)) became proportional to the frequency of the highest frequency ω_+, but not to ω_- as in the usual single-frequency sequence.

For RDX the experimental results are presented in Figure 13.

It is clear from Figure 13a that the FID signal appearing in the first observation window is not observed in the next windows in the case of single-frequency excitation. However, in the case of two-frequency excitation, the secondary NQR signals appear in the next observation windows. The characteristic time of their damping is very close to the transverse relaxation time T_2 on frequency ν_- (after the fifth pulse no NQR signal is observed in the window). Using the NQR transition frequencies one can calculate the parameter m in equations (19)–(24). For RDX we have $m \approx 1.5$, while for $NaNO_2$, $m \approx 1.29$.[30] Therefore, it follows that the secondary echo signals are expected to appear not in the window just after the applied pulse but in the next windows. For instance, equation (21) for RDX shows that the maximal signal has to be in the center of the second observation window, while equation (22) results in a maximum NQR signal for the third observation window just before the next transmission pulse. In the fourth window the tail of the signal given by equation (22) and appearing at the end of previous window may be still observable, while the echo due to equation (23) evolves only in the next, fifth window. However, the signal observed has the form of an FID, which is not expected from our calculations. The reason for this discrepancy is obviously in the relaxation parameters of RDX as well as in the high m value. Therefore, our calculations (equations (19)–(24)) have to be corrected in order to be applicable to RDX. It is interesting to compare these results with NQR measurements on $NaNO_2$ (Figure 14).

It is obvious that our calculations fit better for the $NaNO_2$ signals presented in Figure 14b than for RDX. The signals observed in the second and third observation windows may be explained by equations (21) and (22). Furthermore, as follows from our calculations, the signal in the fourth window is practically absent, while there is a signal in the fifth window. Unfortunately the signal-to-noise ratio is not enough to trace the signal for the windows having higher numbers. But in any case, it is clear that there is better agreement of these NQR experiments with our model calculations, which are in fact expected to be more valid for $NaNO_2$ due to the relaxation parameters of this compound.[31]

Thus, the shape of the two-frequency multipulse signals for $NaNO_2$ and RDX reveals an important effect of the relaxation parameters (see data in Section 3). The secondary signals in RDX show an FID character which is worth studying further. Application of a pulse with frequency ν_- in the sequence studied results in an abrupt decrease in the magnitude of the RDX signals on ν_+ compared to the single-frequency sequence on the same transitions, that reflects polarization transfer to the adjacent transition with frequency ν_-.

(a)

(b)

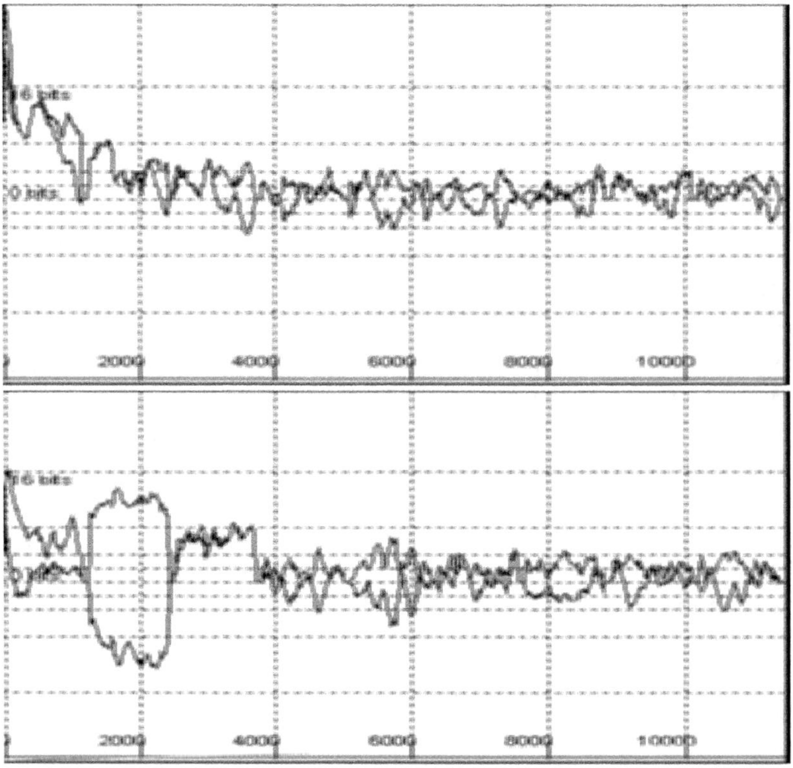

Figure 14. NQR signals in two-frequency multipulse sequence, recorded for $NaNO_2$ on the transition ν_- = 3,600 kHz. Pulse duration of all pulses is $t_1 = t_2 = t_3$ = 200 μs, time interval between pulses 1,680 μs, observation window time 1,280 μs. There are nine observation windows; number of scans is 100. (a) Single-frequency FID signal on ν_- = 3,600 kHz (magnitude and imaginary part of the signal); (b) secondary signals (magnitude and imaginary part of the signal). Tick labels on timescales are in microseconds.

5. Conclusions

We show that using the two-frequency excitation of energy levels of ^{14}N quadrupolar nuclei (I = 1) it is possible to control the signal response of the detected transition. One can use a long saturation pulse to increase or decrease the signal intensity or various combinations of two-frequency pulses to excite secondary NQR signals. The observed secondary NQR signals on one of the excited NQR transitions are similar to the secondary signals on the third non-irradiated transition. It is possible to excite such a two-frequency steady state in the system, which is locked by a continuous pulse series applied to one of the adjacent transitions and detected on another transition due to the secondary signals. In fact we observe a set of echo signals as a result of a "single" pulse on the observed transition. Thus, our studies reveal that sequences similar to that used in our work may have potential for the application of NQR to the detection of various nitrogen compounds, such as explosives, narcotics and illegal substances.

Acknowledgments This work has been supported by the grant for the NATO project CBP.NR.NRSFPP 892836. This work was also supported by the grant No. 106T321 of TÜBİTAK (The Scientific and Technological Research Council of Turkey) and in part by the grant No. 00062.STZ.2007-1 of Ministry of Industry and Trade of Turkey.

References

1. C.P. Slichter: Principles of Magnetic Resonance, Springer-Verlag, Berlin, Heidelberg, New York (1980).
2. R.V. Pound: Nuclear electric quadrupole interaction in crystal. Phys. Rev. 79, 685–705 (1950).
3. M. Emshwiller, E.L. Hahn, D. Kaplan: Pulsed nuclear resonance spectroscopy. Phys. Rev. 118, 414–424 (1960).
4. N. Weiden, A. Weiss: A double resonance NQR experiment in Al_2Br_6. J. Magn. Reson. 20, 334–340 (1975).
5. N.E. Ainbinder, V.S. Grechishkin: Two-frequency nuclear quadrupole spin echo. Izvestiya Vuzov, Radiofizika 10, 186–191 (1967).
6. K.L. Sauer, B.H. Suits, A.N. Garroway, J.B. Miller: Secondary echo in three-frequency nuclear quadrupole resonance of spin-1 nuclei. J. Chem. Phys. 118, 5071–5081 (2003).
7. G.V. Mozjoukhine: The two-frequency nuclear quadrupole resonance for explosives detection. Appl. Magn. Reson. 18, 527–535 (2000).
8. K.L. Sauer, B.H. Suits, A.N. Garroway, J.B. Miller: Three-frequency nuclear quadrupole resonance of spin-1 nuclei. Chem. Phys. Lett. 342, 362–368 (2001).
9. D.Ya. Osokin, R.R. Khusnutdinov, V.A. Shagalov: Two-frequency multiple-pulse sequences in nitrogen-14 NQR. Appl. Magn. Reson. 25, 513–521 (2004).
10. K.R. Thurber, K.L. Sauer, M.L. Buess, C.A. Klug, J.B. Miller: Increasing ^{14}N NQR signal by ^{1}H-^{14}N level crossing with small magnetic fields. J. Magn. Reson. 177, 118–128 (2005).
11. D.Ya. Osokin, R.R. Khusnutdinov: Theory of two-frequency excitation in ^{14}N NQR. Appl. Magn. Reson. 24, 145–156 (2003).
12. A.R. Kessel, V.L. Ermakov: Multiqubit spin: multilevels instead multiparticles. JETP 90, 452 (2000) [transl. from Zh. Eksp. Teor. Fiz. 117, 517 (2000)].
13. V.S. Grechishkin, R.E. Ainbinder: Two-frequency excitation of the quadrupole spin echo. JETP Lett. 5, 87–90 (1967).
14. E.O. Asisov, V.S. Grechishkin, U.M. Uspensky: Two frequency method in NQR spectroscopy of N-14 nuclei. J. Phys. Chem. (in Russian), LIII, 58–59 (1979).

15. V.S. Kirchanov, V.S. Grechishkin, B.V. Dresvankin: Two frequency NQR echo for spin I=1. J. Phys. Chem. (in Russian), LIII, 63–66 (1979).
16. G.V. Mozzhukhin, V.S. Grechishkin, N.Ya. Sinyavsky, T.V. Urepina: Two frequency saturation in pulse NQR of ^{14}N. Russ. Phys. J. 31 (8), 647–649 (1988) [transl. from Izv. Vyssh. Uchebn. Zaved., Fizika (in Russian), 31, 48–51 (1988)].
17. A.G. Landers, T.B. Brill, R.A. Marino: Electronic effects and molecular motion in .beta.-octahydro-1,3,5,7-tetranitro-1,3,5,7-tetrazocine based on nitrogen-14 nuclear quadrupole resonance spectroscopy J. Phys.Chem. 85, 2618–2623 (1981); R.J. Carpowicz, T.B. Brill: Librational motion of hexahydro-1,3,5-trinitro-s-triazine based on the temperature dependence of the nitrogen-14 nuclear quadrupole resonance spectra: the relationship to condensed-phase thermal decomposition J. Phys. Chem. 87, 2109–2112 (1983).
18. R.A. Marino, R.F. Connors: J. Mol. Struct. 111, 323–328 (1983).
19. V.S. Grechishkin, N.Ya. Sinyavsky: Local NQR. Uspehi Fiz. Nauk (in Russian) 163, 95–119 (1993).
20. V.S. Grechishkin, V.P. Anferov: In: J.A.S. Smith (ed.) Advances in NQR, vol. 4. Heyden, London (1980), pp. 71–113.
21. G.V. Mozjoukhine: The two-frequency nuclear quadrupole resonance for explosives detection. Appl. Magn. Reson. 18, 527–535 (2000).
22. S.W. Kercel, R.S. Burlage, D.R. Patek, C.M. Smith: Novel methods for detecting buried explosive devices. Proc. SPIE 3079, 467–477 (1997).
23. S.M. Klainer, T.V. Hirschfeld, R.A. Marino: Fourier Transform Nuclear Quadrupole Resonance Spectroscopy. In: Marshall A. G. (ed.) Fourier, Hadamard and Hilbert Transforms. Plenum Press, New York (1982), pp. 147–181.
24. C.P. Slichter: Principles of Magnetic resonance. Second ed., Springer-Verlag, Berlin, Heidelberg, New York (1980).
25. V.S. Grechishkin, V.P. Anferov, N.Ya. Sinyavsky: Adiabatic demagnetization and two-frequency methods in ^{14}N quadrupole resonance spectroscopy. In: J.A.S. Smith (ed.), Advances in Nuclear Quadrupole resonance, vol. 5. Wiley, London (1983), pp. 1–50.
26. V.S. Grechishkin, N.Ya. Sinyavskii, G.V. Mozzhukhin: A unilateral NQR explosives detector. Russ. Phys. J. 35, 61–64 (1992).
27. P.M. Borodin: Physical Basis of Quantum Electronics, Leningrad University, Leningrad (1985).
28. V.S. Grechishkin, G.V. Mozjoukhin, G.N. Chursin, N.M. Sitnikov: Science-Industry. Ministry of Defence Industry 3, (1987).
29. G.V. Mozzhukhin, A.V. Bodhya, V.V. Fedotov: Effect of ^{14}N nucleus capture in nuclear quadrupole resonance at small relaxation parameters. Russ. Phys. J. 48, 383–386 (2005) [transl. from Izv. Vyssh. Uchebn. Zaved., Fizika (in Russian), 48, 49–52 (2005)].
30. G.V. Mozzhukhin, B.Z. Rameev, N. Doğan, B. Aktaş: Secondary signals in two-frequency nuclear quadrupole resonance on ^{14}N nuclei with $I = 1$, J. Magn. Reson. 193, 49–53 (2008).
31. A.N. Garroway, M.L. Buess, J.P. Yesinowski, J.B. Miller: Narcotics and explosives detection by ^{14}N pure NQR. Proc. SPIE (Substance Detection Systems) 2092, 318–327 (1993).
32. B.V. Dresvyankin, V.S. Grechishkin, T.G. Balicheva, I.V. Pologikh: ^{14}N NQR in hexamethylenetetramine complexes with hydrogen bonding. Theor. Exp. Chem. 11, 231–234 (1976) [transl. from Teoreticheskaya i Éksperimental'naya Khimiya, 11, 276–280 (1975)].

THE DETECTION OF INDUSTRIAL EXPLOSIVES BY THE QUADRUPOLE RESONANCE METHOD: SOME ASPECTS OF THE DETECTION OF AMMONIUM NITRATE AND TRINITROTOLUENE

GEORGE V. MOZZHUKHIN [1*], SERGEY V. MOLCHANOV [1], GALINA S. KUPRIYANOVA [1], ALEXANDER V. BODNYA [1], VLADIMIR V. FEDOTOV [1], HAO GUOXIN [2], JIN YANBO [2], REN TIANLIANG [2], ZHANG GUOJIN [2]
[1] *Quanttex Ltd, Kaliningrad, Russia*
[2] *China Research Institute of Radiowave Propagation, Chingdao, China*

Abstract This paper is devoted to some aspects of the detection of Industrial explosives. The main components of these explosives are trinitrotoluene $C_6H_2(NO_2)_3CH_3$ (TNT) and ammonium nitrate NH_4NO_3 (AN). A study of the behaviour of the nuclear quadrupole resonance (NQR) signals of AN and TNT is reported. A method for attenuating outside interference signals under conditions of an industrial environment are proposed and discussed.

Keywords: industrial explosives, nuclear quadrupole resonance, TNT, AN, detection

1. Introduction

Explosives are probably the most popular tool used by terrorists against civilians. Nowadays there are several groups of explosives for industrial purposes. Explosive mixtures are grouped with respect to the major explosive material as ammonium nitrate-based NH_4NO_3 (AN), nitroglycerin-based $C_3H_5(NO_3)_3$ (NG), hydrazine-based N_2H_4, ammonium perchlorate-based NH_4ClO_4 (APC), and explosives based on processed military pyroxylin and gunpowder.[1] Also, in a number of countries 2,4,6-trinitrotoluene, $C_6H_2(NO_2)_3CH_3$ (TNT) is widely used, and there are considerable reserves of

* E-mail: mgeorge@yandex.ru

TNT in the world. At present, for example, in the USA, TNT is not used in industry due to its toxic explosion products. However, several industrial explosives based on AN are nothing more than mixtures including small quantities of TNT. Many substances among those enumerated above require special storage and transportation conditions. Nitroglycerin, for instance, is sensitive to detonation; explosives based on hydrazine are toxic, while APC is sensitive to rubbing and shocks. The special conditions of storage and transportation complicate the access to these substances for terrorists. Thus, the simplest and most popular industrial explosives for terrorists are those based on AN and TNT.

Methods of explosive detection are an important part of security policy. They include the x-ray method, nuclear (neutron and gamma), electromagnetic (nuclear magnetic resonance and nuclear quadrupole resonance), trace (biosensors, mass spectrometry, luminescence, etc.) methods.[2] One of the fastest developing methods is that of pulse nuclear quadrupolar resonance, which is among the most convenient. The major characteristic detected by nuclear quadrupolar resonance is the frequency spectrum. As the nuclear quadrupole moment in the material depends on the electronic environment, the NQR frequency spectrum is a unique fingerprint of the given chemical compound. At present the equipment can record the NQR spectrum of RDX ($C_3H_6N_6O_6$), PETN ($C_5H_8N_4O_{12}$) and HMX ($C_4H_8N_8O_8$).[3] As the detection of AN and TNT is performed in laboratories and, in the case of considerable quantities of explosive (600 g), in scanned luggage, the development of explosives detection methods is far from complete.

The China Research Institute of Radiowave Propagation (CRIRP) is working on the development of explosives detection in China, paying great attention to the detection of these AN and TNT based explosives, which are produced in China for industrial purposes, as they are the easiest for terrorists to access.

In the present work, we consider the particular qualities of explosives produced in China, and look into the methods for the distinction of signals from noise, which complicate explosives detection. These include the equipment and special methods of signals processing. The results of the present work were applied to equipment for the detection of explosives in luggage, developed at the CRIRP.

2. Detection of industrial TNT and AN by pure NQR

Nuclear Quadrupole Resonance (NQR) is based on the observation of phenomena of the absorption or emission of energy from the radio frequency magnetic field in solids containing quadrupolar nuclei. Asymmetrical nuclei have an electric quadrupole moment. These nuclei are under the influence of the electric field of the surrounding electrons. As a result of electrostatic interaction between valence electrons and the charge of the nucleus there exists a set of energy levels of the nuclear quadrupole interaction. The populations of these energy levels are defined by the Boltzmann distribution. Radiofrequency magnetic fields may excite transitions between these levels. Nitrogen ^{14}N (nuclear spin $I = 1$) is the most important quadrupolar nucleus and is one of the components of known explosives. The main method for the observation of NQR signals is the pulse resonance method. This includes the excitation of the substance by short (10–200 μs) radiofrequency magnetic pulses in the 400 kHz to 5.3 MHz range. The reference frequency of these pulses coincides with the transition frequency of the NQR energy levels. The response after the action of the pulses shows the NQR signal. The observation process is defined by the following parameters: T_1 – the spin-lattice relaxation time; T_2 – the spin-spin relaxation time, and T_2* – the decay parameter of the signal response after the radiofrequency field, $T_2^* = 1/\pi\Delta\nu_{1/2}$ where $\Delta\nu_{1/2}$ is the width of the line at half height. One of the advantages of the pulse NQR method is the use of the radiofrequency magnetic pulse and radio receiver to record the NQR signal. Thus, the structure of the device is analogous to a radio transmitter–receiver system.

Traditionally,[4,5] the most widely used explosive is 2,4,6-trinitrotoluene, $C_6H_2(NO_2)_3CH_3$ (TNT). Trinitrotoluene has a wide range of applications in industry, especially in the mining industry and for military purposes. Industrial TNT is accessible to potential terrorists. However, one can assume that the quality and parameters of military and industrial TNT may differ[6] depending on the manufacturer. The detection of TNT by different methods is an important problem for humanitarian mine-clearing and the struggle against terrorism. Now, there are two methods for the NQR detection of TNT. The first is the pure pulse NQR method[7]; the other is double nuclear magnetic-nuclear quadrupole resonance (double NMR-NQR).[8]

The use of the cross-level double NMR-NQR method for NQR detection of explosives was proposed by Blinc's group.[8] Three double-resonance approaches were developed: (i) detection of NMR signal changes by level crossing in the double NQR-NMR system; (ii) Zeeman polarization-enhanced NQR or the method of increasing ^{14}N NQR signals by ^{1}H–^{14}N level crossing after application of a preparatory static (inhomogeneous) magnetic field; (iii) triple resonance method or quadrupole-quadrupole enhanced NQR. For the application of double NMR–NQR a static magnetic field for proton polarization is necessary. For a field of 0.8 T the longitudinal relaxation time of protons is $T_1 \cong 30$ s.[9] The application of these methods in detection systems is not possible now, because the use of a static strong magnetic field decreases the feasibility, the time of detection and the mobility of the

NQR method. However, the application of this method using a low magnetic field (0–26 mT) is in progress.[10]

The problem of pure NQR for TNT detection is a small signal-to-noise ratio. It is a result of the large number of NQR frequencies in the frequency band up to 0.9 MHz, and the high relaxation parameters[5] (2.2–9.8 s). However, in the detection of explosives it is very important to use multi-pulse sequences. In this case one can detect signals in observation windows between the pulses. The normal time for this sequence is within T_1. The important parameter of this sequence, for instance the spin locking spin echo (SLSE), is $\boldsymbol{T_{opt}}$, the optimal scan time or duration of the SLSE pulse train.[7,11] This parameter strongly depends on the space between the pulses, SLSE. For instance, one can use 15 observation windows for signal acquisition with the following SLSE parameters: pulse space = 1.5 ms, $\boldsymbol{T_{opt}}$ = 30 ms at the TNT frequency of 842 kHz. The next measurement may be made after a delay of 10 s. Moreover, in pulse NQR there are problems related to the relatively long "dead time" of the receiver, after the action of the radiofrequency pulses, and the presence of spurious signals.[11,12]

The samples of Chinese industrial TNT used for the experiment consist of squamuliform TNT, used in open-cast mining. The substance looks like a light yellow scale. The hardening temperature is not less than 80.2°C. The mass fraction of moisture and volatile substances does not exceed 0.07%. Acidity recalculated as a sulfuric acid does not exceed 0.01%. The mass fraction of substances that are insoluble in organic solvents does not exceed 0.05%. The mass fraction of sodium sulfite is negligible. The weight of samples used ranges from 50 to 500 g.

Experimental investigations were carried out on a pulse NQR spectrometer based on the Tecmag "Apollo" console, operated in the frequency range 0.1–100 MHz. In the series resonance circuit of the probe-head we used a variable Jennings capacitor, 5–400 pF, 7.5 kV, 20 A, and a radiofrequency coil. We used a solenoid coil with a volume of 0.5 and 160 l. The probe-head of the spectrometer also includes a low-noise preamplifier and a system to reduce the dead-time of the preamplifier. The preamplifier has the following parameters: gain not less than 30 db; noise not more than 2.7 nV; recovery time approximately 400–500 μs. The variable impedance of the spectrometer is connected with the series resonance circuit. The spectrometer changes of the total Q factor of the circuit during the action of the radiofrequency pulses and the short time-period after pulses. The Q factor is near 140–200 at 400–900 kHz. We used the following sequences of radiofrequency pulses: phase-alternated pulse sequence (PAPS)[13]; multiple pulse spin locking spin echo (SLSE),[4] and combinations of these pulses to reduce the dead-time. RF pulses last between 50 and 300 μs. We used a homemade power amplifier with a large bandwidth (0.4–6.0 MHz) and output power 1.25 kW.

The results of our measurements of the TNT frequencies were already reported in reference.[14] TNT has two forms, orthorhombic and monoclinic, with different sets of frequencies. It is necessary to note that lines concerning the monoclinic form are more convenient for TNT detection by NQR than those for the orthorhombic form. The presence of the orthorhombic form in the sample is visible. However, the basic crystalline form for detection in Chinese industrial TNT is the monoclinic form. We believe that NQR lines at 842 kHz are more convenient for detection. These lines have high intensity and ability to compensate the temperature shift due to the presence of two close lines. The frequencies of these lines are 842 and 844 kHz at room temperature, and have temperature coefficients $\Delta\nu/\Delta T$ of +180 and 120 Hz/K, respectively. The relaxation parameters T_1 are 3.5 s for the 842 kHz line and 4.7 s for the 844 kHz line. Now, the weight of TNT to be detected in a baggage scanner is nearly 400 g. It is very important to detect a small quantity of TNT in a short time by pure NQR. Probably only some kind of double NQR-NMR technique[8] can give an NQR signal in an acceptable time. Figure 1 demonstrates the detection of 25 g of industrial TNT in a 0.1-l laboratory coil.

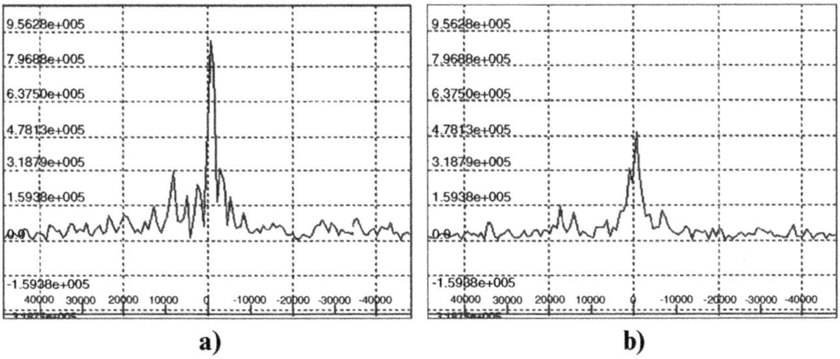

Figure 1. Spectral intensity curve for 842 kHz line obtained for the detection time of 13 min (a) and 6 min (b).

The other type of widely used industrial explosive is ammonium nitrate NH_4NO_3 (AN). AN contains nitrogen ^{14}N nuclei and has the form of colorless crystallites.[15] It has five crystalline forms, of which only three are relevant for explosives detection. The α form is stable below a temperature of −16.9°C (at −25°C the density is 1.71 g/cm^3). The β form is stable from −16.9°C to +32.2°C (at 25°C the density is 1.7125 g/cm^3) and the γ form is stable from +32.2°C to +84.2°C +32.2°C (at 40°C the density is 1.661 g/cm^3). However, the addition of potassium nitrite (about 10%) prevents the phase transitions. AN is a widely used component of explosives for industrial and military applications (ammonites, amatol, etc.). Large quantities of AN are used as fertilizers. NQR frequencies were first detected in reference[16]. However, the application of NQR for detection of AN is very suitable as the result of the favorable relaxation parameters. Moreover, it is possible to detect AN in cars.[17] We can use the following conditions. For instance, the most popular explosive, amatol, is a mixture of AN and TNT. The minimal ratio of AN to TNT is 20:80, the maximum 80:20. Another explosive is ammonal, which contains ammonium nitrate, trinitrotoluene, and aluminium powder mixed in a ratio of roughly 22:67:11. Thus, detection is possible if we use an AN or a TNT NQR frequency.

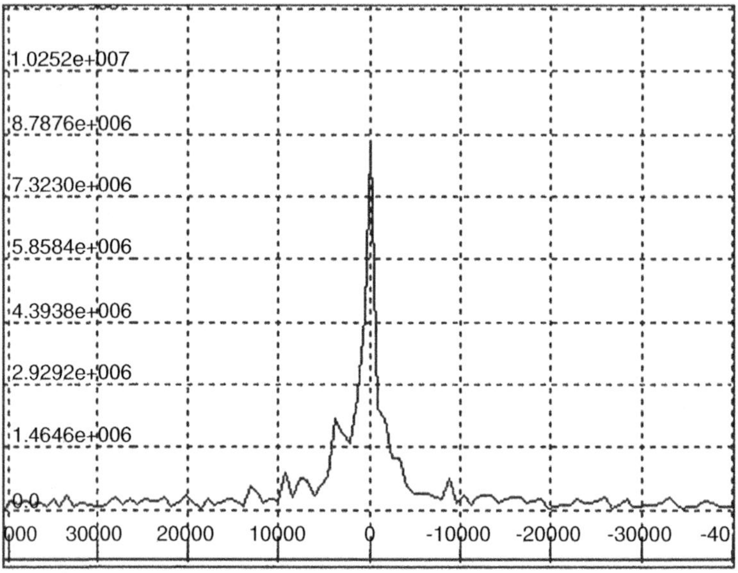

Figure 2. Spectral intensity curve for 424 kHz line obtained in 160 l coil for 125 g of AN.

The most suitable frequency for the detection of AN in ammonal is ν_{-} = 424 kHz at room temperature. This transition has a smallest temperature coefficient, $\Delta\nu/\Delta T$ = +100 Hz/K. For the other transition with a frequency of ν_{+} = 496 K at room temperature, the temperature coefficient is −300 Hz/K. The relaxation parameters T_1 are 13 and 16 s for NQR frequencies ν_{+} and ν_{-}, respectively. The relaxation parameters T_2 are 8 and 6 ms for frequencies ν_{l} and ν_{-}, respectively. However, for the detection of NQR by SLSE the optimal scan time $\boldsymbol{T_{opt}}$ is very important.[12] This parameter for AN is better than for TNT. For instance, $\boldsymbol{T_{opt}}$ is 3 s for the detection of 125 g of AN in the volume of 160 l with the following sequence parameters: pulse time 2.5 ms; number of pulses 1,000 (Figure 2). We display the signal with averages after 8,000 pulses. We used a small repeat time of 8 s between series. In this case we compared the decrease in intensity of the signal and the normal time of detection. In the figure we demonstrate the experimental dependence of the signal intensity on the repeat time for SLSE. Usually one expects to use a repeat time between series of at least $(3 \div 5)T_1$. However, the detection time is a very important parameter of real application. We show the dependence of the signal intensity on the repeat time in Figure 3.

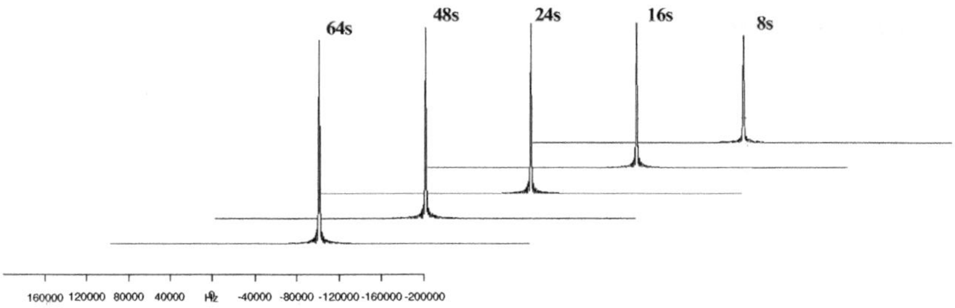

Figure 3. Spectral intensity curve for 424 kHz line obtained by SLSE sequence for different repeat parameters.

Thus, for the detection of explosives which include a mixture of AN and TNT, the detection of AN is the most convenient. The ratio of intensities of the signal of AN at a frequency of 424 KHz and that of TNT at 842 KHz is at least 2:1.

3. NQR Detection under conditions of strong interference signals

One of the important tasks of the recoding of low-frequency NQR signals is to increase the signal-to-noise ratio (SNR) on the background of external noise and interference signals. This task is very critical for remote detection of nitrogen compounds. In this device the receiver coil is partially shielded against spurious external signals. Under the conditions of industrial or other disturbing signals the identification of the small useful signal becomes more complicated. Very strong interfering signals create difficulties for the accumulation and averaging of the NQR signal for the improvement of the SNR.[18]

One of the conditions for wide application of NQR detectors for material detection is that the observation time should be of the order of a few seconds. However, the relaxation parameters of some nitrogen substances do not allow to reduce the detection time. The solution may be to use methods of adaptive two-channel filtering, digital processing of NQR signals, reduction of the noise signal. The application of these methods leads to improvements in data verification of the NQR method.

Distant sources of external signals have a very strong influence on the NQR signals in the case of detection without shielding. The frequency range of a radio station can coincide with the NQR signal frequencies. We display in Table 1 the distribution of frequencies used in Russia.

TABLE 1. Sources of radiofrequency interference signals (for Russia) during the detection of nitrogen compounds by NQR.

Assignment	Frequency, kHz
Radio air and marine navigation	85–526.5
Broadcast radio	526.5–1,606.5
Radiolocation, marine radio, amateur radio	1,606.5–2,173.5
Standard signals, marine radio, distress signals, air communications	2,173.5–3,400
Air communications, amateur radio, standard signals, mobile marine and land radio communications	3,400–5,050

It is probably possible to use two methods. These methods are active and passive attenuation of RFI signals. For instance, there is a gradiometer developed by the Naval Research Laboratory[19] which attenuates the RFI signal by 17–30 dB. The important limitation of this system is low sensitivity and reduction of the radiofrequency magnetic field far from the plane of the coil. The method of active attenuation proposed by Quantum Magnetic[18] includes a coil for NQR detection and an external aerial designed to record only external interference signals. Signals from the aerial pass via an additional receiver to the data system. Additional resources are needed for processing the two signals.

We also used the two-channel system to attenuate the interference signals. However, our two channels do not use independent inputs to the data system, and the subtractor is an analog unit for connection of the data system with the detection system (Figure 4). Channel 1 consists of the transmitter-receiver coil (main aerial), with partial shielding for detection of NQR, and preamplifier, receiver noise and interference signals. Channel 2 consists of the outside aerial with preamplifier, adaptive filter with feedback. The position of the aerial of the second channel excludes the mutual influence of the two aerials. The main task during the tuning of the system is to reach the maximum coincidence of the noise and interference signals in the two channels. Adding the adaptive filter unit with feedback in channel 2 improves the coincidence of the amplitude and phase of the interference signals in the two channels. The interference signal is attenuated by at least 20 dB.

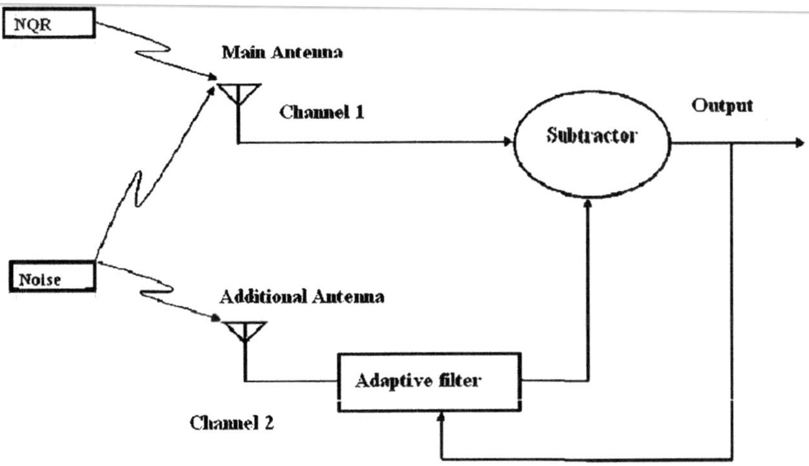

Figure 4. Two-channel system for attenuation of external interfering signals and noise.

We demonstrate the NQR signals of 200 g of TNT in the coil without shielding (Figure 6).

One of the conditions for the application of FFT and traditional filtering methods is a stationary signal for processing. Coherent signal averaging and FFT lead to an increase in SNR. However, the application of wavelet transforms[20] for data processing allows us to analyze the main parameters of noise and NQR signal components of the wavelet decomposition. The most used method of signal de-noising is the Cadsow de-noising algorithm (CDA), based on the decomposition of the time domain signal with the use of the wavelet base function.

We tested this method for TNT on the 0.8415 MHz frequency. Our method has the following steps: (1) record signals in channel 1 without NQR sample; (2) record signals in channel 2; (3) subtract signals from the two channels via subtractor unit; (4) tune in adaptive filter and gain, frequency range and phase to minimize the output signal; (5) excite NQR signals by multipulse sequence (SLSE or others); (6) record NQR signals with use of digital processing. The records of the "single shot" signals for points (1) and (3), (4) are shown in Figure 5.

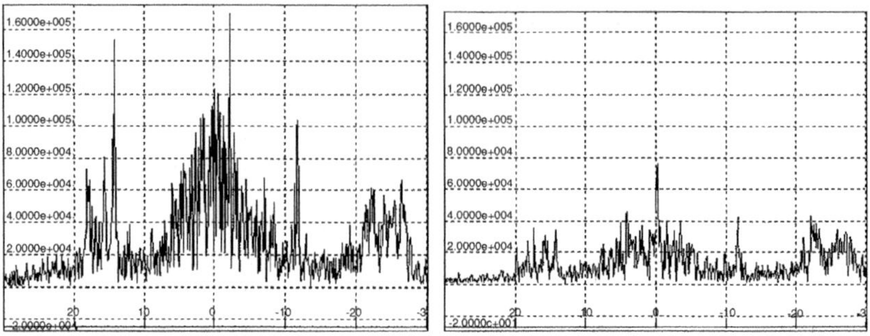

a) b)

Figure 5. 'Single shot" signals after Fast Fourier Transform (FFT) in two-channel system at 0.841 MHz: (a) noise and interference signal in channel 1; (b) noise and interference signal on output. Tick labels on the frequency scales are in kilohertz.

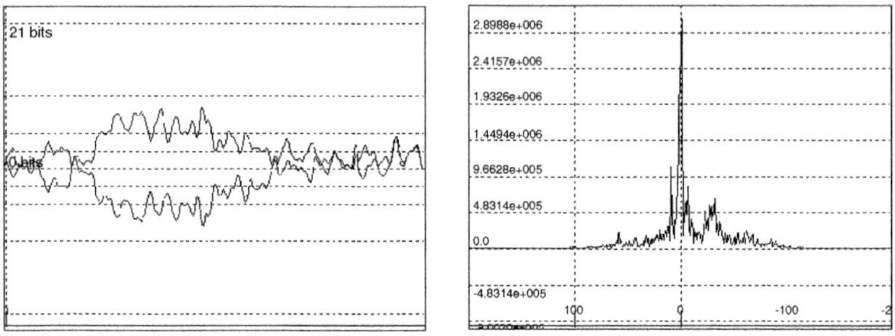

(a) (b)

Figure 6. NQR signal of 200 g TNT without shielding (1,800 scans, observation window 750 μs): (a) signal in time domain, magnitude and imaginary part (low curve) of the signal; (b) signal (magnitude) after Fast Fourier Transform (FFT). Tick labels on the frequency scales are in kilohertz.

A non-stationary signal is de-noised by wavelet analysis as follows: analysis of noise spectrum; the choice of function for thresholding; choice of wavelet basis; choice of decomposition levels. The signal *f(t)* may be described through linear decomposition as:

$$f(t) = \sum_{j,k} C_{jk} \psi_j(t) \tag{1}$$

where j – level and k – number of levels of decomposition, j and k are integer indexes, ψ_i – a set of time domain wavelet functions, C_{jk} – wavelet coefficients of the expansion *f(t)*. We use the additive model of the signal:

$$f(t) = \sum_{j,k} C_{jk}^{s} \psi_j(t) + \sum_{j,k} C_{jk}^{n} \psi_j(t) \tag{2}$$

where C_{jk}^{s}, C_{jk}^{n} are the signal and noise coefficients of the decomposition, respectively. It is necessary to find the operator for transformation of the coefficients of C_{jk}^{n} to zero. The thresholding operator T_{jk}[21] (hard and soft-thresholding) may be used for de-noising.

$$f_t(t) = \sum_{j,k} C_{j,k}^{s} \psi_{j,k}(t) + \sum_{j=j_0}^{n} \sum_{k=1}^{L_j} T_{j,k}\left(C_{j,k}^{n}\right) \psi_{j,k}(t) \qquad (3)$$

where $f_t(t)$ is the estimated function after thresholding, $j = j_0 \ldots j_n$, $k = l \ldots L_j$.

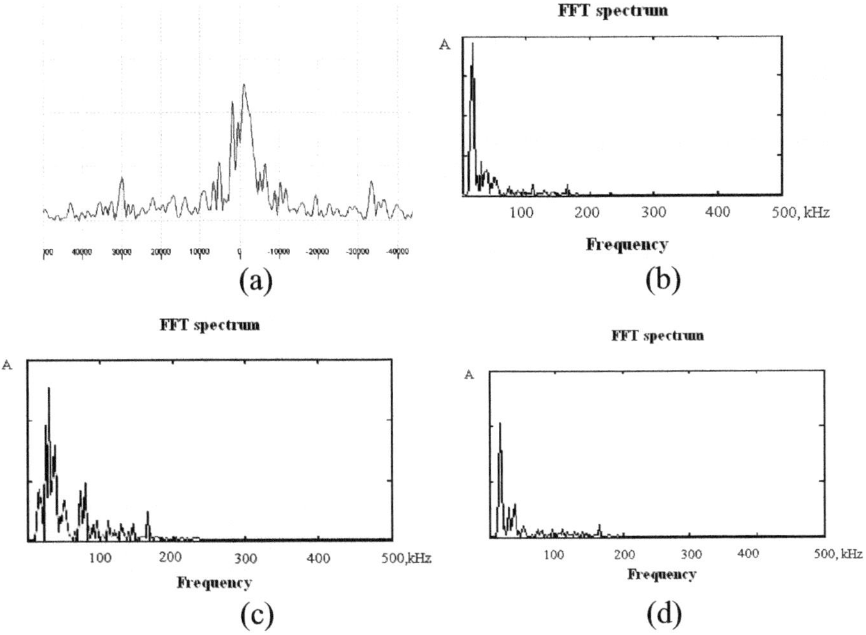

Figure 7. Spectra of 200 g of TNT at 842 kHz: (a) spectrum of NTNMR (Tecmag inc.); (b) spectrum in Matlab Wavelet ToolBox with use of haar wavelet; (c) spectrum in Matlab Wavelet ToolBox with use of sym-wavelet; (d) spectrum in Matlab Wavelet ToolBox with use of coif wavelet.

We used the Matlab Wavelet ToolBox for signal de-noising. We tune the different levels of decomposition to get the best SNR. The zero level of decomposition is the genuine signal. The levels of the decomposition are the wavelet tree. The accuracy of the imaginary part of the signal decreases at low levels of decomposition. However, there a ways for wavelet filtering of the signals and de-noising. For a one-dimensional signal one can use Wavelet Packet 1D or Continuous Wavelet 1D.

We show the results of the application of wavelet transform in Figure 7. The haar and coif wavelets produce the best results for de-noising. However, the presence of the stable interference signal of an external station produces an additional line in the spectrum which may be identified as the NQR signal. In this case the best choice is to subtract this interference signal by means of an independent channel, as was shown in Figure 4.

4. Scanner for explosive detection in passenger baggage created at CRIRP

The CRIRP research group has developed the pulse NQR method for an explosives detection scanner for aviation security in airports. The scanner works independently and allows to detect the following explosives in a volume of 160 l: RDX, HMX, Tetryl, AN, PETN and optionally TNT. The CRIRP scanner is shown in Figure 8.

Figure 8. The baggage scanner developed at CRIRP.

The main advantages of this device are the following:

1. High speed of operation.
2. Automatic detection of explosives.
3. Low false alarm.
4. Scanner may be used in integrated system after some development.

5. Conclusion

NQR technology allows to detect explosives directly via NQR spectra. Industrial explosives have different components, which can decrease the quality of explosives. The detection of most widely used explosives is very important for public security. However, the main problem of this method is the low signal-to-noise ratio under real conditions of an industrial environment. The subtraction of the industrial noise signal from the normal signal-to-noise ratio is a good addition to the usual shielding of the scanner sensor. Wavelet filtering improves detection results. The observed features help to develop NQR detectors.

References

1. B.T. Fedoroff et al.: Encyclopedia of explosives and related items, U.S. Army Research and Development Command (TACOM, ARDEC) Warheads, Energetics and Combat Support Center, Picatinny Arsenal, NJ, vol. 2–8, 1962–1980.
2. C. Bruschini: Commercial systems for direct detection of explosives, ExploStudy, Final Report, École Polytechnique Fédérale de Lausanne (EPFL) & Vrije Universiteit Brussel (VUB), Web: http://diwww.epfl.ch/lami/detec/, 17/2/2001
3. T.N. Rudakov, T.J. Rayner, P.A.S. Hayes, K.L. Russeth: Detection of explosives by quadrupole resonance method: new aspects for security. In: Y. Schuber and A. Kuznetsov (eds.), Detection and Disposal of Improvised Explosives, Springer, Netherlands, 2006, pp. 191–204.
4. Existing and Potential Standoff Explosives Detection Techniques, Committee on the Review of Existing and Potential Standoff Explosives Detection Techniques, National Research Council. ISBN: 0-309-09130-6, 2004, p. The National Academies Press. http://www.nap.edu/catalog/10998.html, R.Meyer, J.Koehler, A.Homburg: Explosives, 5th edn, Wiley-VCH, Weinheim, Germany, 2002.
5. R.A. Marino, R.F. Connors: Orthorhombic and monoclinic TNT: a nitrogen-14 NQR study. J. Mol. Struct. 111, 323–328 (1983).
6. M. Ostafin, B. Nogaj: Temperature dependence of ^{14}N-NQR spectra in commercial 2,4,6-trinitrotoluene (TNT). Mol. Phys. Rep. 29, 64–67 (2000).
7. T. Hirshfeld, S.M. Klainer: Short range remote NQR measurements. J. Mol. Struct. 58, 63–77 (1980).
8. R. Blinc, T. Apih, J. Seliger: Nuclear quadrupole double resonance techniques for the detection of explosives and drugs. Appl. Magn. Reson. 25, 523–534 (2004).
9. M. Nolte, A. Privalov, J. Altman, V. Anferov, F. Fujara: 1H–14N cross-relaxation in trinitrotoluene – a step toward improved landmine detection. J. Phys. D: Appl. Phys. 35, 939–942 (2002).
10. D.W. Prescott, O. Olmedo, S. Soon, K.L. Sauer: Low-field approach to double resonance in nuclear quadrupole resonance of spin-1 nuclei. J. Chem. Phys. 126, 204504 (2007).
11. T.N. Rudakov: Modification of SSFP technique for the effective detection of NQR signals. Physics Letters A 358, 322–326 (2006).

12. G.A. Barrall, D. Gregory, D.K. Lathrop, J.K. Lee, R. Matthews, H.G. Robert, S. Vahidinia, S.A. Vercotter, C. Wu, A.D. Hibbs: Important consideration for NQR detection of TNT, RDX and TETRYL. In: Int. Meeting "The Advances in NQR Detection of Landmines and Explosives". Abstracts, Lubljana, Slovenia, J. Stefan Institute (2000), p. 31.
13. D.Y. Osokin: Pulsed line narrowing in nitrogen-14 NQR. Phys. Stat. Sol. (b) 102, 681–686 (1980).
14. G.V. Mozzhukhin, A.V. Bodnya, V.V. Fedotov, C. Chun, Z. Guojin, G. Huamin: The detection of the trinitrotoluene by pure nuclear quadrupole resonance. Appl. Magn. Reson. 29, 293–298 (2005).
15. Ammonium Nitrate. http://en.wikipedia.org/wiki/Ammonium_nitrate.
16. J. Seliger, V. Zagar, R. Blinc: A new highly sensitive ^{1}H–^{14}N nuclear-quadrupole double-resonance technique. J. Magn. Reson. A, 106, 214–222 (1994).
17. J. Barras, M.J. Gaskell, N. Hunt, R.I. Jenkinson, K.R. Mann. D.A.G. Pedder, G.N. Shilstone, J.A.S. Smith: Detection of ammonium nitrate inside vehicles by nuclear quadrupole resonance. Appl. Magn. Reson. 25, 411–437 (2004).
18. A.N. Garroway, M.L. Buess, J.B. Miller, B.H. Suits, A.D. Hibbs, G.A. Barrall, R. Matthews, L.J. Burnett: Remote sensing by nuclear quadrupole resonance. IEEE Trans. Geosci. Remote Sens. 39 (6), 1108–1117 (2001).
19. B.H. Suits, A.N. Garroway, J.B. Miller: Noise-immune coil for unshielded magnetic resonance measurements. J. Magn. Reson. 131, 154–158 (1998).
20. H.F. Cancino-De-Greiff, R. Ramos-Garcia, J.V. Lorenzo-Ginori: Signal de-noising in magnetic resonance spectroscopy using wavelet transforms. Concepts Magn. Reson. 14, 388–401 (2002).
21. D.L. Donoho: De-noising by soft-thresholding. IEEE Trans. Infor. Theory, 41 (3), 613–627, May 1995.

DEVELOPMENT OF ELECTRIC FIELD NMR SIGNAL ACQUISITION SYSTEM

ROBERT J. PRANCE*, AHMET AYDIN,
CHRISTOPHER J. HARLAND, HELEN PRANCE
Centre for Physical Electronics and Quantum Technology, Department of Engineering and Design, University of Sussex, Brighton, Sussex, BN1 9QT, UK

Abstract An entirely new approach to signal acquisition in nuclear magnetic resonance (NMR) systems has recently been demonstrated experimentally at Sussex. We report results which bring this method closer to realising some significant advantages over the conventional approach. The method uses a novel sensor technology, the electric potential sensor, to detect the electric field signal associated with the precessing nuclear spins. It provides an alternative non-inductive method for acquiring the signal via a weak capacitive coupling to the sample. This paper builds on previously published results and demonstrates the ability to detect spin-echo signals from liquid samples as well as free induction decays from solid samples using this technique. In addition, we demonstrate the intrinsic spatial resolution of the method by using a measurement electrodes with a range of sizes from 180 to 0.2 mm^2, corresponding to coupling capacitance of 1 pF to 1 fF. Conclusions are drawn as to some of the likely application areas for this signal acquisition technique.

Keywords: electric field, electrometer, instrumentation, NMR

1. Introduction

A new approach to signal acquisition in nuclear magnetic resonance (NMR) systems has recently been demonstrated experimentally at Sussex. Initial results were reported at EUROMAR 06[1] and subsequently published in the scientific literature.[2] The method utilises the electric field signal associated

* E-mail: r.j.prance@sussex.ac.uk

with the precessing nuclear spins and provides an alternative to the conventional magnetic field readout method for acquiring the signal.[3,4] The details of the electric field sensor used, the Electric Potential Sensor (EPS), which was invented and patented at Sussex, are published elsewhere.[5,6] The technique is based on simple assumptions about the nature of the signal which exists in magnetic resonance systems. It assumes the existence of an electric field component (E) of the free induction decay signal, which according to Maxwell's equations (at least in the free space limit) must co-exist with a magnetic field (B) of finite frequency. This is given by equation (1).

$$\text{Curl } E = -\partial B/\partial t \tag{1}$$

Work published previously[2] by the authors described a system operating at a Larmor frequency of 1.8 MHz, which was capable of acquiring both the magnetic and electric field components of the free induction decay signal simultaneously. This demonstrated that the two signals were equivalent and also that the crosstalk between the transmitter coil and receiver electrode structure was significantly lower than that for the conventional magnetic signal acquisition system. It also established unambiguously that the data presented were magnetic resonance signals. Our work is now focussing on developing this system to demonstrate some of the inherent advantages of the method, in particular the inherent spatial resolution of the sensors which we believe will lead to new applications within magnetic resonance.

2. Method

The system currently operates between one and two MHz, with either liquid or solid samples using simple pulse sequences such as spin-echo. The maximum available radio frequency power is of the order of 50 W. The EPS used for the electric field signal acquisition may be considered as a near perfect voltmeter with extremely high input impedance. As such it is able to measure electric field, or spatial potential signals, non-invasively using small electrodes which produce minimal perturbation of the field. The input and electrode structures are guarded by a non-magnetic semi-rigid coaxial shield with an outer diameter of 1.4 mm. A combination of carefully controlled positive feedback techniques, including guarding bootstrap and neutralisation, is used to enhance the input impedance of the sensor. In addition a bias circuit is included within the sensor to provide the input bias current required by the sensor for stable operation without drift. The guarded coaxial electrode acquires the signal via weak capacitive coupling to the electric field of interest. This arrangement is shown schematically in Figure 1. The electrode tip is located on the outside of the sample and is either coupled capacitively directly to the sample, in the case of solids, or through the wall of the sample tube in the case of liquids. Typically this coupling capacitance is in the range of 1 pF to 1 fF (10^{-12} F to 10^{-15} F) dependent on the effective cross-sectional area of the electrode tip and the wall thickness of the sample tube. The wall thickness of the tube has been reduced to 50 μm in a small region by blowing a hole through the tube wall and covering this with polyamide (Kapton) tape.[7] This allows much smaller electrode tips to be used than was previously the case, since the capacitance is increased by a factor of ten due to the thinner wall for a given cross sectional tip area.

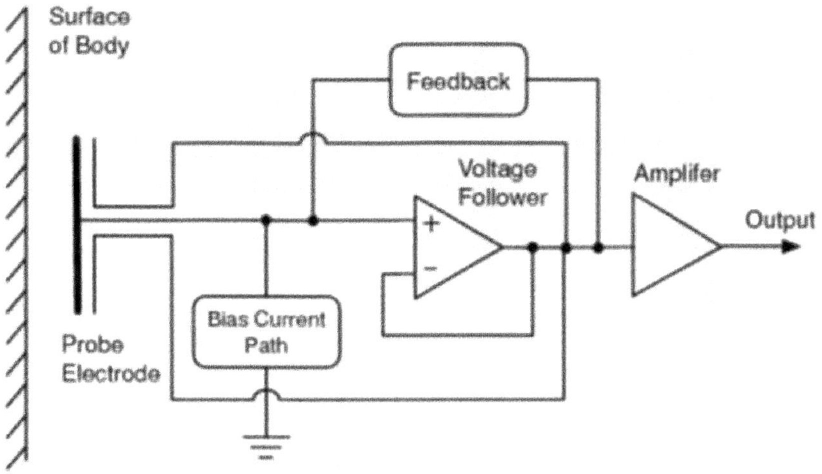

Figure 1. Schematic showing electric potential sensor block diagram with the input electrode capacitively coupled to a source with an air gap as the coupling medium.

One inherent advantage of the electric field method of signal acquisition is immediately apparent, namely the reduced level of crosstalk between the inductive transmitter system and the capacitive receiver. It is not necessary to use any protection circuitry or quarter wave line techniques in order to protect the front end of the EPS from the radio frequency pulses, as is the usual practice with conventional techniques.[8–10] All results reported in this paper were acquired in this way, using a simple gating procedure for the sensor and without front end protection circuitry of any kind on the electric field signal acquisition channel. It is however necessary to use the conventional techniques with the magnetic channel in order to acquire a signal.

The liquid samples used are contained in glass sample tubes 12 mm in diameter with a wall thickness of 0.5 mm. Solid samples are machined to be the same nominal diameter as the sample tubes. The samples are positioned in a screened probe containing a resonant solenoidal radio frequency excitation coil which is fitted with an active Q damping circuit. This probe is inserted into the gap of a conventional copper electromagnet. The field homogeneity of this magnet prevents us from resolving details of structural information from the data. Indeed the linewidths observed correlate well with our measurements of the field inhomogeneity.

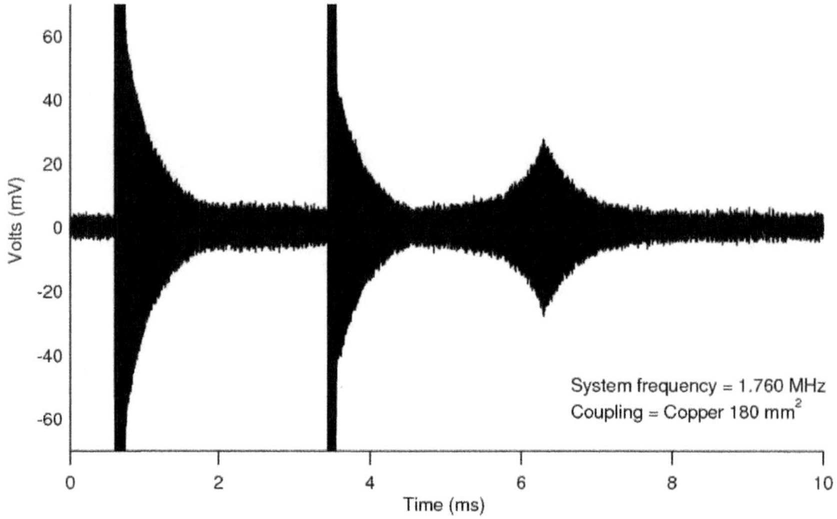

Figure 2. Time domain free induction decay (FID) signal acquired using the electric potential sensor and a pulse-echo sequence for a sample of glycerin. Both the FID and spin-echo signals are clearly visible.

3. Results

An example of a typical time domain free induction decay signal acquired using the EPS electric field signal acquisition technique is shown in Figure 2. For this data the proton resonance of the glycerine sample occurs at 1.847 MHz and the effective area of the EPS electrode was 180 mm^2 consisting of copper tape positioned on the outside of the sample tube. The pulse-echo sequence may be seen clearly with visible free induction decay signals after each pulse. The corresponding Fourier transform for the data in the region of the spin-echo signal is given in Figure 3.

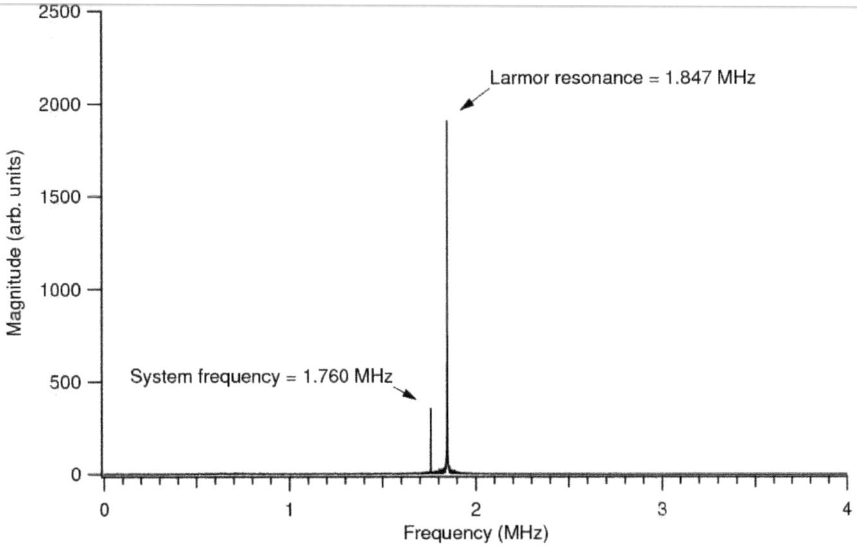

Figure 3. Fourier transform of the data of Figure 3 in the region of the spin-echo signal showing a clear proton resonance signal due to the glycerine sample.

The small secondary peak is a breakthrough signal from the excitation system, set to 1.760 MHz in this case. In order to demonstrate the intrinsic spatial resolution of the electric field technique we have acquired signals using a range of different electrode sizes. As would be expected the signal amplitude is a function of the coupling capacitance and is therefore dependent on the area of the electrode. In all cases the electrode consisting of gold foil positioned on the outside of the Kapton tape window in the sample tube. In Figure 4a the signal has been acquired using a 9 mm^2 electrode, in Figure 4b with 1 mm^2 and in Figure 4c using an electrode with a tip area of 0.2 mm^2. Note that the arbitrary magnitude scales shown in Figure 4 are all the same to allow direct comparison of the relative amplitude of these three signals. It is worth noting that if this signal was the result of simple capacitive coupling to the excitation coil then it would be expected that the amplitude of the signal would be proportional to the coupling capacitance and therefore the area of the electrode for a given wall thickness. Since the ratio of cross sectional areas is $9/0.2 = 45$ this is clearly not the case here. The data of Figure 4 shows that the decrease in amplitude is less than a factor of 2 between the 9 and the 0.2 mm^2 electrodes. This is supporting evidence for our hypothesis that we are detecting the electric field component of the nuclear precession signal rather than the electric field due to the transmitter coil.

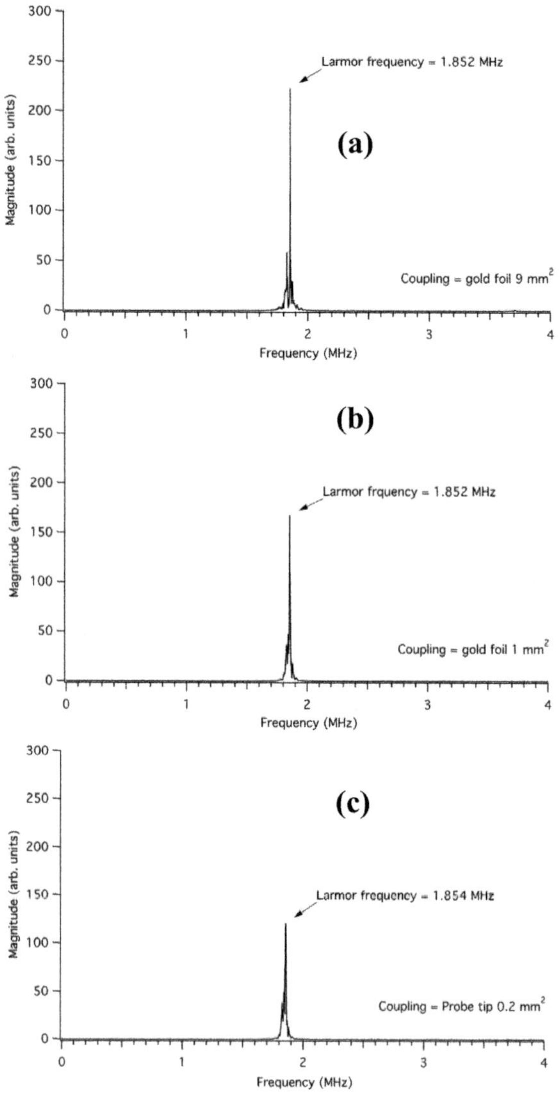

Figure 4. Results showing the effect of varying the area of the measurement electrode of the EPS. In Figure 4a using 9 mm^2, in Figure 4b with 1 mm^2 and in Figure 4c using an electrode with a tip area of 0.2 mm^2. Note that the arbitrary magnitude scales shown are all the same to allow direct comparison of the relative amplitude of these three signals due to a glycerine sample.

As an example of the ability of the EPS to acquire signals from solid samples Figure 5 shows the proton resonance signal due to a sample of solid polythene at a Larmor frequency of 1.855 MHz with the smaller 1.758 MHz signal again due to breakthrough from the excitation system. This signal was acquired using the large area 180 mm^2 electrode.

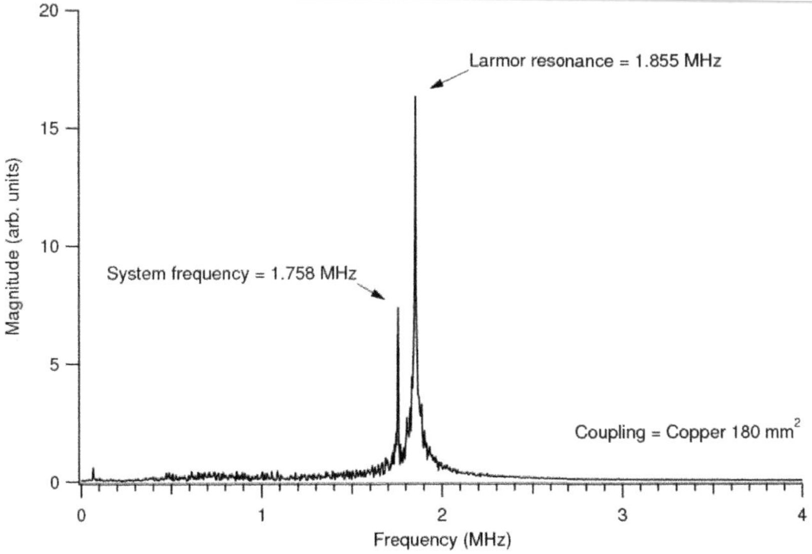

Figure 5. Fourier transform of spin-echo signal due to the proton resonance of a sample of solid polythene. This was acquired using an electric potential sensor with a 180 mm^2 electrode.

4. Conclusions

This paper extends our previously published work on the electric field acquisition of free induction decay signals. We have already demonstrated that this technique is less susceptible to crosstalk between the magnetic transmit and electric receive channels than conventional systems are.

In this paper we show that it is possible to use the intrinsic spatial resolution of the sensors in this application by assessing the performance of a range of different size measurement electrodes. Results were reported for electrodes as small as 0.2 mm^2. The capacitance between the electrode and the excitation coil is proportional to the area of the electrode which should give rise to a large change in the amplitude of the observed signal if this is the mechanism for the observed signals. There is no such change in signal amplitude seen in the data presented in Figure 4. This is interpreted by the authors as supporting evidence for their claim that the measured signal is indeed the electric field component of the free induction decay signal and not due to spurious coupling between the measurement electrode and the transmitter coil.

It is worth noting that in other applications of this EPS technology we have shown that electrode tips may be fabricated and used at the 1 μm resolution level[11] and also that these sensors may be successfully integrated into array formats.[12]

The data presented here from a sensor with a spatial resolution of 200 μm, corresponds to coupling to a small sample volume containing only ~10^{18} spins. Another possible area of interest for the future is low field NMR, since the operating range of the sensors extends from low frequency (<kHz) up to 100 MHz at present.

Acknowledgements The authors would like to thank the Engineering and Physical Sciences Research Council (EPSRC) for supporting this work under grant number EP/E042864/1 – 'The Electric Potential Sensor – a Basic Technology for Measurement Science'.

References

1. Prance RJ, Aydin A, Electric field NMR – a new technique. Frontiers lecture 51, EUROMAR 2006 conference, 16–21 July 2006, York
2. Prance RJ, Aydin A, Acquisition of a nuclear magnetic resonance signal using an electric field detection technique. Appl. Phys. Letts. 91, (2007) doi: 10.1063/1.2762276
3. Torrey HC, Transient nutations in nuclear magnetic resonance. Phys. Rev. 76(8), 1059–1068 (1949)
4. Hahn EL, Nuclear induction due to free Larmor precession. Phys. Rev. 77(4), 297–298 (1950)
5. Prance RJ, Debray A, Clark TD, Prance H, Nock M, Harland CJ, Clippingdale AJ, An ultra low noise electric potential probe for human body scanning. J. Meas. Sci. and Technol. 11, 291–297 (2000)
6. Prance RJ, Harland CJ, Clark TD (2002) patent. WO 03/048789
7. Kapton tape, DuPont, *Electronic Technologies*, US Rt. 23 South & DuPont Road, P.O. Box 89, Circleville, OH 43113, USA.
8. Grannell PK., Orchard MJ, Mansfield P, Garroway AK, Stalker DCA, A FET analogue switch for pulsed NMR receivers. J. Phys. E. 6, 1202–1204 (1973)
9. Hoult DI, The NMR Receiver: A description and analysis of design. Prog. Nucl. Magn. Reson. Spectrosc. 12, 41–77 (1978)
10. Griffin DD, Kleinberg RL, Fukuhara M, Low-frequency NMR spectrometer. Meas. Sci. Technol. 4, 968–975 (1993)
11. Clippingdale AJ, Prance RJ, Clark TD, Brouers F, Non-invasive dielectric measurements with the Scanning Potential Microscope. J. Phys. D. 27, 2426–2430 (1994)
12. Gebrial W, Prance RJ, Harland CJ, Clark TD, Non-invasive imaging using an array of electric potential sensors. Rev. Sci. Instrum. 77(6), (2006), doi: 10.1063/1.2213219

BERRY'S PHASE IN NQR OF POWDERS

NIKOLAY SINYAVSKY[1*], OLGA GLOTOVA[2],
EVGENIY KOROTEY[1]
[1]*Baltic State Academy, Molodiozhnaya st. 6, 236029, Kaliningrad, Russia*
[2]*Department of Physics Adam Mickiewicz University, Umultowska 85, 61-614, Poznan, Poland*

Abstract The results of NQR pattern investigation, modified by the manifestation of Berry's phase upon fast rotation of a powder, are presented. Theoretical and experimental ^{35}Cl NQR spectra, broadened by mechanical sample rotation are compared. NQR splitting, observed for powders with $I = 3/2$, interpreted as Berry's phase manifestation, correlates with adiabatically changing Hamiltonian under the mechanical rotation. Correlation between the accumulation of the geometric phase under sample rotation and dependence of the NQR pattern on the rotation frequency and the asymmetry parameter is shown. Possibilities for selective excitation and nutation NQR spectra ($I = 3/2$, $I = 1$) of powders in circularly polarized r.f. fields are investigated. The results of calculations of ^{14}N NQR signal intensities are described for a case of an excitation of powder explosive substances by an r.f. field with elliptical polarization depending on the field amplitude, the pulse duration and the resonance offset. The manifestation of Berry's phase in NQR under the adiabatic rotation of nuclear magnetization, caused by r.f. pulses with resonance offset, is shown.

Keywords: nuclear quadrupole resonance; Berry's phase; nutation; r.f. field with elliptic polarization

1. Introduction

In designing devices based on NQR for the detection of explosives and narcotic substances the problem of increasing the intensity of the NQR

* E-mail: sinyavsky_physics@bga.gazinter.net

signals and the sensitivity of methods for detecting and identifying substances is critical. At the same time, there are a number of questions which, we feel, have not been adequately investigated in the NQR spectroscopy of powders, which could give new ideas for solutions to some aspects of the above-mentioned problem. This paper is devoted to the manifestation of a geometrical phase in NQR spectra and selective excitation of samples by an r.f. field with elliptical polarization.

The evolution of quantum-mechanical system under adiabatically changing Hamiltonian has important consequences in magnetic resonance. Let us consider the experiment when crystalline sample with nuclear spins I, characterized by quadrupole coupling constant e^2Qq, is rotated at angular velocity ω_R and the nuclear spin Hamiltonian is changing adiabatically. Hence the eigenstates of Hamiltonian, but not the energies change with time. Berry[1] showed that if the Hamiltonian is slowly changed along a cyclic path, a quantum state can acquire a geometric phase. Any observable influence of rotation in NQR is a result of Berry's phase manifestation. So the NQR signal phase has a shift, equal to the accumulated geometric phase. Since the increasing changing of phase is equal to the frequency shift, geometric phase is observable as a frequency shift in spectrum.

The publication of famous Berry's work[1] caused the flow of publications inexhaustible till now. Since then the concept of a geometric phase (Berry's phase) has essentially extended and has penetrated into the various sections of physics. In literature geometric phase can be invoked in discussions of different questions of molecular spectroscopy, atomic-molecular scattering, quantum Hall effect, light spreading in light-carrying fiber, etc. There are many publications, devoted to Berry's phase in NMR and NQR. The broad practical using of Berry's phase in radiospectroscopy is groundless ignored till now mainly because of nontrivial nature of its manifestation in experiment. But the angular dependence of the accumulating geometric phase can provide new structural information.

A first pure NQR experiment was performed by Tycko,[2] who used a crystal of sodium chlorate in which the quadrupole chlorine nuclei were aligned with the crystal's symmetry axis, which is here the quantization axis. He excited these nuclei into a superposition of two spin states by using a pulse of r.f. field. By rotation the crystal about some other axis, the two spin states then acquired geometric phases that increased with time, producing a difference in the frequency of the two states. This shift was detected by a change in the NQR frequency of the nuclei, measured using standard phase-sensitive detection techniques.

The non-adiabatic geometric phase was also observed by Suter et al.[3] using NMR techniques. They studied the system of coupled protons, the quantum system with total spin $I = 1$. The two-level subspace was wade to undergo a cyclic evolution in Hamiltonian by applying time-dependent magnetic field. The geometric phase was measured by looking at interference between these two states and the third, unperturbed level.

Zee[4] offered to use Tycko experiment for rotation not around one, but two axes, and showed the occurrence of a non-Abelian Berry phase. Such experiment was executed by Zwanzinger, Koenig et al.[5] In[6] Zwanzinger reported about the measuring of the geometric component of the transition probability in two-level NMR system by sweeping of r.f. irradiation. Furman[7] showed that in the off-resonance case NMR frequency shift consists of two parts: Bloch–Siegert shift and Berry effect shift. Appelt, Waeckerle and Mehring[8] using optically detected ^{131}Xe NQR, spread Tycko's experiment on nonadiabatic region and for the analysis of Berry's phase, using rotating frame. Berry's dephasing due to diffusion in optically detected NQR of gaseous ^{131}Xe was investigated by Jones and Pines in reference[9]. Nontrivial mixing of the spin states at rotation of NQR-system around two axes and formation of non-Abelian Berry's phase was shown in the work of Asakura.[10] Gopinath and Kumar[11] suggested that the Berry's phase can be used in NMR to implement controlled phase shift gates for quantum-information processing and for strongly dipolar coupled nuclear spins.

Despite of multiple publications about Berry's phase in NMR and NQR, there are not many experimental works, and nobody carried out the experiment with registration of Berry's phase for pure NQR except for Tycko (monocrystal, $\eta = 0$) and the authors[12] (powder sample, $\eta \neq 0$), as far as we know.

In our publications[12,13] Tycko experiment for pure NQR has been extended on cases of powder samples with total spin $I = 1$ and $I = 3/2$ and non-zero asymmetry ($\eta \neq 0$). These works, as a mater of fact, represent the first attempt of practical using of Berry's phase manifestation in pure NQR for measuring of EFG asymmetry parameter in powders with $I = 3/2$. Perhaps the using of the off-resonance impulse consequences for rotation in spin space instead of mechanical rotation of the sample can provide important advantages.

When Hamiltonian follows closed path in parameter space in the time interval $[0, T]$ and the system is initially in its eigenstate then, as was recognized by Berry, system ends up in the state, depends on both the dynamical and geometric phases. If the eigenstates of Hamiltonian has no degeneration (case of NMR) the geometric phase for quantum state with magnetic quantum number m is equal to $\beta = 2\pi m(1 - \cos\theta)$, i.e. it is proportional to the solid angle of simple cone movement.

The influence of Berry's phase on NQR spectrum of powder, which rotated around the axis of r.f. coil, for cases $I = 1$ and $I = 3/2$ is described in reference[13]. For case $I = 3/2$, even with non-zero EFG asymmetry parameter, the spectrum consists of one line and eigenvalues of Hamiltonian have degeneration. If the eigenstates of Hamiltonian are known, anybody can determine Berry's phases for this states, according to Anandan[14,15]:

$$\beta_n = i\int_0^T \langle\psi_n|\frac{d}{dt}|\psi_n\rangle dt\,.$$

However, the simple analytical expressions for Berry's phases in case of $I = 3/2$ come out only for $\eta = 0$[2].

For the first time we calculated the Berry's phases for all four states of spin $I = 3/2$ with $\eta \neq 0$ in dependence of angles θ and φ, and the parameter of EFG asymmetry η in work[13]. We have simulated NQR pattern of rotated powder with $I = 3/2$ for different values of angular velocity. We took into account the Gaussian line broadening of a pattern with mean-square deviation parameter $\sigma = 350$ Hz. In model spectrum we also supposed that $t_w = 3$ μs, $\gamma B_1 = 2\pi\cdot 96$ kHz. Simulated line has three pairs of frequency singularities (ν_1, ν_2 и ν_3), symmetrically relative the center of line, which corresponds to the rest sample. Singularity ν_1 correspond to $\theta = \pi/2$, $\phi = 0$; ν_2 – to $\theta = \pi/2$, $\phi = \pi/2$; and ν_3 – to $\theta = 0$ (ϕ – is free). We have determined the dependence of frequency singularities on the values of asymmetry parameter and the angular velocity of rotation. So if we can measure the singularities we can determine the value of EFG asymmetry parameter of powders in case of $I = 3/2$.

So the analysis carried out in reference[13], let us to propose a new method of EFG asymmetry parameter determination for NQR powders. The method consist in the rotation of the sample around the axis of r.f. coil and registration of the lineshape singularities for nuclei with $I = 3/2$. In comparison with the Zeeman effect, where Helmholtz coils are needed, this experiment can be carried out with standard NMR probe with devise for the sample rotating.

As was reported by Tycko, the rotation of the sample has the same effect as the presence of external magnetic field along the axis of r.f. coil and in may be taken into account by the effective field in Hamiltonian. So the NQR pattern of powder with $I = 3/2$ in the case of Zeeman effect is analogous to the case of sample rotation.

For the pure NQR with $I = 1$ and $\eta \neq 0$ we have determined, that Berry's phases: $\beta_1 = \beta_2 = \beta_3 = 0$ despite on the value of η and angles θ and φ. For case of pure NQR two levels with $\eta = 0$ have degeneration. Degeneration can be replaced by weak external magnetic field and in this case Berry's phases for two states is non-zero, but even with small increasing of the asymmetry parameter the phases becomes equal to zero. So for I = 1 the rotation has the same effect as NQR Zeeman effect, where lines with larger asymmetry parameter η have smaller feeling of weak magnetic field.

In Figure 1 the ^{14}N NQR spectrum of $C_6H_{12}N_4$, is given, in which are observed two closely situated lines. There are no literature data concerning the origin of this phenomena, whether it is caused by small parameter of asymmetry (i.e. $\eta \neq 0$) for all four nuclei of nitrogen in a symmetric molecule, or it follows from two not equivalent positions of nitrogen in a molecule or crystal. The rotation of a sample will give in these cases spectra of a different form (Figure 1c and d). For a spin system with $I = 1$, where there is no problem in determining spectral parameters, the method of sample rotation can be used for the identification of lines and their correlation with the non-equivalent nuclear positions in the molecule.

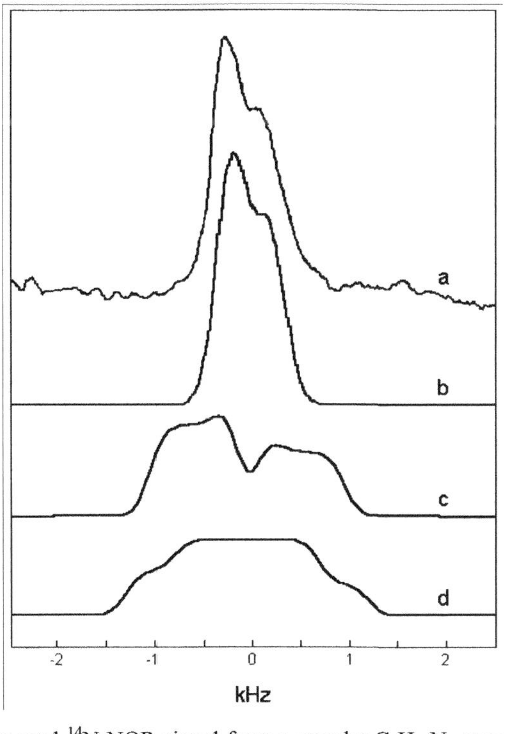

Figure 1. Static experimental ^{14}N NQR signal from a powder $C_6H_{12}N_4$ at room temperature (a). For zero the resonant frequency 3.306 MHz is accepted. The simulated signal for a motionless sample (b). The simulated signals for a rotating sample with $\omega_R = 2\pi \cdot 2$ kHz: all nucleuses are equivalent, $\eta = 0.00025$ (c) and $\eta = 0$, there are two unequivalent positions of nucleuses of nitrogen (d).

The experimental check of theoretical calculations was performed on the NQR spectra of adiabatically rotated samples.[12] ^{35}Cl NQR was observed on powder of $C_6H_4Cl_2$ (*p*-dichlorobenzene) and $C_3N_3Cl_3$ (2,4,6-trichloro-

1,3,5-triazin). The simple single-pulse consequence was used for FID registration. For the frequencies of sample rotation up to 1.5 kHz for both samples only line broadening was observed. Frequency singularities have been found already for 2.0 kHz. For higher frequencies of sample rotation these singularities can be distinguished in spectrum, but the line width increases. So, for the frequency of rotation ν_R = 4.0 kHz the intensities of this satellites become more than intensity of central line. The observed small shift of spectra for smaller frequencies under the sample rotation is caused by sample heating. Perhaps the temperature gradient causes such changes in forms of central line and its satellites.

^{35}Cl NQR spectrum of the static sample of $C_3N_3Cl_3$ for T = 303 K consists of two lines at frequencies 36.261 and 36.279 MHz respectively, and with intensity ratio 2:1. The frequency singularities ν_3 and ν_3' are clearly visible at rotation frequencies higher than ν_R = 2.5 kHz. The resolution of these frequency singularities increases with the rotation frequency, they are visible as satellites of the central line. At rotation frequencies higher than ν_R = 3.5 kHz satellites began mixing. The form of the central lines differ from the form of satellites because the singularities ν_1 and ν_2 (ν_1' and ν_2') are not equal in consequence of the non-zero asymmetry parameter for both lines. ^{35}Cl NQR spectrum of $C_3N_3Cl_3$ with frequency of sample rotation ν_R = 3.35 kHz is shown on Figure 2a. On the same figure simulation of real spectrum (Figure 2b) and idealized spectrum (Figure 2c) for asymmetry parameters η_1 = 0.222 и η_2 = 0.231 are presented for comparison. The figure shows that singularities ν_1 and ν_2 (ν_1' and ν_2') cannot be discerned in the experimental spectrum and only the broad central line is observed. Asymmetry parameters determination was carried out[12] by comparison of experimental ^{35}Cl NQR spectra, exposed to Berry's phase influence, with calculated on computer spectra. The criterion of coincidence for these spectra is the least square method. Accuracy of determination is small and for $C_3N_3Cl_3$ values $\eta_1 \approx \eta_2 = 0.3 \pm 0.1$ has been found. For higher accuracy, first of all, the high stability of frequency rotation is needed. The calculated dependence of the frequency singularities of the NQR powder pattern shows that clearly visible in experiment singularities ν_3 and ν_3' are only weakly dependent on the asymmetry parameter. So for asymmetry parameter determination singularities ν_1 and ν_2 (ν_1' and ν_2'), which strongly dependent on the asymmetry parameter, must be used.

So, our experimental investigations[12] confirm main results of theoretical calculations,[13] concerned with influence of Berry's phase on NQR spectra and demonstrate a capacity for work for a new method of EFG asymmetry parameter determination, which consists of the rotation of a sample around the r.f. coil axis and comparison experimental pattern with simulated line for nuclei with I = 3/2.

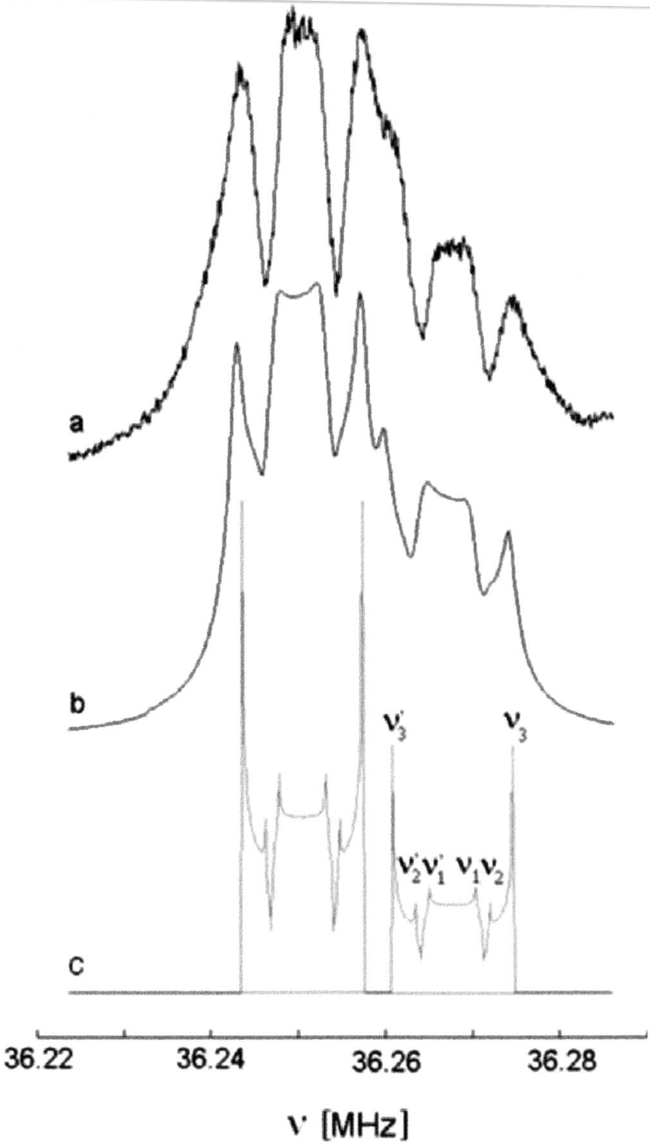

Figure 2. Experimental NQR powder pattern of a sample of $C_3N_3Cl_3$ rotating at $\nu_R = 3.35$ kHz, (b) simulation for $\eta_1 = \eta_2 = 0.3 \pm 0.1$, and (c) idealized calculated spectrum for $\eta_1 = 0.222$ and $\eta_2 = 0.231$.

2. Theory

The geometric phase in NMR two-level system, where cyclic evolution was caused by off-resonance 2π-pulse, investigated in work of Lisin et al.[16] The authors offered to observe Berry's phase influence without macroscopic rotation, but by the way of microscopic rotation of nuclear magnetization in spin space with special r.f. pulse consequences.

Such rotation is represented nutation of magnetization around the effective field, which is dependent on r.f. amplitude of exciting pulses and value of resonance offset. The difference of Berry's phases for different eigenstates of Hamiltonian in rotating frame causes a phase shift, which is dependent on orientation of EFG and its asymmetry parameter. This shift is observed in experiment as frequency shift, i.e. as a rotational splitting for each single-crystal and may be used to receive structural information from powder spectra.

In case of $I = 3/2$, as for all half-integer spin systems, spins in states with $+m$ and $-m$ have equal energy of quadrupole coupling, but opposite precession regarding the electric field gradient. If we use r.f. field with linear polarization to observe NQR, level transitions are caused by both components of field: two fields with opposite circular polarization.[17] One component excites spins with $+m$, while another – spins with $-m$. As a result we observe polarized magnetization, which oscillates along the excited field direction and represents by two components, corresponding spins in $+m$ and $-m$ states, and rotates in opposite directions.

If we use r.f. field with circular polarization, only half of spins (with $+m$ or $-m$) will be excited. On the other site, the using of r.f. field with elliptical polarization, which consists of two fields with circular polarization of different amplitude, give different excitation of spins in states +m and −m. Using r. f. field with circular polarization, as mentioned above, divides spin system on two subsystems with spins in +m and −m states respectively.

The selective excitation of spin system with $I = 3/2$ by r.f. field with circular polarization gives opportunity of individual study of resonance transitions between degenerated levels, contributions from relaxation in NQR signal and accumulating by different states Berry's phases. All cases presented in literature till nowadays are restricted by studying only single-crystal samples with axial symmetry of EFG. At the same time, the case of polycrystalline samples without axial symmetry of the EFG as well as analysis of NQR nutation spectra patterns and singularities by means of an r.f. field with elliptical polarization, have significant practical interest and are important for the investigation of the a Berry's phase manifestation in NQR spectra.

To calculate the observable signal of free induction decay (FID) after the r.f. pulse with elliptical polarization the quantum-mechanical analysis of quadrupole spin system excitation has been used. The procedure of calculation is the same as in work of Pratt.[18] Only one distinguish, which consists in changing interaction representation of coefficients C_n on representation in laboratory frame was used in our works[19,20] to receive a final expressions for $<I_{x,y,z}>$. The response of a spin system with $I = 3/2$, experienced quadrupole interactions in zero-field, on excitation by r.f. pulse has been investigated in terms of time-dependent perturbation theory. In our calculations we suppose that before the application of r.f. pulse, in time interval $0 \leq t \leq t_w$, spin system is in thermodynamic equilibrium with lattice. We have also used the duration of experiment significantly shorter then time of spin-lattice relaxation and our spin system was effectively isolated from lattice. All nuclear spin relaxation effects are considered as unimportant.

The axis of r.f coil, which create pulse of r.f. field B_1, supposed to be oriented with angles (θ,φ) regarding the principal axis system of EFG for examined nucleus. Another r.f. coil with field B_1' is perpendicular to the axis of the first (angles θ' and φ') and produced an oscillations with 90° phase shift between B_1' and B_1. We have described the response of spin system on short pulse with duration t_w, significantly shorter then relaxation time $T_{2\rho}$ in presence of r.f. fields $B_1(t)$ and $B_1'(t)$, i.e. $t_w << T_{2\rho}$.

From the received analytical expressions for nuclear magnetization we have determined that in case of excitation by r.f. field with elliptical polarization each single-crystal has two nutation frequencies, in contrast to case of linear polarized field, where each single-crystal has only one nutation frequency: $\omega_{n+}=2\alpha_+$ и $\omega_{n-}=2\alpha_-$. Here

$$\alpha_{+,-} = \sqrt{\left|\vec{A}\right|^2 + \left|\vec{A}'\right|^2 \pm 2\left|\vec{A}\times\vec{A}'\right|},$$

where instead of vector $\vec{B}_1$ with Cartesian coordinates

$$B_{1x} = B_1 \sin\theta\cos\varphi,\ B_{1y} = B_1 \sin\theta\sin\varphi \text{ and } B_{1z} = B_1\cos\theta$$

vector $\vec{A}$ with coordinates

$$A_{x,y} = \frac{\gamma(\eta\pm 3)}{4\sqrt{3}\rho}B_{1x,y} \text{ and } A_z = \frac{\gamma\eta}{2\sqrt{3}\rho}B_{1z} \text{ is used.}$$

Also instead vector $\vec{B}_1'$ with coordinates

$$B_{1x}' = B_1' \sin\theta'\cos\varphi',\ B_{1y}' = B_1' \sin\theta'\sin\varphi' \text{ and } B_{1z}' = B_1'\cos\theta'$$

we have introduced vector $\vec{A}'$. Signal, induced in first coil of probe, is given by expression:

$$G(t) = -\frac{1}{\gamma B_1}\vec{A}\vec{A}'\left(\frac{\sin 2\alpha_+ t_w}{\alpha_+} + \frac{\sin 2\alpha_- t_w}{\alpha_-}\right)\cos\omega_0(t-t_w) + \\ + \frac{1}{\gamma B_1}\left[\vec{A}(\vec{A}-\vec{F}')\frac{\sin 2\alpha_+ t_w}{\alpha_+} + \vec{A}(\vec{A}+\vec{F}')\frac{\sin 2\alpha_- t_w}{\alpha_-}\right]\sin\omega_0(t-t_w) \quad (1)$$

In analogous, for the signal in second coil we may write:

$$G'(t) = -\frac{1}{\gamma B_1'}\left[\vec{A}'(\vec{A}'+\vec{F})\frac{\sin 2\alpha_+ t_w}{\alpha_+} + \vec{A}'(\vec{A}'-\vec{F})\frac{\sin 2\alpha_- t_w}{\alpha_-}\right]\cos\omega_0(t-t_w) + \\ + \frac{1}{\gamma B_1'}\vec{A}'\vec{A}\left(\frac{\sin 2\alpha_+ t_w}{\alpha_+} + \frac{\sin 2\alpha_- t_w}{\alpha_-}\right)\sin\omega_0(t-t_w) \tag{2}$$

Here we use some definitions: $\vec{F} = \vec{d}\times\vec{A}$, $\vec{F}' = \vec{d}\times\vec{A}'$, where $\vec{d} = \frac{\vec{D}}{|\vec{D}|}$ and $\vec{D} = \vec{A}\times\vec{A}'$.

The analysis of the received expressions has shown, that in case of free angles between two perpendicular oriented coils and the principal axis system of EFG nuclear magnetization will have elliptical polarization. Besides ellipticity parameter will be dependent on both angles of orientation, mentioned above, and asymmetry parameter η. So states with $+m$ and $-m$ will be differently excited, with different selective degree. Even in simple case of single-crystal sample, where coils are oriented along the x and y axis of EFG ($B_1 = B_1'$) without its axial symmetry, magnetization, rotated in xy plane, has elliptical polarization in contrast to work of Weber and Hahn,[17] where magnetization has circular polarization because of $\eta = 0$.

Received expressions for nuclear magnetization for powder have been averaged over all angles θ, φ, θ' and φ'. Nutation spectra of powders, simulated by adding intensities from any single-crystal for the cases of linear and circular polarization of r.f. field and different values of asymmetry parameter, have shown that in case of circular polarized field ($B_1 = B_1'$) nutation spectra of powder consist of two uncrossed lines with six frequency singularities, dependent on asymmetry parameter η and field amplitude B_1:

$$\omega_{n1,2} = \gamma B_1\frac{\sqrt{3}(1\pm\eta)}{2\rho},\ \omega_{n3,4} = \gamma B_1\frac{3\pm\eta}{2\sqrt{3}\rho},\ \omega_{n5} = \gamma B_1\frac{\eta}{\sqrt{3}\rho},\ \omega_{n6} = \gamma B_1\frac{\sqrt{3}}{\rho} \tag{3}$$

In contrast with nutation spectrum of powder, which is registered with linearly polarized field and has only one line with well-defined lineshape, the excitation by r.f. field with circular polarization give nutation spectrum with two lines, characterized by different intensity and lineshape (Figure 3). The low-frequency line is a result of non-selective excitation of spin states $+m$ and $-m$, mentioned above. But high-frequency line is a result of excitation of state, which has the same direction of precession as rotation of r.f. field.

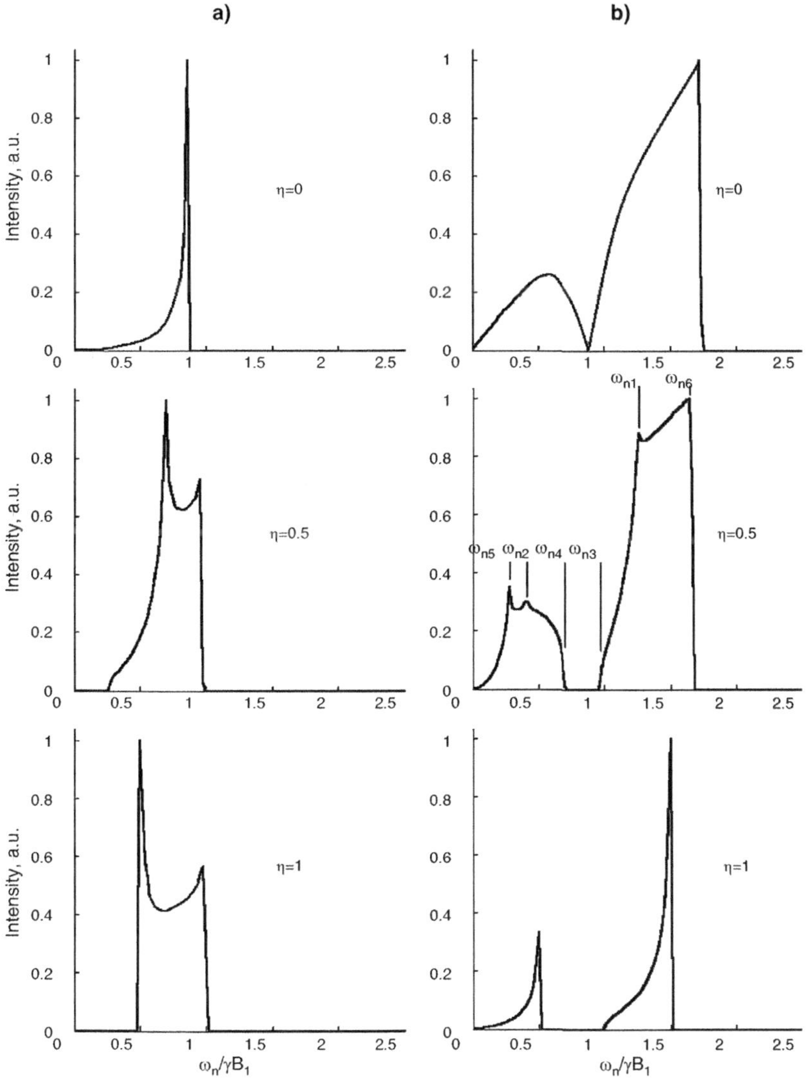

Figure 3. Nutation NQR spectra of powders (I = 3/2), measured with r.f. field with line (a) and circular (b) polarization.

The dependence of singularities on η allows anybody to use them for experimental determination of asymmetry parameter of EFG, because in traditional approach we have no possibility to determine for $I = 3/2$ both asymmetry parameter η and quadrupole coupling constant eQq_{zz} from one measured resonance frequency

$$\omega_0 = \frac{eQq_{zz}}{2\hbar}\sqrt{1+\frac{\eta^2}{3}} .$$

Adding (or subtracting) of signals, induced in both coils, allows to rise the signal-to-noise ratio in √2 times and make frequency singularities precise.

The same analysis, carried out for spin $I = 1$ with resonance offset, has allowed to receive the analytical expressions for NQR signal intensities from powder with elliptically polarized r.f. field:

$$\langle I_x \rangle = \left(\frac{A'_x}{\alpha}\sin 2\alpha t_w + \Delta\omega \frac{A_x}{\alpha^2}\sin^2 \alpha t_w \right)\cos[\omega_+(t-t_w)+\Delta\omega t_w] - \left(\frac{A_x}{\alpha}\sin 2\alpha t_w - \Delta\omega \frac{A'_x}{\alpha^2}\sin^2 \alpha t_w \right)\sin[\omega_+(t-t_w)+\Delta\omega t_w] \tag{4}$$

$$\langle I_y \rangle = \langle I_z \rangle = 0 ,$$

where $\alpha = \sqrt{A_x^2 + A_x'^2 + \Delta\omega^2/4}$, and instead of vector $\vec{B}_1$ with coordinates

$$B_{1x} = B_1 \sin\theta\cos\varphi ,\ B_{1y} = B_1 \sin\theta\sin\varphi \text{ and } B_{1z} = B_1\cos\theta$$

we have introduced vector $\vec{A}$ with coordinates

$$A_{x,y,z} = \frac{\gamma}{2}B_{1x,y,z} \ (\gamma - \text{gyromagnetic ratio}).$$

Also instead of vector $\vec{B}'_1$ with coordinates

$$B'_{1x} = B'_1 \sin\theta'\cos\varphi' ,\ B'_{1y} = B'_1 \sin\theta'\sin\varphi' \text{ and } B'_{1z} = B'_1\cos\theta'$$

we use vector $\vec{A}'$; $\omega = \omega_+ + \Delta\omega$ - spectrometer frequency, not equal to resonance frequency ω_+. The signal, induced in first coil of probe is determined by expression:

$$G(t) = \langle I_x\rangle \sin\theta\cos\varphi + \langle I_y\rangle \sin\theta\sin\varphi + \langle I_z\rangle \cos\theta .$$

The signal, induced in the second coil is determined by the same expression.

The expressions (4) show that despite of excitation with elliptically polarized field nuclear magnetization in this case has linear polarization. Besides the signals, induced in coils are the real and imagine part of one signal, given by (4).

The analysis of case $\eta = 0$ and free values of angles between perpendicular oriented coils and the principal axis system of EFG shows that nuclear magnetization is elliptically polarized and the coefficient of ellipticity is dependent on this angles only and independent on asymmetry parameter η. So states with $+1$ and -1 are excited with different selective degree. Only in simple case of ingle-crystal sample, where r.f. coils oriented along the axes x and y of the principal axis system of EFG and if $B_1 = B_1'$ nuclear magnetization rotates in xy-plane and has circular polarization.

In case of powder expressions (4) must be averaged over all angles θ, φ, θ' and φ' with observance of the condition for perpendicular coils:

$$\sin\theta\sin\theta'\cos\varphi\cos\varphi' + \sin\theta\sin\theta'\sin\varphi\sin\varphi' + \cos\theta\cos\theta' = 0 .$$

The signal, induced in second coil for the case of powder is only shifted in phase on 90° regarding the signal, induced in first coil. Adding (or subtracting) of signals, induced in both coils, if we have two receivers, allows to rise the signal-to-noise ratio because of uncorrelated noises in both channels. The using of r.f. field with circular polarization for case of $I = 1$ in work[21] allows to rise signal-to-noise ratio in $\sqrt{2}$ times, that is very important in solving the problem of explosive detecting by means of NQR.

Taking into account the resonance offset, we have obtained analytical expressions for the intensities of NQR signals for $I = 1$ with powder excitation by an r.f. field with elliptical polarization. It is shown that for arbitrary orientation of the principal axes of the EFG tensor and non-zero asymmetry parameter, nuclear magnetization has (in contrast with the case $I = 3/2$) linear polarization, while in the case of $\eta = 0$ the magnetization is elliptically polarized. The coefficient of ellipticity depends on the orientation of the EFG in this case, but is independent of the EFG asymmetry parameter η. Simulated NQR nutation spectra ($I = 1$) for a powder in the case of a circularly polarized r.f. field have only one well defined singularity, whose position determines the optimal duration of the exciting pulse in the experiment.

The r.f. pulse length dependences of signal intensities induced in one coil in the case of excitation by linearly and circularly polarized fields, are shown in a Figure 4a. The idealized ^{14}N NQR nutation spectra of the powder sample simulated by a method of accumulation of partial intensity for the case of linearly and circularly polarized fields at $\eta \neq 0$ are given in Figure 4b. In the case of a field with circular polarization, the signal intensity is much more sensitive to the exact duration of the $\pi/2$ pulse than for linear polarization.

The theoretical describing of spin system has been carried out[22] in interaction representation, in which spin Hamiltonian is time independent. It has been done for case of resonance offset and with taking into account the accumulating of Berry's phase, caused by excitation of NQR system by strong r.f. field. Berry's phases have been calculated by means of eigenstates, which changing during the r.f. pulse duration. Experimental measuring on ^{35}Cl and ^{14}N have been carried out on NQR spectrometer with Fourier-transformation (0.5–300 MГц). Powders of $KClO_3$, $CCl_3CH(OH)_2$, $C_6H_4Cl_2$

(nuclei ^{35}Cl) and $C_6H_{12}N_4$ (nuclei ^{14}N) have been used as samples. The pulse consequences of Carr–Purcell and strong off-resonance pulses SORS with different resonance offset have been used in experiment. Durations of $\pi/2$ – pulses for nuclei ^{35}Cl from 6 to 15 μs, and for nuclei ^{14}N – from 1.5 to 7.5 μs have been used in experiment. To register the geometrical phase, accumulated during duration of r.f. pulse with resonance offset $\Delta\nu \neq 0$, echo signal (or signal of free induction decay) has been measured between two consequences. It is established that experiment with long pulse consequences is very sensitive to variations in durations of 90° and 180° pulses and value of resonance offset.

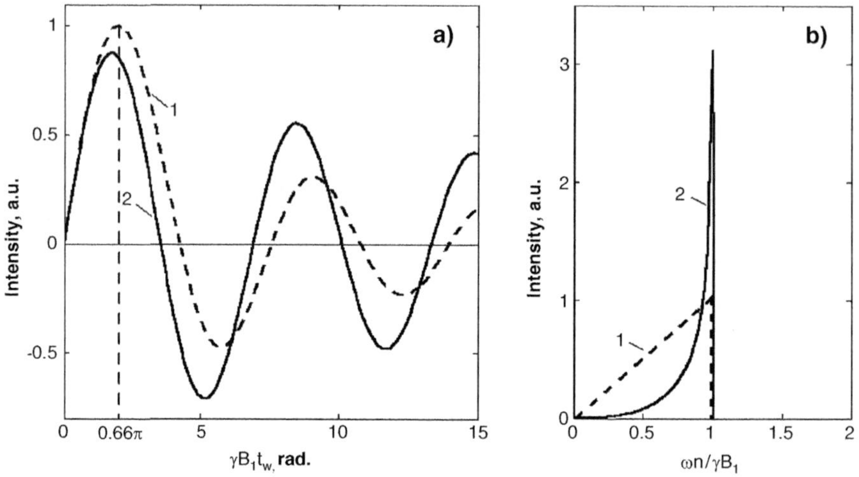

Figure 4. (a) Dependence of a NQR signal ($I = 1$) from duration of a r.f. pulse at excitation of a powder sample by linearly polarized (1) and by circular polarized field (2). (b) Appropriate idealized nutation NQR spectra.

3. Results

The experiments show, that rotation of nuclear magnetization by repeating r.f. pulses with frequency of filling, not equal to resonance frequency, is analogous to fast mechanical sample rotation and causes splitting and broadening of NQR pattern too. Figure 5 illustrates splitting of ^{35}Cl NQR pattern in $KClO_3$ with using SORS pulse consequence (the value of resonance offset was 30 kHz). For splitting measuring it is necessary to take into account scale t_w/τ (because nutation of magnetization takes place only during duration of r.f. pulse). Besides, long pulse consequence causes narrowing of NQR pattern. The resonance offset dependence of NQR splitting for $KClO_3$ in consequence of strong off-resonance pulses is shown on Figure 6.

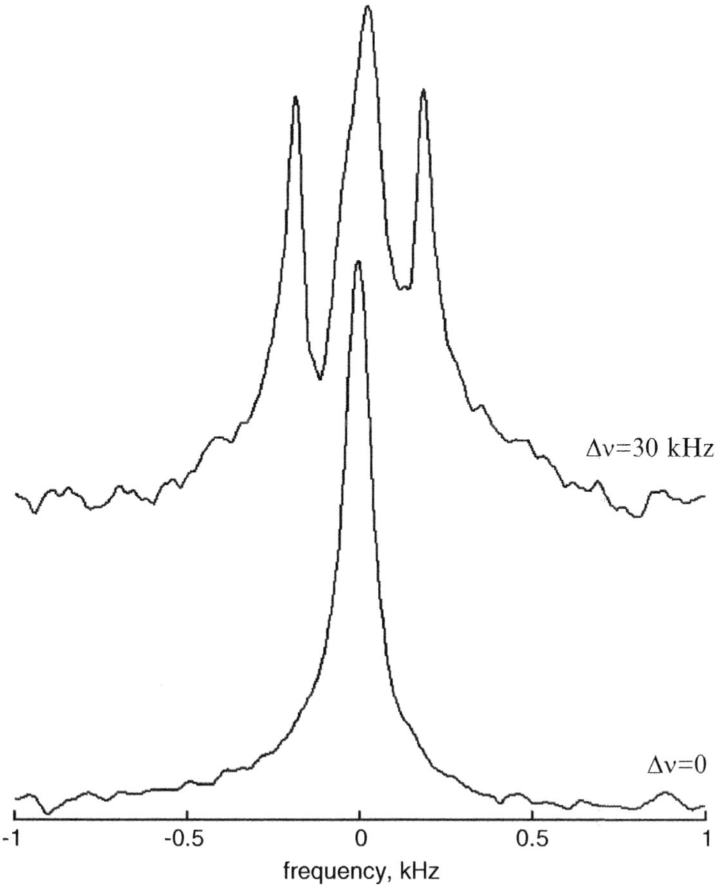

Figure 5. Splitting of ^{35}Cl NQR powder pattern form $KClO_3$ with SORS.

So, the investigations, presented in this work, allow to offer a new method of asymmetry parameter of EFG determination, which consist of sample rotation around the r.f. coil axis and registration of frequency singularities of lineshape for nuclei with $I = 3/2$. In comparison with the Zeeman effect, where Helmholtz coils are needed, this experiment can be carried out with standard NMR probe with devise for the sample rotating. For nuclei with spin $I = 1$, where standard approach allows to determine structural parameters from usual NQR spectrum, this method can be used to identification of lines and their correlation with the non-equivalent nuclei positions in molecule.

It has been shown that NQR (I = 3/2) nutation spectra of powders in case of excitation by r.f. field with circular polarization, consist of two uncrossed lines with characteristic singularities. The asymmetry parameter dependence of singularity positions on nutation spectrum may be used for selective excitation of spin system and for experimental determination of asymmetry parameter from the nutation NQR spectrum, received by applying of r.f. field with circular polarization.

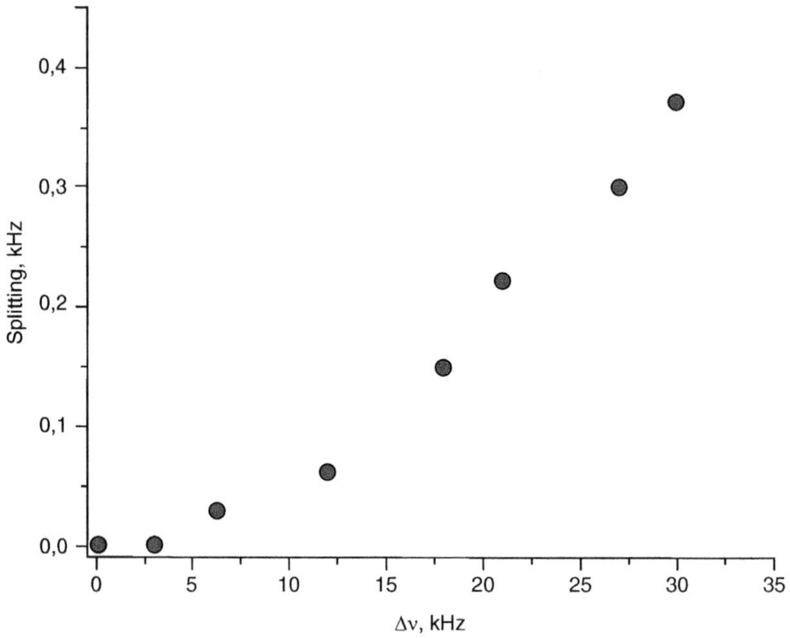

Figure 6. Resonance offset Δν dependence of SORS NQR splitting for $KClO_3$, caused by the accumulation of Berry's phase.

The determination of structural information by means of quadrupole nuclei in solids can be carried out by registration of frequency singularities of NQR lineshape, which are caused by manifestation of geometrical phase under the nuclear magnetization movement along closed path in spin space.

The coherence lost, following free accumulating of Berry's phase, caused by different factors, can be considered as a new relaxation mechanism, which must be taken into account in NQR and NMR experiments.

Acknowledgements Portion of this work were supported by Russian Foundation for Basic Research (RFBR, project nr 08-03-00433).

References

1. M.V. Berry, Proc. R. Soc. Lond. Ser. A 392, 45 (1984).
2. R. Tycko, Phys. Rev. Lett. 58, 2281 (1987).
3. D. Suter, K.T. Mueller, and A. Pines, Phys. Rev. Lett. 60, 1218 (1988).
4. A. Zee, Phys. Rev. A 38, 1 (1988).
5. J.W. Zwanzinger, M. Koenig, and A. Pines, Phys. Rev. A 42, 3107 (1990).
6. J.W. Zwanzinger, Phys. Rev. A 43, 3232 (1991).
7. G.B. Furman and J. M. Kadzhaya, J. Magn. Reson. A 105, 7 (1993).
8. S. Appelt, G. Waeckerle, and M. Mehring, Phys. Rev. Lett. 72, 3921 (1994).
9. J.A. Jones and A. Pines, Chem. Phys. Lett. 245, 215 (1995).
10. H. Asakura, Prog. Theor. Phys. 101, 473 (1999).
11. T. Gopinath and A. Kumar, Phys. Rev. A 73, 022326 (2006).
12. N. Sinyavsky, M. Mackowiak, and C. Schmidt, Z. Naturforsch. 63a, 81 (2008).
13. N. Sinyavsky, E. Korotey, and M. Ma'ckowiak, J. Mol. Struct. 830, 131 (2007).
14. J. Anandan, Nature 360, 307 (1992).
15. Y. Aharonov and J. Anandan, Phys. Rev. Lett. 58, 1593 (1987).
16. W.N. Lisin, G.G. Fedoruk, and E.P. Haymovich, Pis'ma Zh. Eksp. Teor. Fiz. 50, 205 (1989) (in Russian).
17. M.J. Weber and E.L. Hahn. Phys. Rev. 120, 365 (1960).
18. J.C. Pratt, P. Raghunathan and C.A. McDowell, J. Magn. Reson. 20, 313 (1975).
19. A. Sklyar, N. Sinyavsky, M. Ostafin and B. Nogaj, Abstracts of IVth Symposium on nuclear magnetic resonance in chemistry, physics and biological sciences, 24–26 September 2008, Warsaw, Poland, P-24.
20. A. Sklyar, N. Sinyavsky, M. Ostafin and B. Nogaj, Z. Naturforsch A (2009) (in print).
21. J.B. Miller, B.H. Suits and A.N. Garroway. J. Magn. Reson. 151, 228 (2001).
22. O. Glotova, N. Sinyavsky, M. Ostafin, M. Jadzyn and B. Nogaj, Abstracts of IVth Symposium on nuclear magnetic resonance in chemistry, physics and biological sciences, 24–26 September 2008, Warsaw, Poland, P-10.

CONTRIBUTION OF COPPER NQR SPECTROSCOPY TO THE GEOLOGICAL STUDIES OF COMPLEX SULFIDES AND OXIDES

RAMIL R. GAINOV[1*], ALEXANDER V. DOOGLAV[1], IREK R. MUKHAMEDSHIN[1,2], IL'YA A. EVLAMPIEV[1], ANNA YU. ORLOVA[1], IVAN N. PEN'KOV[3], NADEZHDA N. MOZGOVA[4]

[1]*Department of Physics, Magnetic RadioSpectroscopy Laboratory, Kazan State University, Kazan, Kremlevskaya str. 18, 420008, Russian Federation*

[2]*Laboratoire de Physique des Solides, UMR 8502, Universite Paris-Sud, 91405 Orsay, France*

[3]*Department of Geology, Division of areal geology and mineral resources, Kazan State University, Kazan, Kremlevskaya str. 4/5, 420111, Russian Federation*

[4]*Institute of geology of ore deposits, petrography, mineralogy and geochemistry (Russian Academy of Science), Staromonetny per. 35, Moscow 109017, Russian Federation*

Abstract Many energy-related areas such as nuclear waste isolation, continental drilling, fossil fuel recovery, and geothermal energy are directly associated with an in-depth understanding of the earth sciences. Of particular interest is the development of analytical techniques which can augment existing ones in developing a better understanding of mineralogy.

Presently, available instrumental techniques for studying mineralogical problems such as x-ray, electron and neutron diffraction, nuclear gamma resonance (NGR or Mössbauer spectroscopy), electron microscopy and transmission electron microscopy have inherent limitations. These manifest themselves in being unable to characterize mineral samples fully, especially if they are polycrystalline.

Nuclear Quadrupole Resonance (NQR) spectroscopy offers the potential for being able to obtain accurate high resolution spectra. These can then be interpreted to give structural information which can be related to local electronic structure, atomic arrangement, order/disorder phenomena, and crystal phase transformation. In addition, internal dynamics (ionic diffusion,

* Tel.: 7-843-2315175, Fax.: 7-843-2387201, E-mail: ramil.gainov@ksu.ru

J. Fraissard and O. Lapina (eds.), *Explosives Detection Using Magnetic and Nuclear Resonance Techniques*,

metallic behavior, rotations, and so on) in the solid state can be studied. Furthermore, since NQR data are sensitive to changes in temperature and pressure, there is the possibility of obtaining stress/strain information.

As applied to mineralogical and geological problems, NQR can also provide additional information, for example: chemical activity of minerals (genetic and technological aspects) at different hydrothermal conditions, the studies of impurity configurations in ore minerals and their distribution in crystal lattice, and other.

This chapter highlights some NQR studies in copper sulfides, which demonstrate how NQR method can contribute to our understanding of geological problems. Examples are taken primarily from author's investigate groups.

Keywords: ^{63,65}Cu NQR, copper sulfides, oxides

1. Introduction

The significant successes in solid state physics, crystal-chemistry, and physical chemistry provide the new favorable conditions for the development of exploring and technological mineralogy.[1–4] The wide application of the different methods of physical and material science investigation to the mineralogy – electron microscopy and diffraction, x-ray microprobe, IR-spectroscopy, nuclear gamma resonance (NGR or Mössbauer spectroscopy), as well as radiofrequency electron paramagnetic resonance (EPR), nuclear magnetic resonance (NMR), nuclear quadrupole resonance (NQR), and other – permits to reveal the fine details of real mineral structures.[1–4] The main results of mineralogical studies are reduced to the following aspects:

1. Each mineral contains the "traces" of some ore environment, in which this mineral was created and changed. This information is enclosed in peculiarities of mineral morphology and occurrence of manifold crystallographic forms, in appearance of gas–liquid and solid state inclusions in minerals, in crystal lattice disordering, and other. This geological problem is solved by *genetic mineralogy*.
2. A lot of minerals are very sensitive to the slight changes of external conditions of ore formation; therefore even in frame of some separate ore deposit it is possible to reveal the distinctions of mineral crystal-chemistry in different deposit space regions. This circumstance is used in *exploring/estimation mineralogy*.
3. The most minerals are characterized by the micro-scaled inhomogeneity. The investigation of this inhomogeneity allows to mine the additional

information concerning conditions of ore-formation and the nature (forms) of impurity configuration in minerals. These data are necessary for elaboration of extraction schemes in *technological mineralogy*.
4. In the most cases the mineral is originated together with other kinds of minerals. The evolution of these mineral associations (*paragenesis*) obeys the strong rules and describes the ore processing. The paragenic analysis permits to reveal the characteristic minerals, which can serve as indicators of *mineralization processes. Application of such indicators increases the efficiency of geological survey.*
5. The rich variety of the physical and chemical properties of minerals and their analogs is widely involved in creation of laser, radiation, semi-conductive, piezo- and ferroelectric, superionic conductive, thermal, and etc equipment and techniques. The search of new useful mineral properties is the central problem of *mineralogical material-science.*[5,6]
6. The *radioactive mineralogy* has the exceptional and significant twofold importance. The latter is caused by the operability assurance of the nuclear power engineering (the exploring and identification of the uranium, plutonium, and thorium minerals, as well as their by-products) and by the protection of extraction and working up of corresponding ores. In this sense, the nuclear physicists and mineralogists, as well as the ecologists and special services are of great value.
7. Another special tendency in the mineralogy is a *biomineralogical science*. The Earth's interior is not unique source of minerals, but the living substance (including the people) also provides the appearance of solid state materials – biominerals. The biominerals, which arise from living organisms, participate in the construction of teeth, bones, gall-stones, kidney-stones, and other. Therefore, the investigations of composition, structure, properties, as well as the evolution of biominerals are necessary for medical and pharmaceutical purposes. Due to the latter circumstance, *biomineralogical* investigations have the pronounced medical character. On the other hand, in accordance with the nature of this problem the biomineralogy is under consideration of mineralogists, chemists, and physicists, who apply the corresponding methods of studies.[7]
8. Recent trend is nanomineralogy. It was suggested that the development of nano-scaled studies of minerals can stimulate the rapid upsurge of the most sub-areas of mineralogy.[8]

The elucidation of problems mentioned above requires the studies of local (micro- and nano-scaled) properties of minerals, for which the *nuclear quadrupole resonance* (NQR) method appears to be most suitable. Actually, NQR spectroscopy permits the study of some aspects of crystal structures and internal dynamics, which are difficult, inconvenient or impossible to determine by other techniques.[1,3,8]

The NQR method gives the possibility to measure the heterogeneity of internal electrical fields in crystals on quadrupole nucleus site or, more precisely – to study the interactions between quadrupole moment of nucleus and electric field gradient (EFG) of the local crystal environment. The application of NQR spectroscopy in solid state physics/chemistry is based on the extreme sensitivity of the NQR effect to the smallest variations in the electric fields (~0.01%), that makes possible to use the NQR for studying a wide range of problems related to the inter-molecular, inter-ionic and intra-molecular interactions.

The chapter is organized as follows. Section 2 briefly describes some aspects of NQR applications in mineralogical and geological sciences. Section 3 contains necessary NQR and NMR background. In Sections 4–7 we report results and discussions. The paper concludes with a summary in Section 8.

2. NQR in mineralogy and geology

The beginning of NQR in solids dates back to the beginnings of NMR in the late 1940s and early 1950s.[9] The first NQR measurements reported for a solid were by Dehmelt and Kruger using signals from ^{35}Cl in transdi-chloroethylene.[10] An excellent early summary of NQR theory and technique can be found in the 1958 book by Das and Hahn.[11] Several comprehensive summaries can be found listed at the end of this chapter.[12–22]

The pioneering review devoted to NQR in minerals (on the base of own earlier studies) was published by Pen'kov and Safin in 1966 in Russian.[23] One year later the English translation of this paper with small modifications appeared.[24] For the first time these articles show the feasibility of using the NQR method, in some instances, in the study of those features of mineral structure and chemistry which are difficult or impossible to attack by conventional methods, such as x-ray or Mössbauer spectroscopy. By the example of As-, Sb- and Bi-bearing chalcogenides it was demonstrated that NQR gives better picture of the ideal structure and can determine the "degree of defectiveness" and that NQR has a very high sensitivity to crystal-chemical effects. It was also illustrated that NQR can be used as a rapid method of mineral identification. In 1968 Wendling[25] and in 1969 Schultz and Karr[26] also reported that NQR can be applied in studies of mineral chemistry, in particular, for some Cl-bearing compounds (KCl, $BiCl_3$ and other). A comprehensive NQR investigation of the nature of structural impurities, their selective substitutions in some wide-distributed chalcogenides can be found in 1971 work by Pen'kov.[27] In 1980 Marino and co-workers[28] published their report concerning advantages of NQR in studies of some Al-bearing minerals. Few years later, in 1987, Abdullin and co-workers[29] proposed to apply the NQR techniques to the Cu-bearing minerals. In addition to reviews mentioned, there are number of other papers dedicated to NQR in "individual" minerals, but, of course, it is impossible to itemize all of them in this chapter.

3. NQR–NMR background

The NMR–NQR background will be reviewed only for the sake of clarity. The nuclear spins I interact with their electronic environment through quadrupole (i.e. electric) and magnetic hyperfine couplings. In general, the nuclear magnetic resonance (NMR) spectrum of a quadrupole nucleus (i.e. nucleus with spin $I > ½$) is described by the following spin Hamiltonian:

$$H = H_Z + H_Q + H_M \tag{1}$$

Hamiltonian H_Z is the Zeeman interaction of nuclear magnetic moments with the gyromagnetic ratio γ_n with the applied external magnetic field H_0:

$$H_Z = -\gamma_n \hbar H_0 \cdot I \tag{2}$$

Hamiltonian H_Q refers to as the coupling of nuclear quadrupole moment eQ to the local crystal electric field gradient (EFG):

$$H_Q = \frac{eQV_{ZZ}}{4I(2I-1)}\{3I_Z^2 - I(I+1) + \frac{1}{2}\eta(I_+^2 + I_-^2)\} \tag{3}$$

with V_{ZZ} the largest component of the crystal EFG tensor, $\eta=|V_{XX} - V_{YY}|/V_{ZZ}$ the asymmetry parameter showing the deviation of the EFG symmetry from the axial one, i.e. the value of η lies in the range [0,1]. The EFG components satisfy Laplace equation: $V_{XX} + V_{YY} + V_{ZZ} = 0$.

Hamiltonian H_M can be viewed as an interaction between the nuclear spin I and a static plus time dependent local hyperfine field F_L generated by the motion of conduction carriers. The static part of F_L gives rise to a NMR shift expressed by the magnetic (Knight) shift $K = \Delta H/H_0$ due to the polarization of conduction charges. The fluctuating part of F_L is the source of the nuclear spin-lattice relaxation T_1.

The pure NQR spectrum is observed in the case of absence of the external ($H_0 = 0$) and internal ($H_{int} = 0$) static magnetic fields. The remaining H_Q gives rise to doubly degenerate energy levels, between which NQR transitions are induced. The number of NQR lines is defined (i) by the amount of crystallographically nonequivalent positions of quadrupole nucleus in the crystal structure, (ii) by the local geometry (i.e. symmetry) of quadrupole nucleus surrounding, (iii) by the magnitude of nuclear spin I and (iv) by the presence in nature of different isotopes of quadrupole nucleus. In particular, for copper there exist two naturally occurring isotopes ^{63}Cu (69.2% natural abundance, $\gamma/2\pi = 1.128$ kHz/Oe, $Q = -0.22$ barn) and ^{65}Cu (30.8% natural abundance, $\gamma/2\pi = 1.209$ kHz/Oe, $Q = -0.204$ barn) both having spin $I = 3/2$ and thus two doubly degenerate $\pm 1/2$ and $\pm 3/2$ energy levels. Thus, for each isotope a transition between these levels yields a single NQR signal at a frequency:

$$^{63,65}\nu_Q = \frac{e\,^{63,65}QV_{ZZ}}{2h} \cdot \sqrt{1 + \frac{1}{3}\eta^2} \tag{4}$$

It should be noted that the NQR frequency ν_Q, besides the quadrupole moment eQ of the nucleus, depends on the certain arrangement of the surrounding ions through the parameters V_{ZZ} and η. Since ν_Q is defined by two parameters in equation (4), V_{ZZ} and η, it is impossible to obtain both parameters experimentally from the NQR spectra consisting of only one line for $I = 3/2$. Usually it is done from the angular dependence of the NMR spectrum taken for a single crystal or from numerical simulations of NMR spectrum for the unaligned powder. In some cases, for determination of η the 2D nutation NQR studies also could help.

In a semi-empirical approach,[30] it is assumed that components of EFG tensor at Cu nuclei site can be written as the sum of two terms – lattice and valence contributions:

$$V_{ZZ} = (1 - \gamma_\infty) \cdot V_{latt} + (1 - R_{val}) \cdot V_{val} \tag{5}$$

The parameters γ_∞, R_{val} are the Sternheimer antishielding factors, which characterize the effects of charge density distortions induced by lattice electric field and non-fillness of orbitals of central atom. The first contribution arises from all ion charges outside the ion under consideration and can be calculated in a straightforward manner using the model of point-charges (MPC):

$$V_{latt} = \sum_i \frac{q_i \cdot (3 \cdot \cos^2\theta_i - 1)}{r_i^3} \tag{6}$$

where q_i and r_i are the charge and the position of the ith ion, respectively, θ_i – angle between main axis of symmetry and direction into the neighboring ion.

The second term in equation (5) arises from 3d and 4p non-filled shells and non-spherical distortions of inner orbitals of the subject Cu ion. Taking into account only holes in the Cu orbitals, the contributions of 3d and 4p shells can be written as

$$V_{val}^{(3d)} = -\frac{4}{7} \cdot e \cdot \langle r^{-3} \rangle_{3d} \cdot [N_{3d(3z^2-r^2)} - N_{3d(x^2-y^2)} - \\ - N_{3d(xy)} + \frac{1}{2} N_{3d(xz)} + \frac{1}{2} N_{3d(yz)}] \quad (7.1)$$

$$V_{val}^{(4p)} = -\frac{4}{5} \cdot e \cdot \langle r^{-3} \rangle_{4p} \cdot [N_{4p(z)} - \frac{1}{2} N_{4p(x)} - \frac{1}{2} N_{4p(y)}] \quad (7.2)$$

where $N_{3d(x,y,z)}$ and $N_{4p(x,y,z)}$ are the number of electronic holes in different 3d and 4p orbitals, the charge of an electron is given by $-e$.

The computation of total V_{ZZ} constitutes a complex problem, since it requires a detailed knowledge of the structure and the population of the conduction band in metals. Thus, the comparison of ν_Q and V_{ZZ}, deduced from theoretical calculations on the basis of different structural models and approaches (see equation (5)), with experimental values (equation (4)) permits to determine the individual features of local electronic arrangement and peculiarities of chemical bonds[15] and, as consequence, NQR spectra can in general serve as phase-analytical diagnostics of different materials.

The longitudinal (spin-lattice) nuclear relaxation T_1 depends on various and specific sources of field fluctuations in crystal structures of compounds. Therefore the studies of the temperature dependences of nuclear relaxation often allow sensing the lattice dynamics and transport properties of the material.

4. NQR frequencies

NQR has been in many cases the starting point for a detailed resonance investigation. Knowing the NQR frequencies, ν_Q, of, for instance, ^{63}Cu and ^{65}Cu is essential for exact material identification, which can be useful for resolving the geological problems.

In particular, among binary copper-sulfide materials eleven natural minerals and two unstable synthetic phases have been discovered to date: covellite $Cu_{1.00}S$, yarrowite $Cu_{1.12}S$, spionkopite $Cu_{1.40}S$, geerite $C_{1.60}S$, anilite $Cu_{1.75}S$, low-temperature digenite $Cu_{1.80}S$, high-temperature digenite Cu_2S, roxbyite $Cu_{1.8}S$, djurleite $Cu_{1.97}S$, low-temperature chalcocite $Cu_{1.993\text{-}2.001}S$, high-temperature chalcocite Cu_2S and villamaninite CuS_2.[31,32] As is seen, non-stoichiometric compounds prevail in this series. The rich variety of non-stoichiometric minerals in Cu-S system is caused by the fact that even negligible deviations of chemical composition from stoichiometry are accompanied by structural transformations of copper sulfide crystal lattice. These sulfides attract a great interest due to their possible application in different areas of electronic equipment,[5] wide distribution in industrial copper-ore deposits[4] and possession of indicative properties of ore-formation.[32] It is known that the structure and physical–chemical properties of Cu–S minerals depend on the stoichiometry, but the influence of non-stoichiometry on mineral characteristics (structure details, electronic properties, copper valence) is little known. Such problem is defined by the fact that the binary sulfides create the accretions in copper deposits. Moreover, the application of conventional methods (x-ray diffraction and optical spectroscopy) often cannot separate the Cu–S structural phases in natural mixed samples.[1]

The application of ^{63,65}Cu NQR spectroscopy to the studies of some binary sulfides permits to reveal unique properties of these compounds.[33,34] In particular, the low-temperature chalcocite Cu_2S NQR signals were observed at temperatures lower than 290 K.[33] The spectrum consists of 48 lines (24 lines for each copper isotope) in a narrow frequency range (16–29 MHz). According to crystallographic data,[35] the trigonal CuS_3 groups in Cu_2S have various Cu–S distances and various valence angles. The variations cause NQR frequency to be spread within a narrow range. The presence of 24 lines in the chalcocite spectrum for each isotope indicates that the unit cell contains 24 crystall-chemically independent Cu atom sites (see Section 3). This enables the space symmetry group of Cu_2S to be ascertained. Initially, crystallographers[35] defined it as monoclinic ($P2_1/e$) with 24 Cu sites. Later they proposed a monoclinic model for the unit cell but with the space group *Pc* and 48 atom sites (number of the sites is doubled due to lack of an inversion centre).[36] Later crystallographers returned to the unit cell with the space group $P2_1/c$.[37] The NQR data prove the model with the space symmetry group $P2_1/c$ to be more correct.

The comparison of ^{63,65}Cu NQR spectra of low-temperature chalcocite Cu_2S[33] and djurleite $Cu_{1.97}S$[34] shows that these minerals are related. Actually, both minerals exhibit 48 Cu NQR lines at liquid nitrogen temperature (77 K) in a frequency range 16–29 MHz. The latter fact, taking into account the equation (4), demonstrates the similarity of chalcocite and djurleite crystal structures and the same number of non-equivalent copper atoms (i.e. 24). On the other hand, the distinction in chalcocite and djurleite Cu NQR frequencies, especially at high frequency range, points to the evident reorganization of djurleite crystal lattice, which leads to the formation of "new" sulfide.

The ^{63,65}Cu NQR spectroscopy was applied also in studies of copper-poor sulfides: covellite $Cu_{1.00}S$[38–40] and geerite $Cu_{1.60}S$.[41] It was found that copper NQR spectrum of synthetic sample of $Cu_{1.00}S$ at 4.2 K consists of two copper doublets (i.e. the ^{63}Cu and ^{65}Cu isotopes lines): at 1.87 and 1.73 MHz ("low-frequency" doublet, Figure 1a) and at 14.88 and 13.77 MHz ("high-frequency" doublet, Figure 1b). The presence of two spectral lines for both copper isotopes permits to attribute NQR spectra to the two crystallographically non-equivalent sites of copper nuclei in $Cu_{1.00}S$ (so-called Cu(1) and Cu(2) sites).[40]

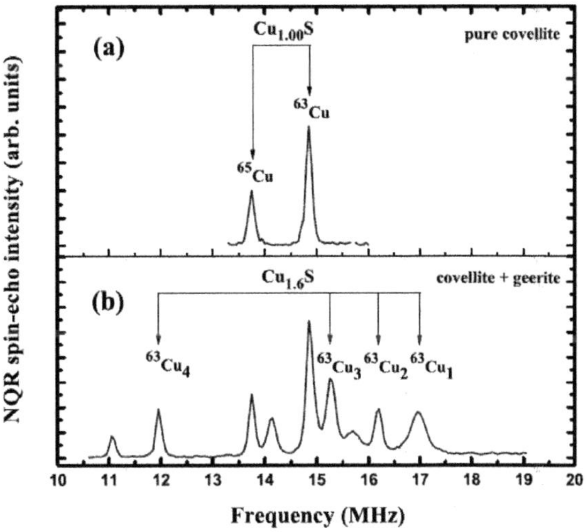

Figure 1. High-frequency copper NQR spectra for pure covellite $Cu_{1.00}S$ (a) and mixed sample, which consists of covellite and geerite $Cu_{1.60}S$ (b) at 4.2 K (solid curves). For details, see Section 4 and reference[40].

The comparison of ^{63}Cu NQR spectrum for $Cu_{1.00}S$ and ^{63}Cu NQR spectrum for geerite $Cu_{1.60}S$[41] can be used for discussions of crystal-chemical relationship between mentioned minerals. Geerite $Cu_{1.60}S$ is characterized by eight $^{63,65}Cu$ lines with NQR frequencies in the range of 10–17 MHz at 4.2 K,[41] which are grouped around the high-frequency Cu NQR signal of covellite $Cu_{1.00}S$. This proximity and analogous temperature dependences of covellite and geerite Cu NQR frequencies provide evidence that the electronic structures of these minerals have the same motive, but are still different. At the same time, the exhibition of eight $^{63,65}Cu$ NQR lines by geerite instead of two covellite $^{63,65}Cu$ NQR lines at high frequency range suggests that the geerite structure can be interpreted as the derivative of covellite structure with the strong deformations.

Thus, in spite of the similarity between chemical compositions of binary sulfides, it is clearly seen that their NQR characteristics are quite different. The investigation of other binary sulfides (synthetic and natural) would permit to find the new details in crystal-chemistry of these minerals and relations with ore-deposits.

5. Temperature dependences of NQR frequencies

The study of ν_Q as a function of temperature (T), pressure (p) and chemical composition immediately provides valuable information since the components of EFG tensor are very sensitive to small atomic displacements or modifications of charge distribution.

For instance, the experimental ratio of the NQR frequencies of the two isotopes in Figure 1a is equal to the ratio of their quadrupole moments (1.081) for both Cu sites in covellite $Cu_{1.00}S$. We have found the same results for the geerite $Cu_{1.60}S$. This proves that no internal magnetic field due to any type of magnetic ordering is present at the Cu sites in these compounds for the temperature range studied. Such a field would split or shift the NQR lines as it does for the Cu signal in anti-ferromagnetic minerals chalcopyrite $CuFeS_2$[42] and cubanite $CuFe_2S_3$.[43]

Although quite informative, studies of $\nu_Q(T)$ and $\nu_Q(p)$ in sulfides are relatively rare. In Figure 2 we have plotted $\nu_Q(T)$ dependence for triangular-coordinated Cu sites in covellite $Cu_{1.00}S$.[30] In general, the quadrupole frequency ν_Q decreases with increasing temperature without any significant anomalies. We thus exclude a major structural transition. However, we focus here on two weak effects: the change of the slope in the ν_Q versus T dependence at 65 K and at 210 K. The analysis of such $\nu_Q(T)$ dependence permits us to conclude that change of slope in the ν_Q versus T dependence at 65 K is related to the strong deformations of CuS_4 tetrahedrons in covellite $Cu_{1.00}S$, whereas the change of slope at 210 K is attributed to Cu valence state fluctuations near average value 1.3+ owing to an electron transfer between two Cu non-equivalent ions.[40]

Another example is the investigation of $\nu_Q(T)$ dependence for Cu in CuS_3 triangular groups of tennantite $Cu_{12}As_4S_{13}$.[44,45] It was found that the dependence is almost linear in the range 60–210 K, but it begins to slow down at lower temperatures (Figure 3). This unusual fact is accompanied by the drastic broadening of the NQR spectrum below 65 K.[44,45] Consistently with the crystal structure, the experimental data can be explained by an occurrence of a magnetic phase transition, which takes place near 65 K.[44,45] The analysis shows that the low-temperature phase is characterized by Cu^{2+} electron magnetic moments freezing in the form of a spin-glass-like constitution.

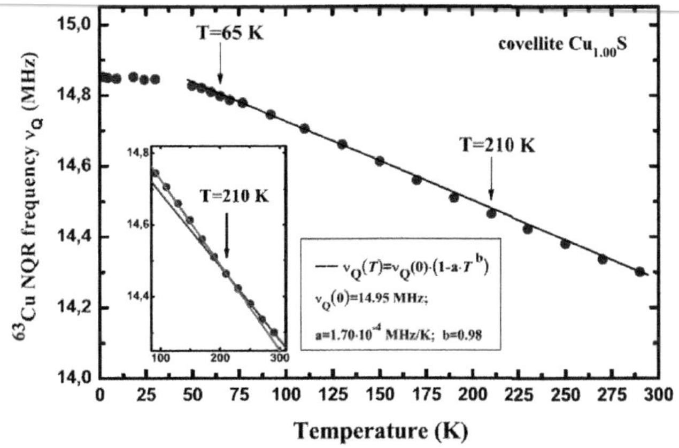

Figure 2. The temperature dependence of ^{63}Cu(1) NQR frequency ν_Q for pure covellite $Cu_{1.00}S$ with the fit of data by equation, depicted in the bottom of figure, within 65–290 K (solid curve) and extracted fitting parameters. Arrows point to the positions of the change of slope in the $\nu_Q(T)$ dependence at 210 K and at 65 K. Inset shows that the $\nu_Q(T)$ dependence can be described by two linear functions in the ranges 65–210 K and 210–290 K. For details, see Section 5 and reference[40].

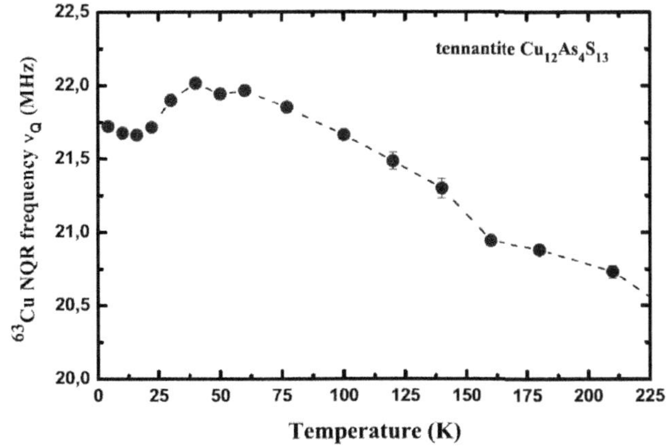

Figure 3. The temperature dependence of ^{63}Cu NQR frequency ν_Q for natural tennantite $Cu_{12}As_4S_{13}$. The dashed curve is guide for the eyes. For details, see Section 5 and references[44,45].

6. Electric field gradients and hole concentrations

The ultimate goal in these quadrupole coupling studies is, of course, the complete determination of the EFG tensor components (see Section 3). Knowing the EFG tensor components, V_{ij} (i,j = x,y,z), is necessary for determination of the local electronic structure details. As to the calculation of EFGs in sulfides and oxides, there are several techniques: the method of pointed charges and pointed multipoles, the Towns-Daily approach, the semi-empirical Mulliken–Wolfsberg–Helmholtz (MWH) method, different cluster and band structure calculations (in particular, the Hartree–Fock–Routhaan procedure), as well as their numerous modifications. Good review of the various techniques calculations can be found in reference[22].

It turns out that the MWH technique within the MO-LCAO approximation adequately describes the charge distributions for Cu sites in complex sulfides and oxides. For instance, the difference between calculated values and experimental data for Cu in Cu_3BiS_3 (wittichenite)[46] is only about 10–15% for all sites. It was shown that EFG (Section 3) is mainly formed by lattice term (90%), and, at less extent, by Cu-3d and Cu-4p orbitals (altogether 10%). In general, the dependence of the value of the ^{63}Cu nuclei quadrupole frequency ν_Q on the geometry of the most typical cases of Cu^{1+} coordination environment in series of copper sulfides was experimentally revealed.[29] In particular, in compounds with triangular CuS_3 complexes the value ν_Q lies within the range 20–23 MHz, whereas for tetrahedral coordination, CuS_4, the ν_Q either equals zero (regular tetrahedron) or is small (distorted tetrahedron), usually less than 3 MHz. Such quite narrow range of the change of ν_Q in these sulfides can be explained by non-significant variations of Cu–S distances, S–Cu–S angles and polarity of chemical bonds.

It is remarkable how the differences in the electronic distribution of related minerals can show up in the corresponding Cu EFG components. The delafossite-based copper oxides $CuMO_2$ (M = Fe, Al, Cr, Ga) may serve as an example for such behavior.[47] It was illustrated by MWH technique that the reduction of only one Cu–O distance in linear CuO_2 units from 2.00 Å ($CuAlO_2$) down to 1.84 Å ($CuGaO_2$) exhibits the decrease of electronic density in Cu-4p_z orbital and increase of the population of Cu-4p_x and Cu-4p_y orbitals, which play the key role in the experimentally observed lowering of total EFG value at the Cu site by about 1.5 MHz (see equation (7.2)).[47]

However, for the Cu sites in triangular complexes CuS_3 of covellite $Cu_{1.00}S$ the experimental values of EFG components are rather low compared to those predicted by MWH technique for Cu sulfides.[40] The origin of this discrepancy is probably that the MWH technique does not fully account for exchange-correlation effects between Cu 3d states of non-equivalent Cu sites.[40] In other words, the "failure" of the Cu EFG calculations in covellite $Cu_{1.00}S$ confirms the exceptional case played by the "triangular" planar Cu and that the Cu–S planes are mainly responsible for superconductivity in this material.

7. Nuclear spin-lattice relaxation

The longitudinal (spin-lattice) nuclear relaxation time T_1 depends on various and specific sources of field fluctuations in crystal structures of compounds. Therefore the studies of the temperature dependence of nuclear relaxation often allow sensing the lattice dynamics and transport properties of the material.

Nowadays it is well-known that among binary Cu sulfides the mineral chalcocite Cu_2S represents the good ionic conductor.[48] The earlier NQR investigation of chalcocite Cu_2S at low-temperature range showed the exponential increase of the nuclear spin-lattice relaxation rate T_1^{-1} with the activation energy more than 0.25 eV which are typical for many ionic conductors at temperatures when the ionic diffusion starts up.[33] Recent studies of chalcocite by precise potentiometric method confirmed this result to be correct.[49]

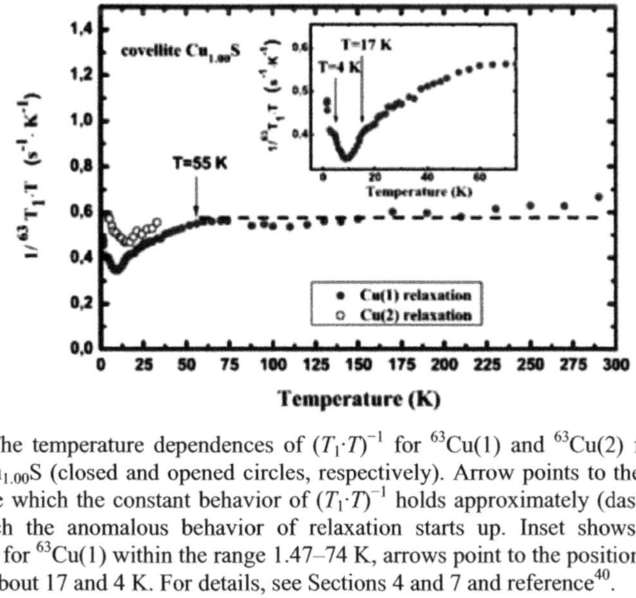

Figure 4. The temperature dependences of $(T_1 \cdot T)^{-1}$ for ^{63}Cu(1) and ^{63}Cu(2) for CuS pure covellite $Cu_{1.00}S$ (closed and opened circles, respectively). Arrow points to the temperature 55 K, above which the constant behavior of $(T_1 \cdot T)^{-1}$ holds approximately (dashed line) and below which the anomalous behavior of relaxation starts up. Inset shows the $(T_1 \cdot T)^{-1}$ dependence for ^{63}Cu(1) within the range 1.47–74 K, arrows point to the positions of bends in $(T_1 \cdot T)^{-1}$ at about 17 and 4 K. For details, see Sections 4 and 7 and reference[40].

At the same time, the metallic mineral covellite $Cu_{1.00}S$ demonstrates another temperature dependence of nuclear relaxation at low temperatures,[40] see also Figure 4. Above 55 K and up to 290 K the magnitude of relaxation rate depends on temperature linearly, i.e. covellite $Cu_{1.00}S$ exhibits the constant $T_1 \cdot T$ behavior. Such behavior is typical for metals.[21,50] It was shown that the conduction electrons in metallic crystals are the main source of this "simple" temperature dependence of nuclear relaxation rate. The T-dependence of the relaxation rate T_1^{-1} in this case is determined by fluctuations of the hyperfine magnetic field or the EFG with the frequencies $\sim 10^{15}$ s^{-1} created by the conducting spin-carrying charges at the nucleus site.[40]

However, below T = 55 K, the relaxation rate in covellite $Cu_{1.00}S$ demonstrates a surprising crossover from constant $T_1 \cdot T$ relation to unusual T-dependent $T_1 \cdot T$ behavior (Figure 4), in spite of the fact that $Cu_{1.00}S$ appears to be a metal down to superconducting transition $T_c \approx 1.6$ K. This anomalous dependence was treated as a "pseudogap" opening.[40] The surprising result of the covellite $Cu_{1.00}S$ NQR relaxation study is that the ratio of the relaxation times of two non-equivalent coppers is constant. Since, in any case, the nuclear relaxation is governed by conduction charges, our result implies that common electronic dynamics must exist for the two non-equivalent Cu ions, i.e. they "feel" not the individual, but the same conduction band. To our knowledge, this is the first observation of such a coupling in $Cu_{1.00}S$.

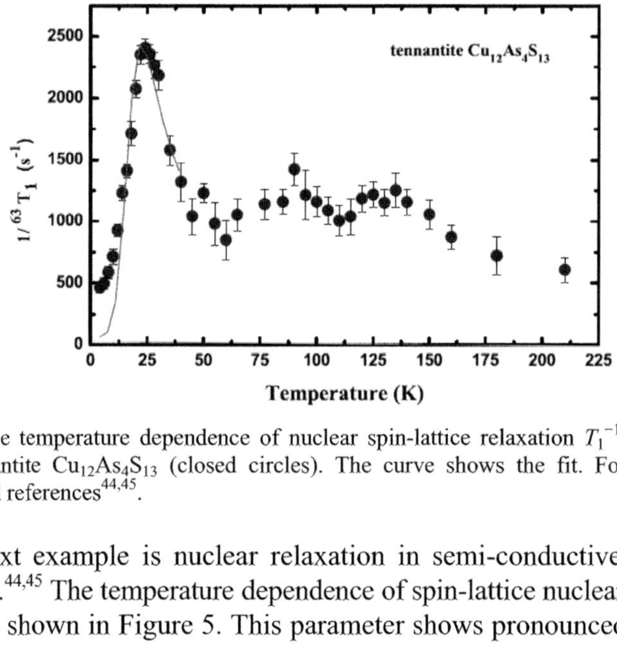

Figure 5. The temperature dependence of nuclear spin-lattice relaxation T_1^{-1} for ^{63}Cu for natural tennantite $Cu_{12}As_4S_{13}$ (closed circles). The curve shows the fit. For details, see Section 7 and references[44,45].

The next example is nuclear relaxation in semi-conductive tennantite $Cu_{12}As_4S_{13}$.[44,45] The temperature dependence of spin-lattice nuclear relaxation rate $1/T_1$ is shown in Figure 5. This parameter shows pronounced peak near 24 K. It was shown that such behavior can be described very well by magnetic or electric field fluctuations with the correlation time τ_C depending on temperature. These data were analyzed from the viewpoint of the inhomogeneous electron distribution in the Cu_6S_{13} cluster.[44,45]

8. Conclusion

Our examples of NQR studies of copper sulfide and oxide minerals have shown how broad and diverse the information that may be obtained is. There are topics typical to many solid state investigations, i.e. phase transitions, magnetic order and the relation of electric field gradients to electronic structure. NQR has proven to be an indispensable tool in this field.

References

1. Marfunin, A.S.: Physics of Minerals and Inorganic Materials: An Introduction. Springer-Verlag, Berlin (1979) [English translation of Russian original edition: Publishing House Nedra, Moscow, USSR (1974)]
2. Vaughan, D.J., Craig, J.R.: Mineral Chemistry of Metal Sulfides. Cambridge University Press, Cambridge, England (1978)
3. Ginsburg, A.I., Kuz'min, V.I., Sidorenko, G.S.: Mineralogical Studies in Geological Survey. Publishing House Nauka, Moscow, USSR (1981)
4. Izoitko, V.M.: Technological Mineralogy and Estimation of Ore. Publishing House Nauka, St. Petersburg, Russia (1997)
5. Abrikosov, N.K., Bankina, V.F., Poretskaya, L.V., Skudnova, E.V., Chizhevskaya, S.N.: Semiconducting Chalcogenides and Their Alloys. Publishing House Nauka, Moscow, USSR (1975)
6. Zakery, A., Elliot, S.R.: Optical properties and applications of chalcogenide glasses: a review. J. Non-Cryst. Solids 330, 1–12 (2003), and reference therein
7. Korago, A.A.: Introduction to a Biomineralogy. Publishing House Nedra, St. Petersburg, Russia (1992)
8. Yushkin, N.P., Askhabov, A.M., and Rakin, V.I.: Nanomineralogy (ultra- and microdispersed mineral states). Publishing House Nauka, St. Petersburg, Russia (2005)
9. Pound, R.V.: Nuclear electric quadrupole interactions in crystals. Phys. Rev. 79, 685–702 (1950)
10. Dehmelt, H.G., Kruger, H.: Kernquadrupolfrequenzen in festem dichloräthylen. Naturwiss. 37, 111–112 (1950)
11. Das, T.P., Hahn, E.L.: Nuclear Quadrupole Resonance Spectroscopy. In: Seitz, F., Turnbull D. (eds.) Supplement 1 of Solid State Physics. Academic Press, New York (1958)
12. Abragam, A.: Principles of Nuclear Magnetism. Oxford University Press, Oxford (1961)
13. Lucken, E.A.C.: Nuclear Quadrupole Coupling Constants. Academic Press, London/ New York (1969)
14. Schempp, E., Bray, P.J.: Nuclear quadrupole resonance spectroscopy. In: D. Henderson (ed.) Physical Chemistry, An Advanced Treatise, pp. 521–632. Academic Press (1970)
15. Semin, G.K., Babushkina, T.A., Yakobson, G.G.: Nuclear quadrupole resonance in chemistry. Wiley, New York (1975) [English translation of Russian original edition: Publishing House Khimiya, Leningrad, USSR (1972)]
16. Grechishkin, V.S.: Nuclear quadrupole interactions in solids. Publishing House Nauka, Moscow, USSR (1973)
17. Safin, I.A., Osokin, D.Ya.: Nuclear quadrupole resonance in nitrogen compounds. Publishing House Nauka, Moscow, USSR (1977)

18. Fukushima, E., Roeder S.B.W.: Experimental Pulse NMR – A Nuts and Bolts Approach. Addison-Wesley, MA (1981)
19. Bouznik, V.M.: Nuclear Resonance in Ionic Crystals. Publishing House Nauka, Novosibirsk, USSR (1981)
20. Buslaev, Yu.A., Kravcenko, E.A., Kolditz, L.: Nuclear quadrupole resonance in inorganic chemistry. Coord. Chem. Rev. 82, 9–231 (1987)
21. Slichter, C.P.: Principles of Magnetic Resonance, 3rd ed. Springer-Verlag, Heidelberg (1989)
22. Kasperovich, V.S., Shelyapina, M.G.: Nuclear quadrupole resonance. In: Chizhik, V.I. et al. (eds.) Quantum Radiophysics, pp. 446–539. Publishing House of St. Petersburg State University, St. Petersburg, Russia (2004)
23. Pen'kov, I.N., Safin, I.A.: Application of nuclear quadrupole resonance in study of minerals. Izv. Akad. Nauk SSSR (Geol. Ser) 12, 41–52 (1966)
24. Pen'kov, I.N., Safin, I.A.: Application of nuclear quadrupole resonance in study of minerals. Int. Geol. Rev. 9, 793–801 (1967)
25. Wendling, E.: La resonance quadrupolaire nucleaire (RQN) et ses applicasions en chimie minerale. Bull. Soc. Chim. France 1, 181–190 (1968)
26. Schultz, H.D., Karr, C.: Quantitative aspects of nuclear quadrupole resonance spectrometry of inorganics and minerals. Anal. Chem. 41, 661–664 (1969)
27. Pen'kov, I.N.: The nature of small structural impurities in some chalcogenides of As, Sb and Bi by data of nuclear quadrupole resonance. Geokhimiya [now Geochem. Int.] 6, 731–742 (1971)
28. Marino, R.A., Wenk, H.R., Apps, J.A., Klainer, S.M.: Proposed applications of NQR techniques to the geosciencies. J. Mol. Struct. 58, 445–457 (1980)
29. Abdullin, R.S., Kal'chev, V.P., and Pen'kov, I.N.: Investigation of copper minerals by NQR: crystallochemistry, electronic structure, lattice dynamics. Phys. Chem. Miner. 14, 258–263 (1987)
30. Kaufmann, E.N., Vianden, R.J.: The electric field gradient in non-cubic metals. Rev. Mod. Phys. 51, 161–214 (1979)
31. Goble, R.J.: The relationship between crystal structure, bonding and cell dimensions in the copper sulfides. Can. Miner. 23, 61–76 (1985)
32. Gablina, I.F.: Copper sulfides as indicators of the ore-forming environment. Dokl. RAS, Earth Science Sect. 356, 657–661 (1997) [English translation: Transactions RAS, 357, 1133–1137 (1997)]
33. Abdullin, R.S., Kal'chev, V.P., Pen'kov, I.N.: ^{63}Cu NQR spectra in low-temperature chalcocite, Cu_2S. Zapiski RMO 1, 99–103 (1988)
34. Safonov, A.N., Pogorel'tsev, A.I., Matukhin, V.L., Gablina, I.F.: Cu NQR spectra and electroconductivity in $Cu_{2-x}S$ compounds. Izv. Vyssh. Uchebn. Zavad., (Energetic problems ser.) 1–2, 100–102 (2006)
35. Evans, H.T.: Structure of low chalcocite. Nature (London), Phys. Sci. 232, 69–70 (1971)
36. Evans, H.T.: Djurleite ($Cu_{1,94}S$) and chalcocite (Cu_2S) new structural studies. Science 203, 356–358 (1979)
37. Evans, H.T.: The crystal structure of low chalcocite and djurleite. Z Kristallog 150, 299–320 (1979)
38. Abdullin, R.S., Kal'chev, V.P., Pen'kov, I.N.: ^{63,65}Cu NQR in covelline, CuS. Dokl. Akad. Nauk SSSR [Sov. Phys.–Dokl.] 294, 1439–1441 (1987)
39. Itoh, Y., Hayashi, A., Yamagata, H., Matsumura, M., Koga, K., Ueda, Y.: Cu NMR and NQR study of CuS. J. Phys. Soc. Japan 65, 1953–1956 (1996)
40. Gainov, R.R., Dooglav, A.V., Pen'kov, I.N., Mukhamedshin, I.R., Mozgova, N.N., Evlampiev, A.V., Bryzgalov, I.A.: Phase transition and anomalous electronic behavior in

layered dichalcogenide CuS (covellite) probed by NQR. Cornell University Library. http://arxiv.org/abs/0809.0560. Accessed 3 September 2008.

41. Anashkin, V.N., Kalinina, T.A., Matukhin, V.L., Pen'kov, I.N., Safin, I.A.: NQR spectra of $^{63,65}Cu$ in coexisting covelline (CuS) and geerite ($Cu_{1,6}S$). Proceedings RMS 5, 59–63 (1994)
42. Kal'chev, V.P., Abdullin, R.S., Pen'kov, I.N.: Nuclear magnetic resonance of $^{63,65}Cu$ in local fields in antiferromagnet $CuFeS_2$. Sov. Phys. Solid State 21, 1801–1803 (1979) [English translation from Fiz Tverd Tela 21, 3132–3134 (1979)]
43. Abdullin, R.S., Kal'chev, V.P., Pen'kov, I.N.: Nuclear magnetic resonance of $^{63,65}Cu$ in local fields in cubanite $CuFe_2S_3$. Fiz. Tverd. Tela 22, 2862–2864 (1980)
44. Gainov, R.R., Dooglav, A.V., Pen'kov, I.N.: Evidence for low-temperature internal dynamics in $Cu_{12}As_4S_{13}$ according to copper NQR and nuclear relaxation. Solid State Commun. 140, 544–548 (2006)
45. Gainov, R.R., Dooglav, A.V., Pen'kov, I.N., Mukhamedshin, I.R., Savinkov, A.V., Mozgova, N.N.: Copper valence, structural separation and lattice dynamics in tennantite (fahlore): NMR, NQR and SQUID studies. Phys. Chem. Miner. 35, 37–48 (2008)
46. Pen'kov, I.N., Abdullin, R.S., Yunusov, N.B., Togulev, N.V.: $^{63,65}Cu$ and ^{209}Bi nuclear quadrupole resonance in wittichenite, Cu_3BiS_3. Izv. Akad. Nauk SSSR Ser. Fiz. [Bull. Acad. Sci. USSR, Phys. Ser.] 42, 2104–2109 (1978)
47. Abdullin, R.S., Pen'kov, I.N., Yunusov, N.B.: Nuclear quadrupole interactions in semiconductors $CuMO_2$. Izv. Akad. Nauk SSSR Ser. Fiz. [Bull. Acad. Sci. USSR, Phys. Ser.] 45, 1787–1791 (1981)
48. Pearce, C.I., Pattrick, R.A.D., Vaughan, D.J.: Electrical and magnetic properties of sulfides. Rev. Miner. Geochem. 61, 127–180 (2006)
49. Berger, R., Bucur, R.V.: Potentiometric measurements of copper diffusion in polycrystalline chalcocite, chalcopyrite and bornite. Solid State Ionics 89, 269–278 (1996)
50. Winter, J.: Magnetic Resonance in Metals. Clarendon Press, Oxford, England (1971)

INDEX

GPSR Compliance

The European Union's (EU) General Product Safety Regulation (GPSR) is a set of rules that requires consumer products to be safe and our obligations to ensure this.

If you have any concerns about our products, you can contact us on ProductSafety@springernature.com

In case Publisher is established outside the EU, the EU authorized representative is:

Springer Nature Customer Service Center GmbH
Europaplatz 3
69115 Heidelberg, Germany

Batch number: 10755284

Printed by Printforce, the Netherlands